Victory and Deceit

ALSO BY JAMES F. DUNNIGAN AND ALBERT A. NOFI

Shooting Blanks

Dirty Little Secrets

Dirty Little Secrets of World War II

ALSO BY JAMES F. DUNNIGAN

From Shield to Storm (with Austin Bay)

How to Stop a War (with William Martel)

A Quick and Dirty Guide to War (with Austin Bay)

How to Make War

The Complete Wargames Handbook

Getting It Right (with Raymond M. Macedonia)

ALSO BY ALBERT A. NOFI

The Alamo and the Texas War for Independence

The Civil War Treasury

Eyewitness History of the Civil War

The Gettysburg Campaign

Napoleon at War

The War Against Hitler: Military Strategy in the West

VICTORY *and* DECEIT

Dirty Tricks at War

James F. Dunnigan
and
Albert A. Nofi

William Morrow and Company, Inc.
New York

It is the policy of William Morrow and Company, Inc., and its imprints and affiliates, recognizing the importance of preserving what has been written, to print the books we publish on acid-free paper, and we exert our best efforts to that end.

Library of Congress Cataloging-in-Publication Data

Dunnigan, James F.
Victory and deceit : dirty tricks at war / James F. Dunnigan and Albert A. Nofi.
p. cm.
Includes index.
ISBN 0-688-12236-1
1. Deception (Military science)—History. I. Nofi, Albert A. II. Title.
U167.5.D37D86 1995
355.4'22—dc20 94-40096
CIP

Printed in the United States of America

First Edition

1 2 3 4 5 6 7 8 9 10

BOOK DESIGN BY RICHARD ORIOLO

FOREWORD

This is a short foreword, so you might as well read it. This book is organized for random access. We have found that that is what our readers prefer, and if you have read any Dunnigan or Nofi books before, you know what we mean.

If you don't know what a "random access book" is, the explanation is simple. Most of the information in the book is self-contained in sections within chapters. Each of these sections tells a specific story without heavy reference to any other part of the book. You might already know some of the stuff in this book. In that case, just go for those areas where you are a little light and expand your knowledge, or whatever.

You can read the book from front to back if you like, and many people do. But you can also scan the table of contents or index and go right for what interests you most.

If you are a little vague on exactly what deception is, then read Chapter 1. It's a good introduction to the subject. That done, you can safely wander.

That's random access.

ACKNOWLEDGMENTS

James J. Bloom, Laura Kramer-Carini, Dennis Casey, Richard L. DiNardo, Valerie Eads, Kenneth S. Gallagher, William Kaiser, Michael Nofi, Marcus Nofi, Dani Shanske, Jay Stone, Brian Sullivan, Ed Wimble, Bob Shuman, Kurt Aldag, Allen Rehm, Al Marchioni, Kirk Schlesinger, and Mary S. Nofi, for putting up with one of us.

CONTENTS

Victory and Deceit

Deception Explained, Described, and Revealed

The most potent weapon in any soldier's arsenal is deception. That you don't hear much about deception in warfare tells you something about how elusive and apparently rare this item is. Yet, as the ancient Chinese adage puts it, "There can never be enough deception in war." In fact, Sun-tzu, the noted Chinese strategist, went further, saying, "All warfare is based on deception."

Seemingly rare, deception is actually one of the most common occurrences in war, and has been from earliest times. But deception, by its very nature, is frightening to soldiers, not to mention politicians and civilians. War is dangerous enough without having to contemplate yet another uncertainty like deception as well. However, like it or not, all soldiers throughout history have used some deception at one time or another, or been the victims of deceptive measures. For example, rare is the soldier who has not been on the losing side in at least one battle. Fleeing defeat, and the uncertain mercies of the victors, tends to concentrate a soldier's mind sharply with regard to deception. Members of a defeated army have lost most of their combat capability, and their ability to defend themselves against a bloody-minded enemy. But what always remains is deception, even if it is as simple as just finding a place in which to hide.

But this common use of deception in defeat shows that most soldiers will attempt deception only if there is little risk to themselves, or they have no other choice. The relatively low risk makes ambush one of the most popular forms of warfare, if you can pull off the deception necessary to make it work. Most soldiers feel more comfortable with low-risk alternatives, and using deception generally involves a bit of risk. Indeed, deception is at once one of the most powerful weapons a soldier has access to, and the one most frequently avoided because of the risk.

Soldiers who can handle risk are considered exceptional warriors, or fools. Because anyone can attempt the risks of deception, but most cannot pull it off well—if at all—anyone proposing deception is often suspected of imprudent behavior. While fortune favors the bold, the flip side is the ever-present risk of calamitous failure when a bold move doesn't come off as intended. In the life-and-death environment of the battlefield, caution is quickly embraced by leader and follower alike.

Many of the examples of deception we provide are from ancient history. These examples are still relevant. The stratagems of biblical prophets and potentates are still used. Remember, if you will, some of the tricks Saddam Hussein used to survive. And survive he did, by using bluff, bluster, false moves, and deceptive moves any biblical warrior would be proud of.

The world is currently full of nations that are militarily weak, but ruled by despots who do not lack for cleverness or the willingness to use deception to maintain and expand their power.

While deception is often a necessity for victory, it is seen by too many troops and national leaders as the handmaiden of defeat—because deception is often viewed as the weapon of the weak. After all, the strong don't need it. Or do they? In fact, deception is equally the tool of the strong, if only because it may make the cost of victory cheaper. In many cases, winning is only possible through the use of deception.

The Many Faces of Deception

Deception is most frequently used by soldiers. It is the desperate situation on the battlefield that prompted soldiers to invent the basic deception techniques. From this crucible, military commanders and national leaders draw their inspiration for deceptions used on a large scale.

Deception comes in many forms: surprise, stealth, misinformation, disinformation, false moves, camouflage, and anything else a desperate or steely-nerved soldier can conjure up to deceive his opponent.

The importance of deception in warfare is often overlooked, often by the soldiers whose lives depend on it. Consider the following items:

- Most aircraft are shot down by an opponent they didn't know was there. This bit of reality is drummed into young pilots, although the lesson doesn't usually take hold until the pilot in question has had his first brush with death. In the air, superior deception is the key to victory. Be where the enemy won't expect you, and shoot him down before he knows where you are. Simple to say, hard to do.

- Naval warfare is largely a matter of finding your opponent before he finds you. While this is particularly true in the twentieth century because of carriers and submarines, it was also true in centuries past. Hiding your moves from your opponent—i.e., deception—is the hallmark of a successful naval commander.

- Ground combat can accurately be described as a series of ambushes. An ambush is one of the more blunt applications of deception.

- Senior commanders consistently believe that the enemy will behave as he would in a similar situation. This happens because the military is an insular organization. Officers rising through the ranks don't get much opportunity to see how other nations deal with battlefield situations. Moreover, an officer on the way up tends to be a true believer in the way his organization does things. By the time the poor fellow becomes a general, he is ripe for deceptions sprung by an enemy commander who has a broader outlook on the way foreigners think.

Not Again!

Some tricks have been used so often, it's amazing that folks haven't caught on. But then, history isn't always a prerequisite for high command, whether military or political. Moreover, we tend to believe what we want to believe. And besides, those stupid yahoos on the other side couldn't possibly come up with a trick that would fool us. War is not an undertaking for the introspective or the humble. A good commander has to have a big ego and a lot of confidence, or he'll probably lack the guts to come up a winner. So unless he's a really devious fellow, he's also likely to overlook the possibility that the enemy may be even smarter than he is. Often, the enemy isn't smarter. But that doesn't mean he may not come up with some clever trick that's so old, no one in his right mind would ever think of using it again. It's worth recalling that history is full of spectacular upset victories which resulted when a supposedly outnumbered, ill-equipped, and poorly trained army led by an amateur soldier managed to put one over on a supposedly more numerous, better-trained, and superbly equipped one led by professionals.

You want examples? How about the ragged forces of both George Washington (American Revolutionary War) and Vo Nguyen Giap (Vietnam wars). Both Washington and Giap used deception extensively to defeat their superior (on paper) opponents. They both used some of the same techniques, many of which were already ancient when they used them, having been mentioned in the Bible. As recently as the Afghan War of 1979–1988, the Iran-Iraq War of 1980–1988, the Persian Gulf War of 1990–1991, Somalia (1993), and the Balkans (the entire twentieth century), the same old deceptions keep getting recycled. There's nothing new under the sun, unless the other fellow has forgotten it and you haven't.

Old Manners Linger On

Despite the do-or-die nature of deception, there's a very human trait that causes most people to avoid practicing it, or even acknowledging its existence. The attitude, conscious or subconscious, is that a real warrior doesn't hide from the enemy, but simply gets out there and kills things. Professionals know better, but most people carrying weapons on a battlefield are not professionals.

Some forms of deception can be taught. But there are many things a soldier must learn, like how to use his weapons. Training in deception, aside from things like camouflage, doesn't often occur. In any event, you need very well-trained, well-disciplined, and confident soldiers to practice deception successfully. Most soldiers in the world are not very well trained. This is nothing new. War is a sometime thing, and peace is actually the norm. Indeed, most soldiers spend most of their time at peace, not at war. In peacetime, there are other distractions and demands on one's time. Somehow, training never gets the attention it deserves, except in wartime. Of course, by then, it is too late for many, and the urge to send ill-trained troops to the front is usually overwhelming. This cycle has been repeated again and again over the centuries.

So if training for the business of fighting is so often slighted, how can we prepare ourselves to fight effectively against deception, and to employ it ourselves? Ultimately, the arts of deception and trickery in war can be learned only through exposure to the experience of history. However, even that is not always sufficient preparation. The great practitioners of deception throughout history have not always been learned soldiers. But they usually have been resourceful, clever commanders thoroughly grounded in the practice of war. Getting leaders with such capabilities is not easy.

For example, the U.S. Army recently decided that deception is a good idea, which was a wise move. The army also has mandated that all operational plans include a deception plan. That sounds good, too. But it isn't. The net result has been some of the lamest tricks and ruses ever to masquerade as deceptions. This lip-service observation of the mandate has caused these ill-conceived attempts at deception to be now included in plans for potential wartime operations. So more work is obviously needed.

Deception is too useful a tool to abandon simply because the troops might not be up to it. During wars, most of the troops who survive their bad training become skilled fighters and often rediscover ancient deception techniques. Officers, who tend to be better trained than their troops in the first place, also become quite adept at deception after a year or so of battle has weeded out the inept among them.

But, invariably, once peace returns, so does the stingy attitude toward thorough training. And the wartime imperatives that fostered the practice of deception are soon forgotten. The nation that develops realistic training for deception before a war begins will have a formidable advantage.

How Do You Touch an Idea?

One major problem with deception is that, except through the lessons of history, deception is also very difficult to demonstrate. Weapons and equipment can be seen and touched, and their use demonstrated in a very tangible fashion. Deception is, well, sort of invisible, at least most of the time. It's not something that one can practice, except at the simplest levels. Just how invisible can be seen from the many ways in which deception has been used over the centuries.

At the lowest level, deception is implicit in the mastery of a soldier's basic skills. Knowing how to patrol, conceal oneself and one's equipment, prepare fortified positions, and move a vehicle at night are all things that professional soldiers practice regularly and do well. But, as noted, training is not a high-priority item in most armed forces. Pilots, for example, don't really appreciate what it takes to "sneak up" on another aircraft unless they have spent a lot of time in the air practicing such maneuvers. Seeing is believing, but pilots have to spend a hundred or so hours in the air each year just to maintain their basic flying skills. Many nations can't even afford that. Most pilots are lucky to get that minimum. You have to fly several hundred more hours each year to get to the point where you realize what airborne deception is, how useful it is, and that you are at the point that you can pull it off. Flying that many hours is hideously expensive, and few pilots do it except in wartime. When civilian pundits declare the 1990–1991 Gulf War to be "a training exercise," they are more accurate than they know. The skills of American pilots in that war were brought to much higher levels because of all the combat flying they were able to do. It made no difference that the opposition was inept. It was the chance to pile up all those hours in the air that gives these pilots an additional edge in any future war.

The same situation applies with navies. Warships spend a great deal of their time tied up in port during peace, as few nations can afford to keep the crews at sea long enough to become truly expert. As has long been the case in warfare, when the fighting starts, it's not a matter of who's better, but who's worse.

Some nations grasp the concept of hard peacetime training making for a more bearable degree of loss in wartime. But this approach is rare. While deception is hardly recognized as a key combat ability, training for deception is even less so.

Learning by Example

The best way to explain deception is to give examples, a lot of examples. A lot of different examples. Well-explained examples. In these pages, we will present over one hundred and twenty examples of military deceptions, ruses, and tricks. We take these examples from history, putting before you successful applications of deception and trickery from earliest antiquity up through the present. Most of these tricks are specifically military, but a few are political or diplomatic. After all, wars are not fought in a purely military environment. Some of the examples lead to a look at other cases of the use of the same, or similar, tricks, in later periods, so the historical sequencing found in this book will sometimes appear rather chaotic, if only to show the essential timelessness of many of the deceptions under consideration. Although some of these examples are from non-Western cultures, we've chosen to focus our attention on cases drawn from Western military and diplomatic history, because the impression exists that the Western "mind-set" lacks the subtlety necessary to effect truly successful deceptions. This is one reason for the surprising popularity of ancient Chinese treatises on the art of war, with their stress on deceptions, ruses, and tricks. In fact, the record demonstrates that from the earliest roots of Western civilization, deceit and trickery were common practices in war.

These examples generally use one or more of the traditional deception techniques, which can be summarized briefly as follows:

- CONCEALMENT: hiding your forces from the enemy using natural cover, obstacles, or simply great distance. Ground troops can be behind a hill, aircraft can fly low to avoid radar, or into clouds to avoid eyes. Concealment was the earliest form of military deception, easily adapted from the technique hunters used to conceal themselves from their prey while out searching for a meal (while avoiding becoming one).
- CAMOUFLAGE: hiding your troops and movements from the enemy by artificial means. Some tree branches and leaves will do,

and have been a staple of camouflage for thousands of years. This is also a heritage of man's past as a hunter.

- FALSE AND PLANTED INFORMATION: letting the enemy get his hands on information that will hurt him and help you, but he won't know that he's being snookered. This involves some espio nage (double agents and the like) as well as understanding your opponent better than he understands you (or himself, for that matter).

- RUSES: tricks, such as displays that use enemy equipment and procedures to deceive, like "false colors" (enemy flags and uniforms) that make the enemy think he is seeing his own troops when, in fact, they are enemy forces.

- DISPLAYS: using techniques to make the enemy see what isn't there. Having a few horses drag branches behind them, to create a lot of dust so that a distant enemy thinks it is a large cavalry force, was a common display technique until recently (and can still be used today, but not as frequently). Lighting many campfires where there are no troops is also an ancient display technique, and one that still has uses today. More modern examples are: fake artillery (painted logs); dummy aircraft and armored vehicles; lots of simulated radio traffic (to represent units that don't exist).

- DEMONSTRATIONS: making a move with your forces that implies imminent action, but is not followed through. In ancient times, this was often something like moving your cavalry back and forth where the enemy could see it (and wonder where it was going to attack). Today, it often means sending an armored unit down a road where it is likely to be spotted by enemy recon aircraft or satellites.

- FEINTS: like a demonstration, but you actually make an attack, or retreat. When this is an attack, it is meant to make the enemy think it is more than it appears to be. It's usually done to distract the enemy while your real main assault occurs elsewhere. Since modern attacks go on for days and involve several waves of troops, a feint would be an attack that lasted only a day and involved far fewer troops than the real thing. But for that one day, the enemy could be deceived into believing that the attack at hand was the "big one." Using a false retreat as a feint is an ancient trick, the usual result being that the enemy troops advance in pursuit in a disorderly fashion.

This breaks up the attackers' battle formation (usually a tightly packed mass of spearmen) and makes the "pursuers" more vulnerable to counterattack. Once the "fleeing" troops see their pursuers break ranks, a signal is given, and the counterattack begins. It usually worked, especially if the "fleeing" force had a fresh reserve of troops to throw into the fray.

- LIES: flat-out lying when communicating with the enemy is something that is timeless. Especially today, opposing armed forces always have easy recourse to radioing messages to their opponent. Often, it's a demand from one side for the other to surrender. These conversations are often full of imaginative fibs, and the side that can lie most effectively under these circumstances often gains an advantage. Nowadays, we also have the media, which can help transmit our little tales to the enemy.
- INSIGHT: the ability of one general to deceive his opponent by outthinking him. This is not usually stated as a form of deception, but it is a common technique. Deceptions are, by definition, largely in the minds of the beholder, and some of the most common deceptions require little beyond one general having a better insight into how his opponent operates. And, of course, it involves the ability to be wise to deception being used against yourself.

What is remarkable about many of the cases that follow is that they demonstrate how little deception has changed over several thousand years. And how short the list of basic deceptions is. The weapons and uniforms have changed greatly; the deception has not. Moreover, many ancient deception techniques can still be used today. The same cannot be said about ancient weapons.

In the Beginning . . .

No one will ever know who first used deception in warfare. It most likely arose as soon as men began slaughtering each other in an organized fashion. Then again, it probably antedates war, in the form of tricks used in hunting. After all, putting out salt for animals to lick (while you take aim with your spear) is not too far removed from positioning a dummy unit to draw the enemy's attention (while you take aim with your artillery). In fact, nature herself deceives, what with animals whose colors enable them to move with relative invisibility through forests or across deserts, and others who have false faces to intimidate possible predators, and still others with built-in lures to entice possible prey. So, perhaps no one consciously invented deception in war; it just was there, grown out of humanity's animal past.

Giving Deception a Little Respect

We in the West regard deception differently from the way most other people do on the planet. In Asia, the Middle East, and other parts of the world, skill at military and political deception is seen as a worthy talent and a more humane alternative to armed conflict.

Take, for example, one of the more successful practitioners of deception in European history, the Byzantines. These were the people who, from their capital at Constantinople (now Istanbul), kept the eastern half of the Roman Empire going after the city of Rome fell in the fifth century. In fact, the Byzantines kept their part of the Roman Empire alive for another thousand years (giving the Romans a two-thousand-year run on the world stage). One of their most successful, and most widely used, tools was deception. All forms of deception, and lots of it. They even incorporated it into standard army manuals, such as *Tactica*, written by Emperor Leo VI (reigned 886–912). So, for example, in addition to useful tactical advice about what some termed "treacherous and cowardly" tactics like night attacks, ambushes, and surprise attacks, Leo suggested some even more devious devices, such as:

- Bribing important officers in the enemy's army to throw a battle. Many military leaders in this period were mercenaries, and thus susceptible to a better offer. The Byzantines knew the best time to make the offer. And the Byzantine intelligence and diplomatic services made it their business to know who was who in foreign armies. Even commanders who were not mercenaries could be tempted by gold, or other favors. There was little sense of nationhood or patriotism in this period. The enemies of Byzantium were either tribal or feudal kingdoms. In both, there were generally plenty of potentially disaffected notables looking for a better deal. As recently as the 1979–1988 Afghan War, the Soviets were able to bribe several Afghan tribal chiefs to stay out of the fighting. Remember, and we'll say it again: The Russians have long considered themselves the heirs to the Byzantines.

- Circulating false reports of distant victories to bolster the morale of one's own men, or reduce that of the enemy. Communications were poor in this period, and the skill with which one could compose and deliver a false message was more important than whether it was true or not. The Byzantines had skilled forgers and experts who could craft convincing false messages. Indeed, the Byzantines prized all forms of showmanship in diplomatic affairs and were very successful at it. In 1982, the British sent one of their feared Ghurka battalions to the Falklands, and made sure the Argentine defenders knew that these knife-wielding Ghurkas were on their way. This had the desired effect of reducing Argentine morale even lower.

- Engaging in protracted negotiations with the enemy, even when all hope of peace was gone, in order to gain time and allow the negotiators to spy out his forces. This is an ancient practice at which the Byzantines were very skilled. Once more, it was usually the case of the practiced Byzantine professional versus the unskilled barbarian negotiator. As always, the pros had a decided edge. You can see this ploy played out in many places today, ranging from Bosnia to Iraq to North Korea.

- Fabricating treasonable letters from important officers in the enemy's army, and making sure their commander accidentally got hold of them. Again, the Byzantine skill at forgery and their superior information-gathering ability enabled them to create convincing deceptions. As the tools of forgery have improved, this deception has become more common in the twentieth century. It was a much-used technique during the Cold War, and ever since.

It was because of this sort of cleverness at deception that Leo VI was nicknamed "The Wise." And he wasn't alone. The Byzantines were consistently very good at this sort of thing. Perhaps too good, as they were usually fighting with various Italian city-states as well as Normans, Slavs, Arabs, Persians, Turks, and anyone else who wanted to finish them off once and for all. For a number of social, political, and even religious, reasons, people in the West came to consider the Byzantines and their deceptions in an unfavorable light.

There were many reasons for this bad reputation in the West. Initially, there were those events that created this "Eastern Empire" in the first place. The Byzantines had a good claim to being the heirs to the Roman Empire. Certainly, when it came to deceptions and ruses, they were true sons of Rome. In fact, they called themselves "Romans." It was a European custom to call them "Byzantines," because the eastern portions of the empire had long spoken Greek as much as Latin. Thus the language of the Byzantine Empire became Greek, and thus different from the rest of Europe (where the Roman Catholic Church kept Latin alive). This did not sit well with those who were left behind in the wreckage of the western part of the empire centered about Rome itself. The Byzantines maintained the efficient government of the Roman Empire while the peoples to the west had to deal with the chaos of the dozens of barbarian tribes that were becoming civilized in fits and starts. It was largely jealousy and insecurity, but the Byzantines were seen as a bunch of snooty know-it-alls. That the Westerners were often on the short end of Byzantine deceptions did little to improve relations. The Byzantines were often undiplomatic with the westerners, feeling, with some justifications, that they were dealing with Italian, German, English, and French barbarians.

Finally, there were religious differences. Minor at first, but by the eleventh century, the Catholic Church in the east split from the Catholics in the west. The eastern ("Orthodox") Catholics no longer recognized the pope, a disagreement that still exists.

As a result, we have today the hoary term "Byzantine politics." It generally means dirty dealing and deception. That's what the Byzantines did. A few people might get killed as a result of such machinations, but it was usually a humane alternative to the wholesale slaughter of war. Always outnumbered and often surrounded by more powerful enemies, the Byzantines learned, as many never do, that foes can be vanquished without bloodshed, or at least without too much of it. Their idea was that if resort must be to war, it should be used sparingly and to good effect. A sound policy it was, and quite parsimonious in terms of lives and treasure, especially Byzantine lives and treasure. Brain over brawn, so to speak. The lesson was not lost on everyone in the West.

Not long after the last remnants of their empire had been overwhelmed by the Turks, and while the Byzantines were still being thought of throughout Europe as big-time, underhanded, bad guys, Niccolò Machiavelli wrote *The Prince.* Meant as a "how to" book for Renaissance rulers, the book presented practical wisdom on how to rule through the use of, well, "Byzantine" techniques. Thus *The Prince,* a generation after Byzantium finally fell to Turkish invaders, gave belated recognition that the last of the Romans knew what they were talking about. And so did many observant students in the West. But we still speak unkindly of those who use "Machiavellian" or "Byzantine" methods. And this makes deception a lifesaving tool that Westerners instinctively shy away from.

How to Deceive, Made Easy

Describing, and understanding, deception is a lot easier than practicing it. But this goes a long way toward explaining why deception is so often ignored, misunderstood, and, all too often, not appreciated. The techniques of deception consist of several different ancient skills. These are described below in some detail, so you'll know what we are talking about when you go through this book's many descriptions of historical deceptions.

Concealment

This is similar to camouflage, except that you are not doing any work other than moving your forces behind a natural obstacle. Concealment is the most ancient deception technique. Armies have long marched all, or part, of their forces behind a hill or into a forest where the enemy could not see them. Ships have long known the advantage of sailing into a fog bank to escape a pursuing force. Aircraft quickly learned the lifesaving trick of flying into a cloud when the opposition got too intense. Concealment is often combined with camouflage, although in a pinch, concealment alone will do the job. Modern sensors like radar have made it more difficult for ships and aircraft to hide in fog or clouds. But ground forces still make good use of what nature has provided. Ground combat is still full of situations where nervous troops approach a tree line that may, or may not, contain enemy troops. There is also strategic concealment—that is, hiding the direction of one's movements on a grand scale or hiding other useful information from the enemy—and political concealment, hiding one's political objectives. Not telling newspapers "the facts" is a form of concealment, eminently valuable on the strategic and political level. As President Roosevelt said shortly after Pearl Harbor, he was not about to tell the American people how badly the Japanese had, or had not, injured us, since it would be precisely what the enemy wanted to know. To which we can only add Winston Churchill's comment that in war, truth is so precious, it must be surrounded by "a bodyguard of lies."

Camouflage

Nearly as ancient a technique as concealment, camouflage is making your own concealment. That is, hiding yourself from enemy view, and it is one of those techniques that man the animal carried forward as human mental skills developed further. Many animals use camouflage to hide themselves from enemies, or prey. In fact, there are few wild animals that do not have camouflage built in. Arctic foxes, for example, have a dark, dull coat in the warm season. This makes it difficult for prey, or enemies, to detect them in underbrush or out in the open. But in the winter, when there is snow everywhere, the arctic fox grows a white coat. This provides excellent camouflage against the nearly universal white background, and is a crucial adaptation of the fox's Arctic environment. Foxes don't hibernate during the winter, but must constantly stalk their next meal. Their white winter coat maximizes their chances of finding enough to eat until spring arrives. Most animals don't change their colors with the season, but they do have a basic color pattern that has evolved as the best form of protection. For soldiers, good camouflage means being able to hide your movements better than the enemy hides his own. While ambushing the enemy is good, being able to move around undetected is even better, and more difficult to accomplish.

Military camouflage was another adaptation of hunting techniques. Early humans quickly learned how to create their own concealment. If there was no shrubbery where they wanted to wait for prey to come by, they realized that they could construct a passable imitation of a natural hiding place. Hunters could cover themselves with the skins of animal species they were hunting to make their prey think there were no predators in the area. The saying "a wolf in sheep's clothing" springs from this ancient practice of the hunter wearing the skin of the animal species he was stalking. When man began stalking man, camouflage techniques became more sophisticated, because the prey was more wary and wise to the tricks of the ultimate predator.

When warfare and armies developed several thousand years ago, camouflage became much less a central part of combat. Before armies came along, the principal form of warfare was raiding (and still is, among primitive peoples). A raid might involve battles, but the object was to ambush your foes, or steal their possessions. Thus the raiding parties had a big incentive to conceal themselves. The raiders were usually hunters and they used the same techniques that worked against animal prey. The raiders would advance carefully, staying out of sight, and would often wear foliage and animal skins. For communication among themselves, the raiders would use animal calls, the same ones they used when hunting to deceive their prey. The only shortcomings with these hunting techniques was that their foes were usually hunters too and knew what to look, or listen, for.

When armies became more the norm in warfare, camouflage became more difficult, and less frequently used. An army, by definition, is a large group of warriors acting together. Such a large group cannot easily sneak around as a dozen or so raiders would. Actually, you could use camouflage with an army, but it was much more difficult, and not many army commanders were up to the task. While the leaders of raiding parties were usually selected from among the most accomplished hunters, army leaders were often chosen from among the most successful politicians. Thus the leaders of raiding parties were much more capable for their tasks than were the commanders of armies. Armies did eventually learn to make use of camouflage, especially when it came to concealing troops waiting in ambush, or even hiding the camps that armies on the march would set up periodically.

A few hundred years ago, armies grew larger and adopted colorful uniforms. This was the antithesis of camouflage, but the uniforms were needed to make it easier to sort out who belonged to which side, and to impose more discipline on large masses of troops. While camouflage was still used by the increasing number of scouts, concealment was not uppermost in the minds of most army commanders. Indeed, until recently, concealment was considered "unsoldierly" and dishonorable. Some of that attitude still persists.

What might be termed "the golden age" of camouflage did not arrive until the twentieth century. By then, armies had become huge (easily, hundreds of thousands of troops each) and numerous (often over a dozen armies per nation). In this century, armies learned to settle along fronts hundreds of miles long, facing enemy armies. Under these circumstances, being so close to the enemy and vast amounts of hostile firepower, camouflage became much more popular among the troops and, eventually, the army leaders themselves. This revealed yet another twentieth-century development: The troops rediscovered the advantages of camouflage before their leaders did. Earlier, army commanders were right out there with the troops, sharing the danger and exposure to what the reality of the situation was. By World War I (1914–1918), army leaders were miles away from the fighting and quite out of touch with what was going on and what needed to be done. Because of this, it took a while before camouflage became part of official army policy, the necessity having to filter back from the front to the rear. However, by the time World War II (1939–1945) rolled around, camouflage was considered a key military technique.

But technology caught up with camouflage in the twentieth century. First came aircraft, which were initially used primarily for scouting. Camouflage that hid you from someone on the ground was usually ineffective when enemy aircraft flew over. Not only that, but camouflage now became a lifesaving measure for troops normally out of sight of the enemy. In addition to attack aircraft, the twentieth century also brought with it long-range artillery. These guns could fire at targets over ten miles away, which put troops far from the front line in danger. Aircraft observers allowed these rear-area troops to be spotted, and accurate artillery fire to be called in. So now, everyone suddenly developed an interest in camouflage. In the past, camouflage was not needed to prevent observation from above. Now it was. Through the middle of the twentieth century, troops developed efficient techniques to hide themselves from aerial observation. This involved more than just putting camouflage nets and foliage over positions and equipment. You also had to, literally, hide your tracks. Trucks and armored vehicles left ruts in the ground when they tried to move cross-country, or on dirt roads, and these ruts were usually quite visible from the air. Even during World War I, the horses used to pull the guns and ammunition wagons left a vivid trail easily seen from the air. The troops soon learned that covering their tracks could be a matter of life or death.

The most ancient camouflage techniques are still very effective. These involve covering the troops, or their positions, with foliage most of the year, or white cloth when there's snow. The use of winter camouflage is a recent development, because winter warfare did not become common until this century. Whatever the season, the principal function of camouflage is to make your troops invisible to the enemy. This invisibility is sometimes achieved, but usually, the invisibility is not complete and the result is that the enemy is never sure exactly what you have and precisely where it is.

In the last eighty years, a vast number of camouflage techniques have been developed. However, technology has rendered a lot of camouflage effort much less useful. The use of infrared (heat) sensors has made it possible to tell quickly what is live vegetation and what has been cut down for camouflage. Infrared can also see through foliage and detect the warm bodies of troops, and the heat from engines and recently fired weapons. Other sensors can detect large masses of metal (tanks and trucks). At the moment, the sensors for seeing through camouflage are not something that every soldier has. These sensors are expensive and require a lot of attention to keep them operational. So the sensors are used primarily by reconnaissance aircraft. For those armies that can afford it, some ground troops haul these sensors around on the battlefield as needed. Eventually, it will be possible to equip patrols of half a dozen troops with infrared and metal detectors. Individual aircraft will also eventually be so equipped, making it easier to bomb camouflaged targets. But, as things stand now and in the foreseeable future, camouflage is still very effective against troops who do not have the latest sensors. Even against high-tech armies, camouflage will hide you a lot of the time, and this can still be an advantage.

False and Planted Information

This is another ancient form of deception, one that became particularly popular once writing was invented and written orders and reports became common tools for generals to communicate with. This form of deception has long taken the form of letting the enemy get his hands on information that will hurt him and help you, but he won't know that he's being taken for a fool. This often involves some espionage (double agents and the like), as well as understanding your opponent better than he understands you.

But soldiers did not have to wait for literacy to use planted misinformation. Bits of clothing or weapons dropped in the right place could give the right bit of misinformation, especially if the troops planting the misinformation wore enemy sandals. Armies have always used scouts, and it was a point of pride to be able to fool enemy scouts. If an ancient general wanted to make it appear that he had more troops than he actually had, or had allies he actually didn't have, he could try to leave signs that suggested greater numbers, such as more litter left behind as his troops pass. An enemy scout looking for signs that troops have passed would note the amount of abandoned or lost clothing along your route of march and make an estimate of the size of your force. Bits of clothing or broken weapons from another tribe, left in the right place, could cause enemy scouts to think this other tribe was marching with your troops. Nowadays, we can also supply our enemies with false or misleading information by feeding it to the press, an increasingly common practice in the twentieth century and one which is likely to remain popular for a long time.

The problem with false and planted information is that you are never sure the enemy will fall for it (See "Richard Meinertzhagen Loses His Dispatch Case," page 145). If your opponent sees through the deception, he can use that as a weapon. Like most forms of deception, this one can cut both ways and only works in the hands of a skillful and resolute general.

Ruses

These are displays that use enemy equipment and procedures to deceive, to make the enemy think he is seeing his own troops when, in fact, he is facing enemy troops. From a distance, all troops tend to look alike. It is only in the details of how they hold their weapons or move across the battlefield that you can sometimes tell who is who. Thus an army a thousand years ago might deceive the enemy by having the troops carry their weapons as did the enemy (for example, with spears pointing skyward, rather than at an angle) and move like the enemy (in a V-shaped formation rather than a straight column). By the time the enemy discovers the deception, it may be too late for him.

These techniques include "false colors" (enemy flags and uniforms) that also make the enemy think he is seeing his own troops when, in fact, they are enemy forces. The effectiveness of this de-

ception can be seen in the centuries-old custom of immediately executing enemy troops found trying to pass as your troops. It's not true that "all's fair in love in war." Well, in love, perhaps, but in war, certain rules have evolved over thousands of years, and wearing the enemies' uniforms has long been considered dishonorable behavior.

In this century, ruses have extended to electronic deceptions. Pilots have been known to switch to enemy frequencies and deliver misleading orders in the enemy's language. Sometimes this works, but often, the accent gives the ruse away. Of course, simply breaking in on the enemy frequency is a ruse, and just giving a wordless cry of pain can be discomforting and distracting to the enemy (who now thinks one of his fellow pilots is in trouble).

Displays

When using techniques to make the enemy see what isn't there, you are not trying to hide your presence, you're simply attempting to make it appear other than what it really is. One of the oldest ploys is to have a few horses drag branches behind them to create a lot of dust so that a distant enemy thinks it is a larger cavalry force than you actually have. Variations on this can still be used today, but not as frequently, because of binoculars and aircraft. Lighting many campfires where there are no troops is also an ancient display technique, and one that still has uses today in guerrilla warfare. More modern examples of displays are: fake artillery (painted logs still work); dummy aircraft and armored vehicles (often inflatable); and a lot of radio traffic (to represent units that don't exist, or units that do exist, being stronger than they actually are).

If effectively used, displays can give you more military power than you actually have. If your display is convincing, the enemy will assign some of his forces to keep an eye on your conjured-up troops. This leaves your opponent with a smaller force for your real units to deal with. In this century, with modern technology making it easier to create effective displays, such practices have made a difference in many battles.

A variant of these techniques is to use such displays to understate one's strength, lighting fewer campfires, doubling troops up in tents, and so forth, in order to entice the enemy to attack where he thinks you are weak, when you are actually strong, or convince him that you are unlikely to attack from a particular quarter.

Demonstrations

These are, literally, demonstrations of your military power, in an attempt to confuse the enemy about exactly what you are going to do with it. Making such moves with your forces attempts to imply imminent action, but does not follow through. In ancient times, this often involved moving your troops back and forth where the enemy could see them. The foe would then have to consider where you were going to attack, and might be encouraged to shift his own forces about. Today, demonstrations often mean sending a naval force near a hostile coast or an armored unit down a road where it is likely to be spotted by enemy recon aircraft or satellites. Anytime armed forces move, the military situation is changing and the other side has to rethink its plans. Cleverly orchestrated demonstrations can drive the enemy to distraction, and often, to make a fatal mistake. These techniques are also used in peacetime. In the year prior to the 1973 Arab-Israeli war, the Egyptians moved their troops to the Israeli border several times. During the last move, they actually used the troops for an attack. Similarly, North Korea has made threatening moves with large numbers of troops in the past, each event requiring South Korean and American troops to respond.

Demonstrations can also be a waste of time, and can turn into a disaster for the side using them. Demonstrations require that you move your forces about, which wears them down and costs you fuel and other supplies. A demonstration also, by definition, lets the enemy know where some of your troops are. This provides the other guy with an opportunity to take a shot at you. Thus demonstrations are not for the faint of heart, as they entail some risk. Demonstrations always provide an opportunity for the enemy, and there's always the risk that the enemy might make use of your demonstrations rather than being deluded by them.

Feints

These are similar to a demonstration, but instead of just showing off your forces, you actually make an attack—or, rather, a "feint." Such an attack is meant to make the enemy believe that this feint is the main attack when in actuality, it is done to distract the enemy while your real main assault occurs elsewhere. Since modern attacks go on for days and involve several waves of troops, a feint would normally last only a day and involve far fewer troops than the main effort. But for that day, and perhaps a little longer, the enemy could be deceived into believing that the attack at hand was the major assault expected. If this deception was successful, the enemy would reorganize his forces to face your feint. This would leave the area where the real main attack was coming rather less well protected. Thus your big attack would have a better chance of succeeding because of a well-delivered feint.

At times, it has been possible to reverse the procedure and deceive the enemy into thinking your main attack is just a feint. This forces the enemy to hold back many of his troops in order to deal with the expected main effort. The most successful example of this ploy was during the Allied invasion of German-occupied France in June of 1944. The Germans were deceived into thinking that the Normandy invasion was simply a feint for the main invasion that was to come later, and farther north. The Germans fell for this, and it was many weeks before German reserve divisions were allowed to enter the Normandy battle. By that time, the Allies were securely ashore and the outcome of the Battle for France was clearly in favor of the Allies.

Lies

The commonest form of deception (in war and peace) is flat-out lying. In wartime, this is usually done to good effect when communicating with the enemy (as in "Joachim Murat Tells a Little White Lie," page 108). This is something that has been going on as long as there has been war. When there is no actual fighting going on, it is common for officers from both sides to "parley." These talks often include a demand for the other side to surrender. These conversations are often full of imaginative fibs, and the side that can lie most effectively under these circumstances often gains an advantage. This was especially true before this century, when "total war" was an alien concept. The side that felt it was about to be trashed in a battle could often negotiate for "terms." While this often meant admitting defeat, the "defeated" soldiers were often allowed to march away with their lives and weapons. The extent to which the lesser army got away intact depended on how good its negotiators were. If the disadvantaged army could convince the other side that their unfavorable situation was only temporary, and that reinforcements were at hand, a more favorable deal regarding the current situation could be made.

In this century, opposing armed forces always have easy recourse to radioing messages to their opponent. This can even be done during a battle. This cuts several other ways, however. The widespread use of radios has made it difficult for commanders to deal with the enemy without first getting permission from their own headquarters. However, there are still those situations where one force is surrounded and has little recourse but to try to lie its way out of the mess. Moreover, the diplomats tend to stay in touch even as their armed countrymen are slaughtering each other on the battlefield. So there's still opportunity for lying your way to victory, even if this job has now been taken from the soldiers and turned over to professional liars.

Insight

Deception is largely a mental game. While camouflage, for example, involves a lot of hard work, its effect is dependent on the enemy believing he is not seeing what is there. Between opposing generals, and to a lesser extent, opposing troops, it's a battle of wits. If one general understands the other better, that provides a huge advantage for some effective deception. The classic example is one general knowing what deceptions his opponent is prone to fall for. This is all sort of like a chess game, where this form of deception is widely practiced.

While it's easier to describe a lot of more tangible deception techniques, all of them depend on how the commanders involved use psychology. The most common cause of a deception failing is that the putative victim simply doesn't fall for it. The examples of deceptions, ruses, and the like in this book describe many instances of tricks that worked, or didn't work, because of the mental abilities of one or both commanders.

Learning from Nature

As noted earlier, camouflage has a natural origin, and it is well to recall that many common deception techniques can be found in use among animals. Concealment is a common animal technique, often accomplished by simply finding a place to hide. False information is used by many predators that leave a tantalizing bit of food in plain sight and wait for some prey to come by and fall for the bait. Ruses are also quite common, with some animals having coloration that indicates another creature that predators would likely avoid. Displays, demonstrations, and feints are all used by one animal species or another as hunting or survival techniques. Lies, of course, are uniquely human, as no other animal has the power of speech.

Animals differ from humans in that their use of deception is programmed in. Their parents might teach them some techniques, but all, or most, are known at birth. Animals don't come up with many new ways to use deception, at least not as quickly as humans.

Humans learned many of their earliest deception techniques from animals, and it is still possible to learn new ones. As mankind increasingly becomes an urban animal, we lose touch with nature and all that it has to teach. Animals are still studied by scientists to learn their secrets, particularly those animals that appear to have built-in means to detect deceptions. Some animals have more acute senses of sight and hearing than humans and it is often possible to duplicate these "antideception" capabilities in electronic sensors. So don't be surprised if you hear of some new deception, or antideception, device developed from existing animal capabilities.

Active Versus Passive Deception

The above deception techniques also fall into two general classes: active and passive. The active techniques are those that require you to have your troops moving around. The safest techniques are the passive ones of camouflage and concealment. These methods depend on trained and disciplined troops for their success, but they have the advantage of being difficult for the enemy to use against you. For this reason, camouflage and concealment are the most common deception techniques. Any commander can use them with minimum risk of his efforts backfiring.

Active techniques (false information, ruses, displays, demonstrations, feints, and lies) are dangerous to use. Warfare is always more dangerous for the army that moves around a lot. For this reason, most commanders prefer passive strategy and tactics. Active deception techniques are even more dangerous, because they are less predictable. Being active with your troops is one thing, as they report to you what they are up to. Active deception techniques usually have a delayed effect, as you won't know until later if the enemy was fooled. Worse yet, if active deceptions fail, the foe may entice you into a deadly trap, using your deceptions against you. Thus military leaders are reluctant to use active deception, and tend to stick with the safer passive methods. It takes a skilled, confident, and experienced commander to use active deception effectively. Typically, there are few experienced military commanders, as armed forces spend far more time at peace than at war. Once a war has been under way and the competent commanders have had a chance to prove themselves, you see more active deception. It's still risky, as the other side also has had time to develop competent commanders. But at least there's a lot less uncertainty borne of inexperience and ignorance. During the two world wars (1914–1918 and 1939–1945), there was a lot more active deception later in the fighting, and quite a lot of it was successful.

Orchestrating Deception

Taken by themselves, none of the deception techniques described above are usually decisive. For almost guaranteed results, you use two or more of these techniques in combination. Perhaps even more important is the intelligence and skill of the commander using deception. Any efforts to deceive the enemy can backfire if the other fellow is cleverer than you are. It's a mental game more than anything else. It takes a fair bit of mental acuity to successfully orchestrate two or more deception techniques at the same time. If your opponent is equally adept or even sharper than you are, he can turn your own deception against you.

It's important to remember that the most effective deceptions involve the enemy fooling himself. He has to learn what you want him to learn, and it has to conform to his expectations or his hopes or fears as to what is actually going on. At times, a deception has failed merely because the enemy never learned about it. For example, the Germans laid on an elaborate deception plan in anticipation of their invasion of Russia in 1941. Unfortunately, it didn't fool the Russians, who were too busy fooling themselves that Hitler would never attack. So the fact that the German attack was enormously successful had nothing to do with their deceptive measures. Which reminds us that this book is about deception, not self-deception, and will confine itself to tricks played by one side against another, and ignore instances in which armies and commanders fooled themselves into disaster, without any help from the enemy.

Most successful deception requires that several tasks be performed nearly simultaneously and, for the deception to work, all the tasks must come off successfully. This alone scares off most who would consider using deception. Commanders who become adept at deception usually have a knack for it, and generally start off with simple subterfuges and, if successful, work their way up to more ambitious undertakings. The problem with becoming adept at deception is that it does require practice as well as talent. Commanders don't practice making war as much as they should (it's expensive, for one thing), and this leaves them to learn under fire.

Yet, every war produces commanders who are masterful deceivers. Such leaders are not only more successful, but generally achieve their victories at less cost to their own troops. Deception is more than battlefield cleverness; it's also a lifesaver.

The Wisdom of the Ancients

Deception and Trickery in Warfare from Earliest Times to 1648

The earliest written references to deceptive measures in warfare more or less correspond to the earliest written references to warfare itself. The available evidence suggests that by the time people got around to writing about deceptions and such, subterfuges were already old hat. The Roman military engineer Sextus Julius Frontinus, who lived during the first century (already a couple of thousand years into the history of organized warfare), devoted a quarter of his book *Strategematicon* (stratagems) to deceptions, feints, and other ruses of war. And Frontinus was by no means the first writer to deal with deception in warfare. For example, the Bible, which is considerably older, mentions military deceptions in many places, and there is plentiful archaeological evidence that military deception is at least as old as military operations themselves. Many written accounts of military deception in ancient, medieval, and Renaissance times have come down to us, and some of the more informative ones are included here.

Muwatallis Puts One Over on Ramses the Great

About the oldest documentary treatment of a military campaign is that of Kadesh, way back about 1288 B.C. It also includes one of the earliest descriptions of the very effective use of military deception, namely false information and concealment. The campaign began when Egypt's young, ambitious pharaoh Ramses II decided he wanted to be known as "the Great." To accomplish this aim, he embarked upon a series of military campaigns in Asia for the purpose of capturing southern Syria from the Hittites, a powerful nation centered in the heart of what is now Turkey. By the fifth year of his long reign (roughly 1292–1225 B.C.), Ramses had secured control of most of the coast of what is now Israel and southern Lebanon. The next step was to cross over the Lebanon mountains and take Kadesh, a major city on the Orontes River. In 1288 B.C., he undertook an expedition against Kadesh with an army of about twenty thousand men. Ramses' army moved rapidly and by the end of May, were within a day's march (about a dozen miles) of the city, camped at a place called Shabtuna, where there was a ford across the Orontes River.

At Shabtuna, two Arabs were brought in. The men claimed that they were messengers from their sheik, who was an unwilling ally of the Hittites. In their words (as passed on to us on the walls of one of Ramses' many monuments): "We will be subjects of Pharaoh, May He Live Forever! and we will desert the vanquished King of the Hittites . . . who now sits in the land of Aleppo." This came as very good news to Ramses, who had hitherto believed that the Hittites were around Hama, about ninety miles to the north, for Aleppo lies much farther northward. So the Egyptian army made camp for the night.

It was not unusual for representatives of local leaders to seek an audience with the head of an invading army. The locals suffered a good deal from looting and general mistreatment by troops passing through, so they often tried to limit the damage by cutting a deal with the invaders' head man. It was thought prudent to bring gifts to this meeting, and in this case, the two Arabs brought offers of subservience and information about the enemy.

The next morning, Ramses took his bodyguard and went ahead on his own, with the army's four divisions struggling behind. About midmorning, Ramses and his escort were a mile or so to the west of Kadesh. As the army was by now strung out over some dozen miles of road, with one, and possibly two, divisions still on the wrong side of the Orontes, Ramses decided to make camp. About midafternoon, his first division came up and its troops also began seeking places to set up their cooking fires and bed down for the night. The second division was by then plodding along the road, while the third and fourth were still disentangling themselves from the ford and some nearby woods. At about this time, some of Ramses' scouts came in with two prisoners, Hittites whom they had captured in the act of observing the Egyptian camp. This was disquieting, as scouts didn't usually operate that far from their army, and the Hittites were supposed to be over a week's march away. Upon being "vigorously" interrogated, the two revealed some very bad news. "The king of the Hittites, and his many allies, is stationed . . . drawn up for battle behind Kadesh." The Arabs had lied, and had only served to confirm the faulty information the Egyptians already had. It was a trap, and the captured scouts were no doubt part of the Hittites' last-minute attack preparations. Ramses suddenly began issuing orders, but it was too late.

Even as Ramses learned of their presence, the Hittites struck, charging across the Orontes and right into the flank of his second division, which was strung out along a road. The onset of the Hittite assault shattered the second division and overran the Egyptian camp.

The Hittite success had been the result of a clever trick. The two Arabs who had so cheerfully offered the information that the Hittites were near Aleppo were plants. Muwatallis, the king of the Hittites (reigned 1306–1282 B.C.), had sent them for the deliberate purpose of deceiving Ramses as to his whereabouts. He guessed correctly that Ramses had such a high opinion of himself that he would not energetically double-check the false information the Arabs brought him.

But that was not the only deception the Hittites used. As the Egyptian army neared Kadesh, Muwatallis had kept the city between his army and Ramses', thus engaging in a useful bit of concealment. The result was a tactical disaster for the Egyptians, and one which could easily have proven fatal. However, Ramses, at least by his own testimony, managed to rally his troops and, with elements from the rapidly arriving third division, was able to turn defeat into a marginal victory.

So Ramses won the battle. But it was a near thing, and it was almost decided by Muwatallis' little tricks. It's worth remembering that no amount of clever scheming can overcome success in battle. The Hittites, in any event, risked little with their deception. If the Egyptians did not fall for the false information, the Hittites still had their "hide behind the city" trick. Moreover, the fact that the Egyptians caught two Hittite scouts indicates that the enemy forces were more energetic in their scouting than the Egyptians. If the Hittites had not achieved surprise, they could have withdrawn and tried something else. As it was, their deception almost gave them victory and was a sobering experience for the Egyptians, after barely avoiding defeat.

Egyptians as Tricksters

What is particularly surprising about the success of Muwatallis' little deception in 1288 B.C. (discussed above) is that the Egyptians had been in the trickery business for a long time. During the Campaign of Megiddo (Armageddon) in 1469 B.C., the first recorded battle in history, King Thutmose III (reigned 1504–1450 B.C.), who also liked to call himself "The Great," had the option of taking three routes over the Lebanon mountains into Syria, and deliberately chose the most difficult and dangerous, a narrow pass very suitable for an enemy ambush. Prompted by Thutmose's insight into the enemy's mind ("Those mummy lovers are too smart to take that route!"), the Egyptians apparently made use of a feint. Enough troops seem to have been left around for the enemy scouts to see, while most of the Egyptians hastened off along the route the enemy was sure would not be used. Since his enemies assumed he would not take such an insane risk, they failed to guard the pass, which allowed Thutmose to get into their rear and inflict a severe defeat.

Other forms of trickery were commonly used by the Egyptians. For example, during an earlier campaign up the coast from Egypt, a Captain Thut in Thutmose's service developed a scheme to take the city of Jaffa (also known as Yafo in modern-day Israel) without resorting to a siege. Taking two hundred select men, Thut embarked by ship for Jaffa while the royal army marched up the coast. When the ship neared Jaffa, some of the men were sewn into great sacks normally used for grain. Upon docking, the "crew" of the ship began unloading its cargo while the "skipper," Thut, took care of port duties and such. That night, Thutmose's army came up and stormed the city, which fell when the defenders were taken in the rear by Thut's commandos.

The difference between Thutmose's successful use of deception and Ramses' failure to see through a relatively transparent Hittite trick must certainly have lain in the personalities of the two men. At Kadesh, Ramses believed what he wanted to hear. And he certainly saw only what he wanted to see, for he completely failed to notice that the site selected for his camp had been used within the previous few days by a considerable army, and one with numerous horses!

Captain Thut's ruse, concealing troops in cargo, has been used with considerable frequency in history, and even in literature (in *Ali Baba and the Forty Thieves*, the thieves are hidden in great jars of grain; Tom Clancy has Soviet commandos taking Iceland with a similar trick in one of his novels). The most recent notable use of the trick occurred in April 1940, when, in anticipation of their invasion of Norway, the Germans thoughtfully "prepositioned" several battalions of troops in Norwegian harbors by concealing them in the holds of merchant ships that ordinarily called at those ports.

Joshua Takes Ai

Moses led the Hebrew Exodus from Egypt not long after the age of Ramses II. He, of course, never made it to the "Promised Land," and the leadership of the Hebrews fell to Joshua, his principal aide. Joshua proved an able commander, undoubtedly aided by the fact that the Hebrews, having spent forty years in the desert, were a pretty hard lot compared with the sedentary inhabitants of "The Land of Milk and Honey," not to mention the occasional bit of divine intercession the Chosen People received. But the Hebrew conquest was certainly not all hard fighting and heavenly assistance. There was a lot of good thought behind their victories as well, as can be seen from time to time in the Bible. One of the most interesting and clever operations was the capture of the city of Ai, shortly after the fall of Jericho, as recounted in Joshua, Chapters 7 and 8 (if your authors must steal, we shall steal from the best).

After the fall of Jericho (in what would today be considered about 1200 B.C.), Joshua sent a reconnaissance party to look over nearby Ai. Upon their return, the scouts reported that Ai was a meager place: "Let not all the people go up; but let about two or three thousand men go up and smite Ai; . . . for they are but few." Taking this advice, Joshua dispatched three thousand men. Perhaps from overconfidence, these men were roughly handled by the men of Ai, who chased them away after a brief skirmish in which thirty-six of the Hebrews were killed. It was the first reverse the Children of Israel had in their conquest of Canaan, and a humiliating one at that. Obviously, something had to be done, if only to restore their morale and military reputation. Consulting with the Lord, Joshua came up with a plan.

In secrecy, a column of select troops—the Old Testament says they numbered about five thousand, a highly plausible figure—was dispatched to a position behind Ai, to the west, while the main body of the army, some twenty-five thousand men, advanced boldly against the city from the north, offering battle before the main gate. "And it came to pass, when the King of Ai saw it, that . . . the men of the city went out against Israel to battle . . . and Joshua and all Israel made as if they were beaten before them, and fled by the way of the wilderness." That is to say, after a brief clash, the Israelites, who, it will be recalled, had broken easily in their first attack against the city, conducted a feigned retreat, fleeing the field toward some rough terrain to the east. Heartened, the king of Ai ordered a pursuit, committing all his troops, so that there was not an able-bodied man left in the vicinity of the city.

As the men of Ai pursued the Israelite main body, the Israelites in ambush behind the city made a sudden descent on the place, stormed through the gates, took it, and put it to the torch. At this moment, Joshua counterattacked, pinning the army of Ai to his front, even as its ranks were becoming disordered by word that the city was in flames. Their ranks grew even more disordered when, its arson done, the Israelite ambush column sortied from the city into their rear, so that "they were in the midst of Israel, some on this side and some on that side." The rest was a slaughter, fully twelve thousand of the people of Ai perishing, either in the battle or in the subsequent massacre, so that the city was destroyed completely.

About That Trojan Horse

The Trojan War is one of the earliest recorded events in European civilization, on the cusp, so to speak, of myth, legend, and history. The extent to which the surviving documentation of the war is historically accurate—embodied in the Homeric epics the *Iliad* and the *Odyssey*, plus a scattering of references in other ancient works, supplemented by archaeological evidence—remains a matter of scholarly debate. And your authors will not bore you with the gory details of scholarly battles over whether Homer really existed, or was maybe a woman, or maybe was several people, or maybe even someone else named "Homer," rather than the traditional one—battles in which much ink and even a little blood have been shed. What is certain is that the city of Troy did exist, and that it was destroyed by war, albeit about a century earlier than the period handed down by tradition; roughly the mid-thirteenth century B.C., rather than the twelfth.

The story of the war is relatively familiar. As a result of the kidnapping, or elopement, of Helen, queen of Sparta, by Prince Paris, a younger son of the king of Troy, the Greeks undertook a war against the Trojans. The suggestion that the war was merely a struggle for control of trade routes is reasonable; but don't dismiss the notion that Helen played a major role, especially if, as is hinted several times in the epics, the Spartan throne was matrilineal. The war dragged on for ten years, with many heroic deeds on both sides, until Troy finally fell to a ruse, the "Trojan Horse."

The story of the Trojan Horse is found scattered through the *Odyssey*, but may be summarized simply.

After ten years of war, the Greeks, with a little divine inspiration, constructed a huge wooden horse, in which they concealed fifty of their bravest warriors. Then, setting fire to their camp (a common practice when abandoning a siege), the rest of the Greeks set sail for the nearby island of Tenedos. Seeing the fire, the Trojans emerged cautiously from their city, to discover that the Greeks had gone, leaving behind this huge wooden horse. They also found one Greek, Sinon, who had "accidentally" been left behind. Upon questioning him, he told them that, despairing of ever taking the city, the Greeks had decided to abandon the siege, and had left the horse as a propitiatory sacrifice to the gods of the place. A debate ensued among the Trojans as to what to do with the horse. Some were for throwing it from a nearby cliff, others for burning it, and still others, for hauling it into the city and consecrating it to Athena.

In the midst of the debate, two vast serpents appeared from the sea and destroyed the priest Laocoön and his sons, who had argued that the horse was a ruse. This prodigy convinced virtually all of the Trojans that Sinon's tale was true—except Aeneas, who saw that the gods were meddling in the business—and he quietly removed his people to nearby Mount Ida. The rest of the Trojans, thinking their toils and dangers were over, broke down a part of their walls and dragged the horse into the city, where they indulged in great festivities, spreading banquets, singing, and drinking through the night, with no thought to their security. Before dawn, when all the Trojans were asleep or drunk, Sinon, the "lost" Greek, raised a torch from the walls, as a signal to the other Greeks, who were at Tenedos. They returned, while the warriors descended from the horse—and the city was taken.

Now, a modern rationalistic explanation of the tale suggests that the story of the Trojan Horse refers to the use of some sort of siege engine in the war; possibly a battering ram sheltered by a sturdy shed or a belfry, a tower which was shoved against the wall and from which the Greeks emerged to storm the walls by means of assault ramps. This is not an unreasonable interpretation. After all, Homer was working from oral tradition at least three, and possibly four, centuries after the war, more than enough time for the memory of the events to grow dim indeed. Even some ancient Greek commentators suggested this interpretation. However, that's not what is important about the tale.

The acceptability of the story of the Trojan Horse demonstrates that deceptions were a commonplace of warfare in Europe at the very dawn of history. The reaction of both Laocoön, who obviously smelled a rat, and Aeneas, who saw in the fate of Laocoön that the game was fixed, certainly emphasize this. And it can also be seen from other bits and pieces drawn from the tale of Troy.

For example, on one occasion during the war, the Trojan hero Dolon wore a wolfskin by way of camouflage while on a reconnaissance mission; and on another occasion as a ploy to get inside Troy itself, the great Odysseus had himself flogged and dressed in rags, so as to appear to be a deserter. And still another ruse mentioned in the *Iliad* has had a long and honorable career: the story of Patroclus and Achilles' armor.

The pivotal plot device in the *Iliad* is that Achilles, the greatest of the Greek heroes, had a falling out with Agamemnon, chief commander of the expedition. In consequence, Achilles refused to fight again until the latter apologized. This gave the Trojans the opportunity to launch a devastatingly successful attack on the Greeks, who were driven back to their fortified camp on the coast, where the Trojans even managed to set fire to some of the Greek ships. In this moment of desperation, Patroclus, Achilles' greatest friend, went to the hero and begged him the loan of his armor and permission to lead his people, the Myrmidons, into battle. Achilles assented. The sudden attack of Patroclus at the head of the Myrmidons totally discomfited the Trojans, who knew Achilles was unbeatable. Soon the Trojans were driven back to the very gates of their city. There Hector, eldest son of the king of Troy, and the city's greatest hero, confronted the false Achilles, despite a prophecy that he was fated to be slain by Achilles. Much to his surprise, Hector kills the warrior whom he, and everyone else, believed to be Achilles. With that, the Myrmidons and other Greeks lost heart in turn, and fled again to the safety of their camp, pursued by the now-elated Trojans. In the end, the battle was a draw.

Although this is its first mention in literature, the wearing of another's armor in battle is by no means an unusual event. It occurs several times in European history (e.g., Henry V at Agincourt), and also in that of other civilizations as well, so this aspect of the story of Troy is certainly very reasonable. As are the consequences when the impersonator is slain.

So while the tale of the Trojan Horse may be based on a misinterpretation of a very mundane event, it certainly suggests the extent to which deception was practiced even in relatively primitive cultures.

Gideon Smites the Midianites

In about 1100 B.C., a people called the Midianites began troubling the northern Israelite tribes. The Midianites were apparently a desert tribe seeking to migrate into Canaan. Their raids and incursions becoming too much for the local residents to handle. Gideon, a proven young warrior, called upon the northern Israelite tribes for volunteers to fight the interlopers. The response was heartening, a thoroughly unexpected thirty-two thousand men coming forward. As these were too numerous, Gideon set up a series of physical and moral tests and reduced this host to ten thousand, the best and bravest of the Israelites, sending the rest home. From this army, Gideon selected an elite band of three hundred. It was with these that Gideon planned to spring an elaborate trap on the Midianites, whose entire tribe was encamped in the Valley of Jezreel.

Gideon prepared his attack carefully. He deployed the bulk of his army in ambush, covering the Jordan River fords at the eastern end of the Valley of Jezreel. Meanwhile, Gideon divided the select three hundred into three companies, equipped them with swords, trumpets, and crocks in which they carried burning combustibles, and led them on a night raid against the Midianite camp.

Approaching the unsuspecting enemy, Gideon's three companies appear to have each "attacked" a different face of the Midianite camp. Loudly blowing their trumpets, they hurled their incendiaries into the camp. This caused great confusion among the Midianites, who thought themselves beset by a great host. So desperate were some to escape, that they turned their weapons against their fellows. Since the camp was not under attack from the eastern side, the fugitives fled in that direction. And as the Midianites streamed away from their camp, they ran into the balance of Gideon's army blocking the Jordan fords. The result was a great slaughter, with only remnants of the Midianites managing to escape east of the Jordan, to be pursued and destroyed over the next few days.

Gideon's "commando" raid was a classic example of a special operation. It was well organized, Gideon making certain that he had selected his men with care and provided them with special equipment. And it was of more than tactical importance, for he also gave some thought to turning what could easily have been a mere harassing raid into a decisive victory.

The Battle of Gibeah: Turning Stupidity into an Advantage

Some time after Gideon smote the Midianites, a Levite (one of the priestly clan of the Israelites) was subject to a violent insult in Gibeah by some members of the tribe of Benjamin. When the Benjaminites would offer no redress, the other eleven of the tribes of Israel mustered their armies to exact vengeance. The result, as recounted somewhat confusingly in Judges, Chapters 19 and 20, was a hard and bloody battle at Gibeah, which was not far from Jerusalem. The Benjaminites numbered 26,700 men (including 700 from Gibeah), a figure by no means improbable. It is interesting to note that Scripture accords the Benjaminites 700 slingers, which suggests that this was an unusual weapon, adding that all of them were left-handed, an interesting notion indeed (now, southpaw swordsmen might be an interesting trick in war, but left-handed slingers are no more dangerous than right-handed ones). The other Israelites are said to have fielded 400,000 men, which is surely excessive. Considering that modern commentary suggests that only the tribe of Ephraim may actually have taken the field against Benjamin, perhaps with token contingents from the other tribes, the figure may have been a tenth of that given. Of course, even 40,000 men was quite a respectable host, so the odds were certainly in favor of the attackers. However, sheer numbers are not always a guarantee of success in war.

On the first day of the campaign, the Israelite army advanced boldly against the city. The Benjaminites offered battle before the city. They appear to have met the Israelite assault with a coordinated counterattack, using slingers and swordsmen in an unexpected fashion, so that they drove off their attackers, who apparently were relying heavily on their superior numbers. As a result, the Israelites were defeated, suffering a reported twenty-two thousand casualties, which we may reasonably reduce to perhaps a tenth that number. After suitable prayers and devotions, the Israelites renewed their attack on the second day. Once again, the Benjaminites defeated them, inflicting another eighteen thousand casualties (eighteen hundred?). At this point, Phineas, the high priest and guardian of the Ark of the Covenant, appears to have taken command, after consultation with the Lord.

So on the third day of the campaign, Phineas ordered ten thousand of the Israelite force to lie in ambush not far from Gibeah, and possibly another contingent of equal or greater size to lie in wait in a "wilderness" at some distance to the rear. He then boldly advanced directly against the city with the main body of the army. Having by now twice defeated their kindred, the Benjaminites once again sortied from Gibeah. When the Benjaminite army attacked his own, Phineas fell back. Greatly heartened by this apparent confirmation of their military prowess, the Benjaminites pressed the attack, and Phineas' army fell back farther, in a classic feigned retreat. As the two armies drew away from the city, Phineas sprang his trap. While the troops concealed in the "wilderness" leaped into action against the Benjaminites' flanks, the ten thousand men who had been lying in wait near Gibeah stormed the city, which turned out to be undefended. Rapidly capturing the place, the troops put it to the torch.

Suddenly, the Benjaminites, who thought they were driving their estranged kinfolk before them, found themselves beset on their flanks, and with their city engulfed in flames to their rear. As they attempted to break off the action to go to the rescue of the town, Phineas ordered the main body to counterattack, catching the Benjaminites in disorder. Then the Israelite troops who had taken Gibeah sortied into the Benjaminite rear. The result was a great slaughter of the Benjaminites, virtually all of them perishing, so that only six hundred survived, a figure by no means unreasonable, given the nature of their defeat.

The campaign of Gibeah demonstrates the use of ambushes and deception on both sides. It also suggests that when arrayed for war, the Israelite tribes lacked an overall command. Had not Phineas usurped that role, the Benjaminites might well have gone on winning. But once in command, Phineas deliberately repeated the pattern of the two failed assaults on Gibeah, in order to lull the Benjaminites into a false sense of superiority, while laying an elaborate trap for them.

Samaria: the Ruse That Wasn't

Deception and trickery in warfare were extraordinarily common in ancient times. And thereby hangs a tale.

During the reign of King Jehoram of Israel (c. 852–841 B.C.), the Syrians invaded the land. At the head of a large army, King Benhadad of Damascus (c. 860–840 B.C.) marched right for Samaria, the capital of Israel, and began a siege. This set the stage for a clash between Jehoram, impious and sinful, and the prophet Elisha, with a resulting lesson in faith for all. However, for our purposes, the tale, as told in II Kings, Chapter 7, has another lesson, one which reveals much about the military practice of the day.

Benhadad's siege of Samaria was long and arduous, and the inhabitants suffered much from famine. Then, as a result of divine intervention (itself called forth by Elisha in the course of his clash with Jehoram), the Syrians were made to hear "a noise of chariots, and a noise of horses, even the noise of a great host: and they said one to another, 'Lo, the King of Israel hath hired against us the Kings of the Hittites, and the Kings of the Egyptians, to come upon us.' Wherefore, they rose and fled in the twilight, and left their tents, and their horses, and their asses, even the camp as it was, and fled for their life."

Of this development, the people of Samaria knew nothing. But it happened that there were several lepers who had been sheltering between the walls of the city and the Syrian lines of investment. They were there because throughout history, it has been a common, if cruel, practice for "useless mouths" to be expelled from a besieged city. Often, the besiegers refuse to let the fugitives pass their lines, lest they encourage the besieged to rid themselves of still more residents. Lepers, of course, would have been just about the most "useless" of "useless mouths." In any case, desperate and starving, the lepers reasoned that if they remained where they were, they would soon die of hunger, and since they could not reenter the city, they might as well try their luck with the Syrians, who at worst would give them a swifter death than would famine. So they approached the Syrian siege lines. And, of course, found them empty. After eating and drinking their fill, the lepers returned to Samaria, bearing what they thought was good news.

When King Jehoram was brought word that the Syrians appeared to have fled, he smelled a rat. "I will now show you what the Syrians have done to us. They know that we be hungry; therefore are they gone out of the camp to hide themselves in the field, saying, 'When they come out of the city, we shall catch them alive and get into the city.' " So rather than throw open the gates, the king ordered a reconnaissance. Of course, the Syrians were nowhere to be found, and the siege was lifted.

We have stripped this tale of most of its religious content, which in the Bible was emphasized, because of its considerable military interest. Jehoram's reaction to word that the Syrians have fled contains an important lesson concerning the frequency of ruses, deceptions, and other tricks in warfare. Seen in this light, the king's wariness becomes understandable.

Of course, it is also possible that when the Syrians heard "a noise of chariots, and a noise of horses, even the noise of a great host," the Lord had a little assistance from human instruments. But if that was the case, we leave it to the theologians to elucidate.

Zopyrus' Nose for Victory

Darius I (reigned 521–486 B.C.) was one of the greatest of the kings of Persia. Having gained the throne against several potential rivals through a little harmless trickery involving a stallion and a mare in heat (the curious may consult Chapter III of Herodotus' *The History* for details), he kept it by his skills in the arts of war and statecraft, and through his ability to inspire the utmost devotion from his followers.

In 518 B.C., Darius was still new to the throne, and there was some resistance to his authority. The most notable holdout was Babylon, in Mesopotamia (present-day Iraq), a great and rich city, possibly the greatest in the world at the time, as it was the center of "global" banking and commerce. As the city was blessed with enormous walls, taking it presented a considerable problem. Nevertheless, Darius laid siege to the place. Much blood was shed in efforts to take the city, all to no avail. Then one of Darius' most loyal supporters, one Zopyrus, hit upon a clever scheme. He would desert to the enemy forces, gain their confidence by convincing them that he had become an inveterate foe of Darius, rise to high rank among them, and, at the propitious moment, betray the city to his erstwhile lord. Of course, convincing the good burghers of Babylon of his sincerity would take some doing, but Zopyrus appears to have thought of everything. Including not informing the king of his scheme until Darius could not turn it down.

In great secrecy, Zopyrus arranged to have various interesting and disfiguring mutilations inflicted upon his person, the most notable of which was the amputation of his nose. This done, Zopyrus presented himself to the king and explained his scheme. Considering that Zopyrus had already given a great deal in his service, Darius was pretty much forced to go along. (Note that there is a parallel tradition that says Darius talked [or, in some accounts, coerced] Zopyrus into taking part in the ruse.)

So one morning, who should show up at the gates of Babylon, but the horribly mutilated Zopyrus, dressed in rags and apparently nearly dead from abuse. Brought before the city fathers, Zopyrus explained that Darius had so punished him for failing to capture the city in a timely fashion. Swearing eternal enmity toward his erstwhile sovereign, Zopyrus asked that the Babylonians allow him to take up the sword in their defense. Surprisingly naive for a bunch of bankers, the good burghers swallowed Zopyrus' tale hook, line, and sinker. Zopyrus rose rapidly in their ranks. Given command of some troops, he led apparently successful sorties against the besiegers (who had been ordered by Darius to offer but feeble resistance). Given still more responsibility, he proved even more useful to the defenders. As a result, it didn't take long for Zopyrus *Rhinotmetus* (i.e., "no-nose") to rise so far that, one fine night, he was able to betray the city into Darius' hands.

Zopyrus' ruse, albeit a bit extreme, is a reminder that deceptions can assume the most extraordinary forms.

Epaminondas Invents the Deceptive "Oblique Order"

In ancient Greece, a highly stylized form of warfare provided an opportunity for a clever general to use a simple, effective, and unexpected deception.

By the mid third century B.C., Sparta was the principal military power among the Greek city-states. Having eliminated Athens as a competitor for primacy in the latter portion of the previous century, the Spartans were now concerned with the growing power of Thebes, the principal city of the region of Boeotia. In July of 371 B.C., an army of twelve thousand Spartans and their various subject allies from Lacedaemon marched into Boeotia, with the intention of teaching the upstart Thebans a lesson. The army soon reached the small city of Leuctra, a few miles southwest of Thebes. There, across the narrow (1,500 meter) valley which stretched northward from the heights occupied by Leuctra, the Spartan king Kleombrotos found a smaller Theban army, about eighty-five hundred men under Epaminondas, the principal Boeotian general, prepared for battle.

Kleombrotos deployed his 10,000 hoplites (heavy infantry) in two divisions, each of about 12 men deep (about 12 meters) with a front of about 320–350 meters, on an overall frontage of about 700–750 meters. On the extreme right of his line, Kleombrotos took up station with 700 Spartans, the best troops in the Greek world. On each flank he posted some 500 light infantry and sent about 1,000 light cavalry into the valley to his front.

The Theban general Epaminondas, aware of the numerical odds against him, posted his forces in a radically different fashion. He divided his hoplites into three divisions, deployed on a front of about 1,000–1,500 meters. However, although each division occupied some 200–220 meters of front, their composition was by no means uniform. The center and right divisions numbered perhaps 1,500 men each, deployed to a depth of about 12 men, while his left-hand division totaled about 3,000 men with a depth of 50 men. In addition, Epaminondas "echeloned" his front somewhat from the left, so that his extreme left was perhaps 200 meters farther forward than his extreme right, where he himself took his station. As had Kleombrotos, Epaminondas divided his 1,500 light infantry between his flanks, and put his 1,000 light cavalry, which was both more numerous and technically superior to that of his foes, into the valley.

This may not seem to be a very clever deployment, but it was. The Greek city-states practiced war according to some rather sturdy customs. One of these was that the "best and bravest" troops deployed on the right of the line. There was a very practical reason for this. Nearly all soldiers (then and now) were right-handed. The Greek heavy infantryman (the hoplite) used his left arm to hold his heavy shield and his right to hold his spear or sword. This meant that the men on the right of the line were exposed on their right hand, so it seemed reasonable to place the bravest men on that flank. Moreover, in action, the troops tended to move to their left, pushing with their shield and hacking with their weapon. This also meant that when two armies faced each other, the best troops on each side confronted the worst (usually) on the other side. The battles came down to whose "worst" (left side of the line) troops could hang on the longest, or whose "best" (the right side) could cut through the enemy line. Battles were decided when portions of one side broke and ran. The battles were essentially gigantic shoving matches, as the heavily armed hoplites crashed into each other, whereupon men in the front ranks hacked and poked at the enemy with sword and spear while men in the rear ranks literally pushed the entire mass forward. It was a messy endurance contest. The "worst" troops were the more likely to break, so it was up to the "best" troops on the right of each army to break through first. Whoever broke through first would win.

The Theban general Epaminondas turned this conventional wisdom on its head. He put his best troops against the enemy's best (on the Theban left side). But more important, he more or less doubled the strength and quadrupled the depth of his right to boot.

The battle opened with a cavalry skirmish, in which Epaminondas' Thebean horse handily routed Kleombrotos', driving the Spartans against the Lacedaemonian left wing, impeding its advance. So as the main bodies of the two armies began to move, the Lacedaemonians' right quickly outdistanced their left. The battle came down to a clash between the Lacedaemonian right and the Theban left. And the overwhelmingly superior numbers on the Theban left (spearheaded by the so-called Sacred Band under Pelopides) crushed Kleombrotos' right. Some four hundred Spartans, including Kleombrotos himself, and six hundred other Lacedaemonians fell, causing the line to break, and the Thebans went on to win the battle. Thus a clever deception, and the exploitation of traditional battle tactics, allowed the Thebans to crush (literally) the normally unbeatable Spartans.

Leuctra marked the end of Spartan primacy in Greece. But Thebes proved unable to maintain her own supremacy for very long. Soon the Greeks were squabbling among themselves again, until they all lost their independence to one who, as a young lad, had been a hostage in Thebes and an ardent student of the military reforms of Epaminondas. This diligent young fellow was Philip II of Macedon, father of Alexander the Great, who learned several lessons, not least of which was to hit the enemy where he least expected it.

Since he invented it, Epaminondas' "oblique order" has been used several times down through the ages. Perhaps the most famous practitioner of this deceptive tactical deployment was Frederick the Great of Prussia (reigned 1740–1786). His victory at the Battle of Leuthen (December 6, 1757) was a masterful demonstration of the use of oblique order, nearly two thousand years after it was first noted in written records.

Frederick, with about thirty-three thousand men, was confronted by an Austrian army of some sixty-five thousand, well deployed behind a five-mile-long ridge, with their reserves behind their left wing in anticipation of a flank attack, since their right rested on a swamp. Frederick approached this position in four columns, the outer two of cavalry and the inner pair of infantry. Taking advantage of the cover that the ridgeline offered, when within some hundreds of yards of the Austrian front, Frederick suddenly marched his army obliquely to the right, his left-hand column of cavalry becoming the rear of the marching columns, and skirmishing with the enemy right, which drew the Austrian reserves thither. Meanwhile, with his right-flank cavalry column now his advanced guard, Frederick's two infantry columns marched in parallel right across the Austrian front, at about a 45-degree angle to their original line of advance. As soon as his leading battalions began to overlap the Austrian left, Frederick opened up with all of his artillery, which had been posted to the rear, and, turning his infantry to the left, began to fall on the Austrian line, enveloping and crushing it.

Leuthen was a victory won against great odds, by taking a great risk. The Austrians expected Frederick to repeat his tactics of several previous battles, which he did, but not before tricking them into thinking he was going to hit the opposite flank. In addition, by using the oblique order he brought the bulk of his numerically inferior army to bear against a small portion of the numerically very superior foe. Altogether not a bad day's work, and very much in line with a trick already at least two thousand years old.

Sun Pin Lights a Few Fires Too Few

Sometimes the simplest deception works wonders, especially if you have nothing to lose.

In 341 B.C., the Chinese kingdom of Han was attacked by two of its rivals, Wei and Chao. The king of Han urgently appealed to his colleague the king of Ch'i for aid, and the latter ordered General T'ien Chi to the aid of his ally. T'ien Chi rapidly collected an army (of, by tradition, one hundred thousand men) and was soon on the march westward, to undertake a counter invasion of Wei. Now, it happened that the army of Ch'i had a poor reputation in Wei and Chao, and even in Han. So poor, indeed, that the commander of the Wei, P'ang Chuan, went home on leave for several days, despite the imminent arrival of the Ch'i army.

On the march from Ch'i to Wei, T'ien Chi had the benefit of having as his principal advisor Sun Pin, a descendant of the famous Sun-tzu, author of what is perhaps the earliest treatise on warfare, appropriately Englished as *The Art of War.* Sun Pin suggested that the attitude of their enemies (and indeed, even of their allies) concerning the military prowess of the Ch'i could be used to their advantage. Since an army on the march generally loses strength—to the rigors of the march, to privation, and, of course, to desertion—he suggested that T'ien Chi create the impression that his army was melting away at an extraordinary rate, thereby further strengthening the contemptuous attitude of the enemy, and putting him off guard. Considering this advice, T'ien Chi thought it reasonable and issued orders accordingly.

As a result, each night as T'ien Chi's army made camp, the troops lit fewer and fewer campfires. Tradition has it that on the first night after entering Wei, the army lit a hundred thousand fires; on the second, fifty thousand; and on the third, only thirty thousand (these figures are probably greatly exaggerated, but the general idea remains the same regardless of actual numbers). The Wei commander P'ang Chuan, by then back from his leave, found this development very pleasing, remarking, "I have always believed that the men of Ch'i were cowardly, and here is the proof. Only three days after entering our territory, they have already lost more than half their number through desertion." Taking heart from this, P'ang Chuan decided to strike quickly, leaving behind his heavy infantry and supply train, advancing by forced marches with only his light infantry and cavalry.

Calculating the rate of advance of P'ang Chuan's advanced guard, Sun Pin suggested that T'ien Chi (who obviously knew his own limitations and the strengths of his advisor) lay an ambush for the Wei army at a place called Ma Ling, in what is now northeastern Shantung. At that point, the main road passed through a series of defiles, in terrain eminently suitable for ambushes, which the men of Ch'i promptly proceeded to lay. In order to provide a spring for his trap, Sun Pin left a message for P'ang Chuan, carved on the side of a tree, reading: P'ANG CHUAN DIES UNDER THIS TREE.

As Sun Pin had calculated, P'ang Chuan arrived at the tree in the dark. Seeing that something was carved upon it, he called for torchlight. And as soon as the torches were lit, ten thousand crossbowmen cut loose, followed by a heavy infantry attack. As his army was being cut to pieces about him, P'ang Chuan cut his own throat, dying beneath the tree, as Sun Pin had predicted.

Because of Sun Pin's stratagem, the Ch'i victory at Ma Ling was complete. Not only was a major part of the enemy army destroyed, but the heir to the throne of Wei was captured, leading to a speedy end to hostilities. It was a simple trick, indeed almost a commonplace one, which has worked many times throughout history. Sun Pin merely took advantage of his enemy's preconceptions.

Philip of Macedon Snookers Some Amateurs

Philip II of Macedon (reigned 359–336 B.C.) was one of the most remarkable commanders of the ancient world. A fine field officer, he was also a military innovator of note, creating the vaunted Macedonian Phalanx, an organizational and tactical system that would dominate the Mediterranean world for nearly two centuries. As a strategist of almost global perception, he envisioned nothing less than a union of all the petty states of the Greeks into a great confederacy which would destroy forever the Persian threat. He was also a deceiver of the first rank. His skills at deception and trickery in war were often demonstrated, but perhaps never so effectively as at the Battle of Chaeronea (338 B.C.).

In a series of wars from 355 through 338 B.C., Philip had gradually established Macedonia, a state in northern Greece that was only by courtesy considered Greek, as the champion of Hellenism. But in 338 B.C., Athens and Thebes, formerly among the leading Greek states, rose in revolt to challenge his authority.

Philip could only muster about thirty-two thousand men against the combined forces of the rebels, about fifty thousand men. However, the allied host included many mercenaries, as was by then common in Greek armies, supplemented by an extraordinary mustering of the traditional levy of the citizenry, something which had not been done for many years. In contrast, Philip's army consisted almost entirely of his Macedonian regular troops, one of the first systematically recruited and trained armies in history, in which men served a portion of their time on active duty before passing into a reserve. After a bit of maneuvering, the armies met at Chaeronea, not far from Thebes in central Greece. At Chaeronea, Philip used two tricks to great effect.

Aware that the Theban and Athenian citizens were unused to field service, but avid for action, he delayed making any aggressive moves, taking an interminable time to get his forces into position. And once the action had begun, he deliberately prolonged it. This had the effect of leaving the enemy to stand in the hot sun for hours on end, sapping both their spiritual and physical resources.

At the same time, aware that the enemy had occupied some very favorable ground, he undertook to advance his shoulder-to-shoulder phalanx very slowly. Once contact had been made with the enemy, he fell back, as if under extreme pressure. The overly enthusiastic, if tired, Athenians and Thebans pressed forward, thereby abandoning the favorable ground which they had previously occupied. Philip kept falling back slowly. At this point, perhaps we should let Polyaenus, a Greek military writer of some four centuries later, tell the tale, from his *Strategica,* composed about A.D. 165: "As soon as he had by this maneuver drawn them from their advantageous ground, and gained an eminence, he halted; and encouraged his troops to a vigorous attack, made such an impression on the enemy, as soon determined a brilliant victory in his favor,"

A victory was sealed by a cavalry charge led by Philip's son Alexander.

The Battle of Chaeronea ended in a smashing Macedonian victory, the Theban and Athenian casualties amounting to almost half their strength, some twenty-two thousand men, killed, wounded, or captured. The rebellion against Philip soon fell apart. At that, he began laying plans for his great crusade against Persia, an undertaking that he would not live to see, but which his son, later surnamed "The Great," would carry through.

Alexander the Great Deceiver

Having, as Polyaenus would later write, an "ambition to unite all mankind to him," Alexander III of Macedon (reigned 336–323 B.C.) very nearly accomplished his goal. An effective field commander, Alexander was also a master of ruses and deceptions. Indeed, Polyaenus devotes more space to Alexander than to any other commander, listing nearly three dozen stratagems, including many ruses and deceptions; and Frontinus, who wrote a century earlier, mentions a few that Polyaenus missed. Some examples will give a fair notion of Alexander's abilities in this regard.

- On a campaign against Thebes, which had risen in rebellion upon the death of his father, Alexander sent off a portion of his army under the command of Antipater and ordered it to conceal itself near the city. He then assaulted the strongest sector of the city's defenses. When the battle for the walls was at its hottest, Antipater emerged from concealment and assaulted the city from the opposite quarter. As the walls there had been stripped of men to cope with the threat from Alexander's main body, Antipater's troops rapidly broke through and was shortly in control of the city.

- At the battle of Arbela (October 1, 331 B.C.), finding that the Persians had sown a portion of the ground between the two armies with caltrops (metal devices like pickup jacks, such that whichever way they fall they always present one spike upward), Alexander arranged his deployment so that the affected ground would help cover one of his flanks from attack by the numerically very superior enemy.

- While operating in what had been the eastern frontier regions of the defeated Persian kingdom, in 327 B.C., Alexander found the enemy in occupation of a narrow pass which proved impossible to force. He pulled back his army several miles and made camp. A search among the local peasantry turned up a herdsman who knew a secret path through the mountains into the rear of the pass. Giving orders for the normal number of cookfires to be lit for that night, Alexander took a portion of the army off in secret, negotiated the path, and fell upon the unsuspecting defenders from the rear before dawn the next morning, in coordination with an assault from the front by the balance of the army.

- During his campaign in India (spring 326 B.C.), Alexander arrived at the Hydaspes River, across which lay the territories of King Porus, the most powerful Indian monarch. Unfortunately, King Porus and a considerable army were operating behind the river, making a crossing an extremely risky business. So Alexander marched upstream for some days and made preparations for a crossing there, only to find Porus in occupation of the opposite bank once more. So he marched back to another possible ford, and once more, there was Porus. Alexander and Porus danced this step several times, until many believed that the Macedonian had no intention whatsoever of crossing the Hydaspes. As a result, Porus' reconnaissance became lax. Meanwhile, in great secrecy Alexander had gathered boats, inflated hides, and rafts. And while a portion of the army made motions as if to cross the river in one place, thereby attracting Porus' attention thither, the bulk of the army made a sudden march to where these had been cached, and proceeded to cross the river unopposed.

- Having captured a young woman of extreme beauty, the betrothed of a prominent chief of a hostile tribe in Central Asia, Alexander carefully protected her from abuse and promptly returned her to her beloved, an act of generosity which caused the tribe to lay down its arms.

Alexander undoubtedly had a knack for both tactics and deception. And it certainly ran in the family, for he was a true son of his father, Philip II.

Gaius Claudius Nero Fools the Great Hannibal: 207 B.C.

The ancients realized that against a highly talented commander, deception could even the odds. Thus the Romans ultimately conquered Hannibal, whom they themselves considered one of the greatest commanders of all time.

One of the most critical struggles in the history of European civilization was the long duel between Rome, a rising Latin city-state in central Italy, and Carthage, her far more prosperous Punic (i.e., Semitic Phoenicians from the present-day Lebanon) rival in modern Tunisia. The two cities fought three wars, known as the Punic Wars. In the First Punic War (264–241 B.C.), the Romans, by heroic efforts and despite horrendous losses, secured command of the seas and inflicted a severe defeat upon their maritime rival. But, under the leadership of the Barca family, Carthage recovered, and over the next generation built an enormous empire in Spain. In the Second Punic War (218–201 B.C.), Hannibal Barca brought an army of mercenaries overland from Spain into Italy, and soon proved himself one of the premier captains of history. After three successive massive defeats (the Trebbia, Lake Trasimenus, and Cannae), the Romans decided to ignore him and fight the Carthaginians where they were weak, which was everywhere Hannibal wasn't.

For more than a decade, Hannibal ravaged and plundered in Italy, winning an occasional battle and taking an occasional city, while constantly being harassed by Roman forces. Hannibal's army was unbeatable on the battlefield, but he could not capture and hold enough of the fortified cities in Italy, nor could he be everywhere at once. The Romans just would not quit and Hannibal was not as able a diplomat as he was a general. Gradually, the Romans bottled up Hannibal in the heel of Italy, neutralized (at great expense), but not eliminated.

Meanwhile, the Romans proceeded with the conquest of Spain, thereby cutting Hannibal off from his base. Hannibal needed reinforcements if he was to continue his campaign in Italy. Blockaded in the heel of Italy, Hannibal sent for reinforcements from Spain. This set the stage for the decisive engagement of the war, the Battle of the Metaurus, in 207 B.C., which saw the army attempting to reinforce Hannibal destroyed. The road to the Metaurus began with a brilliant campaign to reinforce Hannibal, led by his brother, Hasdrubal. Eluding the Roman armies in Spain, Hasdrubal marched an army over the Pyrennees, across southern France, over the Alps, through the Po Valley, and down the east coast of Italy.

As Hasdrubal marched south, Hannibal endeavored to move north. But the Roman consul Gaius Claudius Nero (known as Claudius) outmaneuvered him, keeping his troops confined to southeastern Italy. Then, by a stroke of fortune, Claudius intercepted a message from Hasdrubal in which the latter outlined to Hannibal the full details of his proposed line of march. With this information in hand, Claudius decided upon a daring plan. He would march one of his two legions north to reinforce the two legions confronting Hasdrubal, under his coconsul, Marcus Livius Salinator, while the balance of his forces remained behind to keep Hannibal bottled up. But how to execute this bold maneuver without risking defeat at the hands of the resourceful Hannibal? Sextus Julius Frontinus, a Roman public official and soldier who lived during the latter portion of the first century, tells us in his *Strategematicon* what Claudius did: "Desiring that his departure should go unnoticed by Hannibal, whose camp was not far from his own, Claudius . . . gave orders to the commanders whom he left behind that the established pattern of patrols and sentries be maintained, that the number of campfires lighted be the same each night, and that the existing appearance of the camp be maintained, so that Hannibal might not become suspicious and attempt to attack the small number of troops left behind."

Claudius was soon on the march with six thousand infantry and one thousand cavalry, while the rest of his troops did twice their normal duties, thereby misleading Hannibal, who continued to believe that the whole Roman force was still before him. Meanwhile, Claudius and his troops marched some two hundred and forty miles in seven days, one of the more heroic forced marches in history, to join Livius' army along the Metaurus, a small river on the eastern side of Italy. And once again Claudius resorted to deception. As Frontinus observes, "When he joined Livius in Umbria, Claudius forbade any enlargement of the Roman camp, lest this give some indication of the arrival of additional forces."

Thus the Romans used two deceptions: one to keep Hannibal from realizing that half the Roman troops he was facing had marched off; and the second to deceive Hannibal's brother Hasdrubal and the reinforcing army as to the size of the Roman forces they were facing.

Within a day or so of combining their forces, Claudius and Livius offered battle.

Although he had a river at his back, Hasdrubal, an able and experienced commander, deployed his forces rather well. Holding his cavalry in the rear, Hasdrubal placed his most reliable troops, his Spanish infantry, on his southern, or right, flank, with his less reliable Ligurian infantry in the center, and his formidable, but brittle Gallic mercenaries on the left, where the terrain was most favorable to the defense. Across his front he posted his light infantry and about a dozen elephants. The Romans posted some cavalry on their right (northern) flank, facing Hasdrubal's Gauls. To the left of these were Claudius and his legion from Apulia, while the Roman center and left were held by Livius' two legions and their supporting allied contingents, with the balance of the Roman cavalry posted on the extreme left, and their light infantry spread across their front.

The battle began in a series of small clashes between the light forces. The Carthaginian elephants soon panicked, fleeing into a gully, where they were all captured. Meanwhile, the battle quickly became very intense in the center and on the southern end of the lines, where the Roman left became heavily engaged with the Carthaginian right. At the other end of the front, however, Claudius found the terrain extraordinarily difficult, so that it proved impossible for him to come to grips with the Gallic troops to his front. Reasoning that if the ground prevented him from getting at the enemy it would also prevent the enemy from getting at him, Claudius decided to take his troops and reinforce the Roman left. Leaving his cavalry to screen the right, he made a rapid march with his legionaires behind the rear of the entire Roman troops, passing beyond their southern flank before turning and deploying so as to attack against Hasdrubal's exposed left. The resulting attack demoralized the enemy. After a vain attempt at rallying his men, Hasdrubal plunged into the midst of a Roman cohort and died fighting. His army perished with him, perhaps ten thousand dying and the rest being taken or fled. Only two thousand Romans fell.

The Metaurus was a brilliant victory, but the war was not yet won. With only a day or so's rest, Claudius' weary troops were once again on the march, heading south, where Hannibal was still inactive. After another grueling march, Claudius and his army returned to their original camp, near the modern Canosa di Puglia. Hannibal had been so completely fooled that the first news he received of the disaster on the Metaurus came when his brother's head was hurled into his camp. As Frontinus put it, "By the same plan, Claudius stole a march on one of the two sharpest Carthaginian generals and crushed the other."

After the Metaurus, it was only a matter of time. Without reinforcements, Hannibal was unable to break out of Apulia and eventually had to abandon his campaign in the Roman heartland. Meanwhile, Roman armies mopped up the remnants of Carthaginian control in Spain. Within a few years, a Roman army was operating in North Africa, and Hannibal was recalled to defend the capital, in a battle which he lost. The Carthaginians sued for peace, and the Romans imposed severe terms. Roman fear and hatred of Carthage lingered, however. Finally, the Romans provoked the Third Punic War (149–146 B.C.), in which they destroyed their rival forever.

Hannibal, Master of Deception

Despite the fact that Gaius Claudius Nero hoodwinked Hannibal quite neatly during the Campaign of the Metaurus, the Punic commander had a well-deserved reputation as a trickster himself. Indeed, all of his greatest victories over the Romans (the Trebbia, Lake Trasimenus, and Cannae during 218–216 B.C.) were the result of clever thinking combined with overconfidence on the part of his opponents. At the Trebbia, for example, he lured away the Roman cavalry and then hit the rest of their army from the rear, while at Lake Trasimenus, he enticed the Romans into an ambush by a feigned flight. And at Cannae, his greatest victory, he posted his weakest troops in his center and his strongest on the flanks, so that when the center collapsed under the Roman onslaught, his flanks enveloped virtually the whole of Rome's army, upward of six legions (over forty thousand troops) perishing.

On several occasions during the long war, Hannibal captured cities using a surprisingly simple ruse. He attired some of his men who were fluent in Latin in Roman togas. Accompanied by men uniformed as lictors (i.e., bearer of *fasces,* the bundle of rods and ax which was the symbol of Roman magisterial authority), and a reasonably appropriate train of other escorts, these "Romans" would then approach the main gate of a town and demand entry in the name of the "Senate and People of Rome." As refusing to comply with such a request was likely to have dire consequences, the townsfolk often took the bait, to their great misfortune, for the false Romans would quickly seize the gate and admit Hannibal's troops. This trick has been used often in history. For example, the Grimaldi family seized control of Monaco in 1297, when Francesco Grimaldi, a Genoese merchant prince, gained entry to the town by clothing himself and some of his men as Franciscan monks.

One of Hannibal's cleverest tricks occurred during his investment of Tarentum (Taranto, in southern Italy) in 212 B.C. Even before the siege, Hannibal appears to have been in secret correspondence with a local citizen, Cononeus, a Tarentine Greek unhappy over Roman domination. The two evolved a clever plan to allow the Carthaginian to infiltrate troops into the city.

Cononeus, the turncoat Greek, convinced the Roman commander to allow him to lead nocturnal sorties from the city for the purpose of gathering supplies by means of hunting. As a result, Cononeus began to lead hunting parties out of the city each night, taking them right into Hannibal's camp, where the latter supplied him with boars and other trophies of the chase, which his troops had run down during the day. One night, after Cononeus' nightly comings and goings had become commonplace, Hannibal dressed some of his best troops in hunting togs, loaded them with game, and introduced them into Cononeus' party. Perhaps the guards at the gate noticed that more men were returning to the city with Cononeus than had left with him, but they didn't live long enough to do anything about it. As the hunting party entered the gates, the disguised troops threw down their burdens, drew their swords, and cut down the guards. Forcing open the gate, they admitted Hannibal, who had been lying nearby in ambush with a large body of his troops, and the city was soon his, save for the citadel, to which the Romans clung with great tenacity throughout the war, benefiting from their command of the sea.

So Hannibal was no amateur when it come to trickery.

Hannibal and Fabius: Duel of the Tricksters

Hannibal's wiliest Roman foe was Quintus Fabius Maximus. Fabius earned the nickname *Cunctator*, "the Delayer," because, aware that his tactical skills were no match for Hannibal's, he avoided open battle, relying instead upon ambushes, night attacks, guerrilla operations, sieges, and similar techniques to wear down the enemy. The two clashed many times, each trying to outsmart the other.

On one occasion in 217 B.C., Hannibal had been forced to retreat by some clever maneuvering on Fabius' part. Outnumbered, with supplies running low and night approaching, he found himself confronted by some difficult terrain, which would impede his movement so much that Fabius might be able to pick off a piece of his rearguard. To keep Fabius at arm's length, Hannibal had torches tied to the horns of cattle and turned the terrified animals loose in the direction of the Roman army. As the cattle fled through the countryside, the torches spread the flames to the surrounding brush. When the Romans saw the moving flames, they at first thought they were witnessing a supernatural apparition. They were soon disabused of this notion as their scouts came in. Learning the true nature of the phenomenon, Fabius decided that it might well be a trick on Hannibal's part to set him up for an ambush, so he pulled his troops back to their camp.

Later that same year, Hannibal was confronted by two Roman armies, one under Fabius and the other under his chief subordinate, Marcus Minucius Rufus. Aware that his two opponents were of very different characters—Fabius, wary and clever, Minucius, eager and unthinking—he resolved to separate the two commanders and defeat Minucius.

Boldly advancing his army so that he interposed it between his two opponents, Hannibal concealed a portion of his troops in ambush. Then he sent a small force to seize a hill near Minucius' camp. Minucius took the bait and led his army out to crush these troops, only to fall into Hannibal's ambush. Things would have gone badly for Minucius were it not that the wily Fabius intervened, forcing Hannibal to retreat. Fabius, while careful, was basically using deception.

He had convinced Hannibal that he would not move boldly on the battlefield. Thus when Fabius swiftly intervened to rescue Munucius, Hannibal was unnerved and withdrew.

Fabius once put one over on Hannibal quite nicely. Many of the more gung-ho Romans thought that Fabius' "Fabian" tactics were rather perfidious, and there was some talk that he was in sympathy with the enemy, which was why he was avoiding battle. Realizing this, Hannibal ordered that when raiding Roman lands, his troops were to avoid damaging Fabius' property, in the hope that Fabius' enemies would charge him with treason. When Fabius learned that his estates (which were pretty vast) were being mysteriously spared, he immediately donated them to the Roman state for the support of the poor, thereby neatly turning Hannibal's attempt to discredit him into an image-enhancing gesture. Even in the days before newspapers and television, it was important to project the proper image to the public. Then, and now, this image was an important component of a commander's arsenal.

Deceptive Defeat in the Battle of Ching-hsing, 205 B.C.

The ancients were very fond of the "feigned retreat." This was a dangerous tactic, since what started out as a false retreat could easily turn into a real one—and disaster. For this reason, false retreats were rare enough in ancient warfare to actually have a chance of working. The ancients much admired any commander who could pull off this sort of thing. Here is the story of how one such ancient Chinese general did it.

In the fall of 205 B.C., Liu Pang, founder of the Han Dynasty, sent his general Han Hsin on campaign against the kingdom of Chao. The king of Chao and his general, Ch'en Yu, the lord of Ch'eng-an, concentrated an army of about twenty-six thousand men near the mouth of the Ching Gorge, which was the only practical route between the Chih River and the mountains of Shansi, establishing a fortified camp in an area near the modern town of Shih-chia-chuang, nearly two hundred miles southwest of Beijing. Perhaps overconfident, the Chao general allowed the Han army to move out from the gorge. The battle that followed resulted in a victory for the Han, won as much by guile as by hard fighting.

Han Hsing deployed the bulk of his approximately twenty-two thousand men in two mixed divisions, each of about ten thousand men, mostly spear- or bow-armed infantry, with some cavalry and chariots, plus a detachment of two thousand light cavalry. These troops were posted with their backs rather close to the Chih River, partially in order to inspire overconfidence in his enemies and perhaps also to remind his troops that defeat meant annihilation. Meanwhile, Han Hsing sent his light cavalry detachment to make a secret flank march into the enemy's rear, in order to storm his camp. Details of the Chao deployment are unclear. However, they too probably were posted in two divisions of roughly equal strength, twelve and a half thousand men each—a mixture of spearmen, archers, light cavalry, and chariots, plus a small guard of perhaps a thousand men to guard their camp. The security of the camp was very important, as armies in this period tended to carry most of their portable wealth with them. In addition, thousands of servants, wives, and even some children went along to take care of the soldiers. There were also thousands of pack animals to carry the considerable amount of gear. In all, a military camp was an enormous concentration of wealth in ancient times. The ones who won the battle would find most of their reward when they plundered the enemy camp.

The battle opened with a Chao attack on the leading Han division. When the pressure on this division became too great, Han Hsing caused it to fall back upon his second division. As the battle intensified, apparently by arrangement Han Hsing's troops executed a feigned retreat. As the elated Chao swarmed forward, Han Hsing's light cavalry column emerged in their rear. Falling upon the virtually deserted Chao camp, the Han light cavalry quickly overran it and hoisted Han flags. Seeing the enemy apparently in control of their camp, the Chao began to lose heart, despite still being numerically superior to their foes. Han Hsing also saw that his troops had overrun the Chao camp, and he immediately renewed the attack with his main body. The result was that the Chao host disintegrated, and the Han achieved a signal victory in their struggle to establish a new dynasty in China, which was confirmed in 202 B.C.

Han Hsing's little trick was a simple one. In fact, the loss of their camp was primarily a psychological blow to the Chao. They could easily have overcome Han Hsing's army and then turned to dispose quickly of the Han troops in their camp. But it was an effective trick, and one which had already been used by Phineas at Gibeah, about eight hundred years earlier; by Han Hsing's contemporary Hannibal, in a dimly reported action at Locri, in southern Italy in 208 B.C.; and one which would be repeated several times through history, perhaps most famously by Bonaparte at Arcola, nearly two thousand years later.

Quintus Lutatius Catulus Eludes the Marauding Cimbri

The Romans were masters of deception through psychology. They strove to learn all they could about their enemies. Armed with that knowledge, they could then craft strategies, tactics, and deceptions designed to capitalize on their opponents' customs. Indeed, our only knowledge of many ancient peoples comes from Roman writings of their enemies, whom the Romans usually proceeded to exterminate.

In the late second century B.C., the Roman Republic faced an enormous tide of Germanic barbarians, primarily the Teutones and the Cimbri tribes, swooping down toward Italy around both ends of the Alps. Seemingly unstoppable, Gallic tribes and Roman armies both were crushed by the oncoming hordes. On one occasion, in 102 B.C., a Roman army was saved only by a clever trick.

Having been defeated by the Cimbri in a battle near the present Adige River in northeastern Italy, Quintus Lutatius Catulus and his Roman army found their retreat blocked by a small river, the opposite bank of which was held by their German foes. Unwilling to attempt a river crossing with his weakened and demoralized army, Catulus decided to attempt a deception. He concentrated his troops on a convenient mountain, such as those the Romans were wont to build their nightly fortified camps upon. He then ordered the men to erect a few tents, collect firewood, light fires, and even begin erecting a rampart, as if they were making camp. But he also instructed them not to remove their armor or set down their packs, and to maintain their ranks. Seeing this activity, the Cimbri concluded that the Romans were making camp, and decided to do so themselves. The Cimbri felt that the Romans were going to make a last stand in their fortified camp, and that on the following day, the Cimbri could begin a siege that would eventually result in the destruction of this Roman force. In short order, the Cimbri, who were much less well disciplined than the Romans, were scattered widely about the countryside, gathering materials for their camp.

Judging the moment propitious, Catulus suddenly ordered his men to reform their ranks and advance. They quickly crossed the river, and then made a sudden attack on the partially completed and weakly held enemy camp. Having accomplished his objective, and having disrupted the enemy, Catulus resumed his retreat. He had saved his army, and in the following year, he used it in conjunction with that of the great Gaius Marius to inflict so crushing a defeat on the enemy that they vanished from the pages of history.

Sertorius Hoodwinks Pompey the Great

Quintus Sertorius, a native of Nursia, in Italy's Sabine country, was one of the greatest Roman commanders. A partisan of the so-called Popular party, Sertorius had been a supporter of the great Gaius Marius during the first bout of civil wars to beset the Roman state (88–82 B.C.). Entrusted with command of the Marian forces in Spain, Sertorius was able to keep resistance alive against the so-called Aristocratic party for a decade (82–72 B.C.) after the triumph of the latter with the fall of the Eternal City to Lucius Cornelius Sulla in 82 B.C.

The loss of a great many historical documents from the period of the late Roman Republic makes it difficult to piece together a coherent picture of Sertorius' career. Despite this, it is clear that he was a wily, clever commander, with a healthy appreciation for trickery and deception. One of his finest tricks was played against Gnaeus Pompeius Magnus, better known as Pompey the Great.

Pompey, a youngish fellow who had proven an extremely capable commander, had been assigned the task of securing Spain for the Republic, which essentially meant overcoming Sertorius. In 76 B.C., Pompey's army was encamped at the town of Lauron, near Valencia, on the east coast of Spain. Good commander that he was, Sertorius' army was camped nearby. Because Sertorius controlled the interior, there were only two areas in which Pompey's troops could find forage with relative ease: one quite near to their camp, and the other at some remove. Sertorius gave strict instructions that whenever Pompey's foraging parties visited the nearer area, they were to be harassed with cavalry and light troops. However, under no circumstances were his men to interfere with Pompeians who attempted to forage in the more distant area, despite the fact that they were fully capable of doing so. After a while, Pompey stopped sending foraging parties to the nearer area and concentrated instead on gathering supplies from the more distant one. This was precisely what Sertorius wanted.

One day Sertorius, who had an army of five or six legions plus supporting light troops (say, 30,000–35,000 men), was reliably informed that Pompey had sent a large foraging party to gather supplies in the more distant area. He immediately dispatched some two thousand cavalry, plus five thousand men equipped as Roman heavy infantry and an equal number armed as Spanish light infantry. The troops deployed in a wooded area athwart the Pompeians' line of march, with the light troops in front, the heavy infantry in support of them, and the cavalry to the rear. About 15 percent of the latter were instructed to take up a position between the ambush site and the Pompeian camp. The troops—who were mostly Spaniards stiffened with a few Roman officers and centurions—were carefully instructed to make no move or sound until "the third hour of the morning," which was the Roman way of saying "nine-ish."

Sure enough, at just about "the third hour of the morning," who should march down the road but Pompey's foragers, heavily burdened with goodies. Sertorius' trap now sprang shut. His light infantry attacked, driving the foragers before them in a rout. This assault was followed up by the heavy infantrymen, who ensured that the Pompeian legionaries would not have a chance to rally. As the Pompeians began to break, Sertorius' cavalry went after them, to harry them from the field. And as they fled, the two hundred and fifty cavalrymen whom Sertorius had thoughtfully detached earlier cut them to pieces as they attempted to reach Pompey's camp.

As soon as word of the disaster reached Pompey, he dispatched a legion of about five thousand men to march to the relief of the supply column. When these troops encountered Sertorius' cavalry, the latter fell back, drawing them toward Sertorius' infantry. Soon the reinforcing column was engaged with the Sertorian infantry. Given a respite, the Sertorian cavalry regrouped on their right flank and returned to the fray, passing around Pompey's legion to hit it in the rear. Thus beset, Pompey ordered the balance of his army to the support of the beset legion, but Sertorius countered by deploying the balance of his army in position so obviously superior as to threaten Pompey's whole army with total destruction. Pompey backed down, preferring to lose a part of his army, rather than the whole of it. In the end, Pompey's loss amounted to ten thousand men, while Sertorius suffered no more than a fifth of as many casualties.

Caesar Has a Bad Afternoon: The Sambre, July of 57 B.C.

In 58 B.C., Gaius Julius Caesar, an ambitious and unscrupulous Roman politician, went to Gaul with the express purpose of conquering it so that he could return to Rome a richer and more powerful man. The task required eight years and untold casualties. One of the most hard-fought battles of the war was against a coalition of Belgic tribes, in July of 57 B.C., along the banks of the River Sambre, where now sits the village of Haufmont, about five miles southeast of modern Maubuege in northern France.

It being late afternoon, Caesar's army was preparing its customary nightly fortified camp. He had some 38,000–40,000 troops (about 34,000 Roman legionaries, plus 4,000–6,000 allied cavalry, archers, and slingers), about a quarter of whom were green recruits. Under Caesar's direction, the men of his six veteran legions were already engaged in erecting the camp, on a site on the top of a ridge running along the northwestern side of the Sambre. Most of the men had put off their armor to work better in the heat, and some of them had wandered somewhat away from the site of the camp to cut timber, fetch water, and the like. Caesar's other two legions, composed of recruits, were still on the march, about four miles to the northwest, escorting the army's baggage train.

In his memoirs of the war, *Commentarii de Bello Gallico,* Caesar asserts that the unified forces of the three allied tribes (the Nervi, Atebrates, and Viromandui) had some seventy-five thousand men, but we can safely consider this an exaggeration. It is likely that the Belgians totaled between thirty and forty thousand. The Gauls deployed under cover of the dense forest on the southeastern side of the Sambre River, which in that season was only about three feet deep at that point. The center of their line was only about a mile from Caesar's partially completed camp. The Gauls must have taken some care in deploying, as a large army makes a lot of noise, which can be detected even at some distance; the sounds of men, animals, and of metal clanking against metal all melt together into a low, but audible buzz. The Gallic front was about two miles wide. The troops were in four groups, each of two divisions, one behind the other, with the Nervi, about 50 percent of the total, constituting the left wing, the Vermandui to the right of center, and the Atrebantes on the extreme right. The Belgic light cavalry appears to have been on the flanks.

Precisely how Bogduognatus, who commanded the Belgians, decided that Caesar would camp in that place will never be known. It seems likely that he had some idea that Caesar would choose that particular site, for the Gallic army was already encamped nearby and deploying for battle even as Caesar marched up. Caesar reports that some Gauls who had been prisoners with his army escaped and informed the enemy of his plans. He also says that the Gauls hampered his movements by constructing several abatis, tangled obstacles of felled trees, to block the road. As it would have taken some time to clear these, it's possible that Bogduognatus, knowing the invariable Roman habit of erecting a fortified camp each evening, calculated that by impeding their advance just enough, he would lead them to an ideal campsite, right where he wanted them. Whatever is the case, Bogduognatus clearly laid a trap for the Romans, relying on Caesar's overconfidence and Roman military custom to enhance whatever advantages he would derive from the element of surprise.

The Gallic attack was quite swift, and as they swept down the wooded slopes into the more open area along the banks of the Sambre, they quickly overwhelmed some cavalrymen whom Caesar had posted there. Then they started splashing across the river as the Romans began to notice them. Their onset led to a storm of activity in the Roman camp. Caesar ordered the call to arms to be sounded and began getting his men into line of battle. As the men gathered to face the onrushing Gauls, Caesar (at least by his own account) ran from legion to legion, uttering encouraging words while shouting orders to the officers. The ensuing battle was a chaotic one. The Gauls struck the Roman lines when they were still ill-formed, with many of the troops only partially armed. Caesar himself repeatedly found it necessary to stiffen the resolve of his men, and even to fight in the front ranks. In the end, Roman discipline and experience, not to mention the timely arrival of the two legions of recruits, overcame Gallic courage and ardor. But it was a near thing, and the Gauls might easily have won.

Josephus Defends Jotapata

In A.D. 66, in consequence of many years of Roman misrule, there broke out in Judea what is commonly called the "First Jewish War." It was a brutal affair, with no quarter asked or given by either side. Our knowledge of this struggle is remarkably complete, due in no small way to the fact that one of the principal Jewish commanders in the early part of the war, Joseph, son of Matthias, later became a Roman collaborator, and, eventually, under the name Flavius Josephus, wrote a surprisingly detailed account, which has survived. Not given to modesty, Josephus has left us a particularly good account of his command of the Jewish forces in the Galilee and at the siege of Jotapata during the first two years of the war.

In the face of a massive Roman invasion of the Galilee, Josephus found himself bottled up in Jotapata from about the beginning of June A.D. 68. It was a small city, perched on a prominent rocky rise, virtually impossible of access save from one side, and well fortified, with a strong garrison. Within days, a Roman army under Titus Flavius Sabinus Vespasianus (later, the emperor Vespasian, A.D. 69–79) marched up. Vespasian initially blockaded the town with a strong force of cavalry, and as soon as his legionary and auxiliary infantrymen came up, he determined to take the place by storm. The Romans made several efforts to break into the place, being rebuffed each time by fierce resistance. After five days of this, Vespasian decided to use earth instead of lives. He began a formal siege, setting his engineers to build a massive ramp which when completed would permit his troops to swarm over the walls.

Josephus conducted an active defense, undertaking many raids and sorties, setting fire to the works, often with bloody consequences. Meanwhile, he had the walls of the town raised by thirty feet and strengthened with sturdy towers. Of course, since there was no Jewish army in the field to march to its relief, given time, Jotapata would have to surrender. But Josephus was determined that the city hold out as long as possible, since it pinned down a major portion of the Roman forces in Judea, thus gaining time for resistance elsewhere in the country.

Now, there was plenty of food in the town, but water supply was a problem. Jotapata lacked a well, being dependent on rainwater collected in cisterns. It being high summer, no rain could be expected, a matter that disheartened the defenders as much as it heartened the Romans. To convince the Romans that there was no water shortage in the city, Josephus ordered his men to do their laundry and spread it out to dry on the city walls. The spectacle of so much water being wasted convinced Vespasian that waiting until the city ran out of water might be a bad idea, so he essayed a new series of assaults, all of which were beaten off. Soon the Romans returned to a less active investment. During Roman assaults, the Jews employed a number of tricks to make the work of the attackers more difficult, such as pouring boiled vegetable slops on the ramps of their assault towers to make their footing insecure; hurling hot oil at them to set them afire; and so forth. When the Romans were not making assaults, Josephus subjected them to continual harassment by small raiding parties. All of this was very much traditional stuff in sieges.

Josephus employed a number of other unusual measures in his defense of the city. To prevent panic among the noncombatants, he ordered most of them confined to their houses. Until the Roman siege lines fully encircled the city, he was able to maintain communications with the outside world by sending couriers camouflaged with sheepskins through a particularly torturous ravine, instructing them that on their return, they were to bring some foodstuffs with them. So that he could concentrate his best troops at the points of greatest danger, whenever he noticed the Romans preparing an assault, he posted old men and the walking wounded to guard the least damaged portions of the walls. Since the numbers of these proved inadequate to create the impression of a large garrison, he dressed women in armor and posted them on the walls as well, under strict orders not to hurl their spears, lest by the clumsiness of their throws they reveal their gender, and thus the inadequacy of the defense.

By the fifth (or perhaps seventh) week of the siege, the defenders had taken a lot of casualties. A deserter brought word of this to Vespasian, stressing that the remaining fighting men in Jotapata were not only few in number, but weak from a lack of food, water, and sleep. The man suggested that a predawn assault might catch the sentries off guard, and pointed out a way that a small body of men might scale the walls and gain the citadel. Although Vespasian smelled a rat, he decided to make the attempt, on the grounds that if it was a trick, the number of men put at risk would be small. His own son, Titus (emperor, A.D. 79–81), led the scaling party, which proved an overwhelming success. With no losses to themselves, Titus and his little band secured the citadel and admitted the army. Jotapata fell with great slaughter.

And Josephus? With a few others, he for a time concealed himself in a cave. Since the other fugitives were determined upon death rather than surrender, Josephus proposed a lottery to determine who should slay whom, leaving the last man alone to incur the sin of suicide. He then manipulated the lots so that in the end he was alone with one other man, whom he convinced that survival was better than death. So by a series of deceptions and ruses, Josephus survived not only the siege of Jotapata, but also the zealous determination of his comrades to die rather than suffer further subjugation by the Romans. Thus Josephus ended life as a Roman citizen, indeed a personal friend to the emperor, with a large pension and a fine estate, and the leisure with which to write history.

The tricks that Josephus employed during the siege of Jotapata should not necessarily be considered unique. Rather, his account is one of the few we have from the ancient world that gives details of this sort of thing. He was very likely merely following the customary practice of the age. It is worth recalling that when the Jewish deserter offered to point out a way into the city, Vespasian suspected a trick.

Chu-ko Liang Plays a Lute

The Chinese have a long tradition of developing, using, and rewarding "clever" commanders. It was the Chinese who first explicitly stated that it was better to win victories via clever deceptions rather than bloody battles. In the following example, a noted Chinese commander uses a string of deceptions to defeat a much larger enemy army.

The collapse of the Han Dynasty in China inaugurated the "Three Kingdoms" period (A.D. 221–280). This was an era of chaos and war, as the principal successor states to the Han (Shu, Wu, and Wei) struggled for mastery of China. Wei controlled the region north of the Yangtze; Shu held the inland regions to the southwest; and Wu, the coastal regions to the southeast.

One of the greatest men of the age was Chu-ko Liang, sometimes known as K'ung Ming. For over a decade (A.D. 223–234), Chu-ko Liang served as the chief minister and principal general of the Shu emperor, Liu Pei. Chu-ko Liang was so successful in all of his endeavors—whether in war or government or scholarship—he eventually became the subject of numerous legends. This makes it difficult to determine which stories about him are true and which fabricated. One such tale relates how he defended the city of Hsi from an attack by Wei.

The Wei invaded Shu with considerable forces under an officer known as the Ssu-Ma (it more or less means "field marshal"). They proceeded to carry all before them, taking several cities. Chu-ko Liang had only about ten thousand men with whom to oppose the enemy, whose numbers were great, albeit probably not the more than one hundred thousand that tradition suggests. Learning that the enemy was approaching the city of Hsi, he hastened thither. Hsi commanded a valley through some rugged mountains. As a result, if Chu-ko Liang could hold the city, he could prevent the enemy from securing easy access to the interior regions of Shu, forcing the Ssu-Ma to make a difficult and time-consuming march through the mountains.

Arriving at Hsi, Chu-ko Liang dispatched scouts to apprise him of the enemy's movements, while trying to gather an army from among the local nobility. It soon became clear that the main enemy force was quite near and far outnumbered his own. Many of his officers began to lose heart. Not, however, Chu-ko Liang. Selecting two thousand of his best cavalry, Chu-ko Liang placed them under a couple of his best subordinates and sent them off with the mission of impeding enemy movements in the mountains. Then he turned to the defense of Hsi proper.

Since his forces were small, and the city walls poorly maintained, Chu-ko Liang resorted to deception. He ordered that nothing was to suggest the presence of an army in Hsi. The banners that normally flew from the walls were removed, the watchtowers were left unguarded, drums unsounded, and even the city's four gates were left open and unattended. He placed most of his troops in hiding in buildings, ordering them not to emerge on pain of death. But Chu-ko Liang ordered some troops to assume the clothing of ordinary workmen and peasants, even women, and to occupy themselves with mundane tasks, such as sweeping up trash, sprinkling the streets to lay the dust, hawking vegetables in the market, chasing stray dogs, and the like. Then, as the enemy army came into view over some distant hills, Chu-ko Liang called for his favorite pipa, a seven-stringed lutelike instrument, and a colorful robe such as a gentleman might wear when relaxing, saying "With these arms I shall face the enemy." Accompanied by several young pages, bearing his sword, his staff of office, and some refreshments, Chu-ko Liang mounted the ramparts of the city. Finding a sunny spot, the general made himself comfortable. As incense burned to sweeten the air, he partook of some fruit, cakes, and wine. Then he took up his pipa and proceeded to pluck out a tune, singing softly in accompaniment, while the pages stood guard beside him.

Chu-ko Liang's actions did not go unnoticed by the enemy, whose scouts were by this time quite close to the city. These men soon reported back to the Ssu-Ma that the city seemed unusually quiet, with the people going about their ordinary business, the gates ajar, and Chu-ko Liang carelessly diverting himself with a pipa. No fool, the Ssu-Ma smelled a rat. After a personal reconnaissance to confirm the reports of his scouts, he became convinced that Chu-ko Liang was planning some clever trap. Soon the Ssu-Ma had his army on the march northward, to take the more arduous road through the mountains, a road he believed to be undefended. Of course, once in the mountains, the Wei army came under attack from the cavalry that Chu-ko Liang had earlier dispatched for that precise purpose. With his foe entangled in the mountains, Chu-ko Liang was able to concentrate a larger army at Hsi while evacuating the populace and putting the place in order for a proper defense. By the time the Ssu-Ma realized he had been tricked, it was too late and he was forced to withdraw.

At Hsi, Chu-ko Liang used his own reputation for trickery and ingenuity as a weapon. The Ssu-Ma was defeated because he believed that a general of such resourcefulness and imagination as Chu-ko Liang must have had some wily trick up his sleeve. He did. However, it was not a trap for the Wei army, merely one for the Wei commander. Of course, clever generalship is not something that can be taught.

With the passing of Chu-ko Liang, the fortunes of Shu declined. By A.D. 264, it was incorporated into the kingdom of Wei, under the rule of a son of the too-clever Ssu-Ma.

Belisarius at Daras

Although the Romans were able to conquer nearly everyone whom they came across, they never had much success against their eastern enemies, the Parthians or their successors the Persians, based in modern Iran. On the other hand, neither the Parthians nor the Persians ever had much real success against the Romans, either. This was not for lack of trying, for the two powers were at war off and on for nearly eight hundred years, until the Persians were overthrown and the (by then) eastern Romans were very nearly vanquished by the Arab conquest in the seventh century. There were many great generals in these interminable wars, and they won many notable victories, none of which had any lasting effect on the strategic situation. One of the greatest of these commanders was Belisarius, a Grecianized Slav who commanded a largely mercenary army.

Daras was one of the great Belisarius' most impressive victories. In A.D. 530, Daras (about three hundred miles northwest of Damascus) was the most advanced Roman post on the Romano-Persian frontier. It was just fifteen miles from Nisibis, which the Persians had held for over one hundred and fifty years. In an ill-advised move, the Romans had decided to erect a new fortress very close to the frontier. So close, in fact, the Persians argued that the place violated the terms of a long-standing truce. Belisarius, in command of the Roman forces, ignored this protest. So the Persians decided for war. After some preliminary skirmishing in which the Romans came off second-best because Belisarius was on the sick list, the Romans fell back on Daras, permitting the Persians to raze the new frontier fortress. As Emperor Justinian soon reinforced Belisarius, the Persians decided to forestall a Roman offensive by undertaking one of their own. In June, they advanced boldly into Roman territory, only to find Belisarius deployed for battle against them before Daras.

As Belisarius had only about 2,000 archers, 4,000 spearmen, 6,000 light cavalry, and 12,000 *kataphractoi* (cataphracts, or heavy horse archers), many of whom were green, he decided that a defensive stance gave him the best chance of victory. But, of course, a little trickery wouldn't hurt. He deployed immediately outside the gates of Daras. However, although the front extended for perhaps a thousand meters, Belisarius cleverly "refused" the central portion, pulling it back about a hundred meters to the rear. In addition, he had the entire front covered by a ditch, but left a few crossing points. His deployment was rather complex, and may best be illustrated by a diagram:

```
                  -------------------Persians-------------------

1.      -ha-...........                          ..........-ha-
2.             -cavalry-: --- horse archers --- :-cavalry-
                         :                      :
                         :                      :
                         :......................:
3.                       - - - - Main Body - - - -
4.                               --Reserve--
```

KEY: 1. Horse archers posted off to the flanks
2. First line of resistance
3. Belisarius' Main line of resistance
4. Reserve line
ha = horse archers
····:······· Belisarius' ditch

Belisarius' position was about a hundred meters or so deep, such that archers could deliver fire from his refused center to support the troops roughly on the forward line. The Persians, about 10,000 infantry and 20,000 light cavalry under one Firouz, deployed in two identical lines, the second about a hundred meters behind the first. Each line consisted of about 15,000 men, with 5,000 infantry in the center and 5,000 cavalry on each flank, with Firouz apparently in the second line.

The battle took two days. On the first day, there was a series of skirmishes between light forces, mostly small affairs of no consequence save to men who were killed. On the second day, the Persians attacked in force.

The Persians advanced their entire force directly at Belisarius' army, and quickly came into contact with his first line. The Persian cavalry, on their flanks, became involved with the Roman heavy cavalry on this line, fighting across the ditch. The Persian center made up of infantrymen, pushed back the thin line of horse archers covering the center of Belisarius' first line, and advanced against his refused center. But as Belisarius' horse archers retired across the ditch to join his main line, the Persian infantry came under heavy fire from Belisarius' archers on that line. Meanwhile, the Persians managed to get across the ditch on the Roman left. They began pushing back the defenders, but were in turn taken in the flank by horse archers from the Roman left flank and reserve. Then the troops whom the Persians had driven back counter-attacked, which threw the Persians back across the ditch. Over on the Roman right, the Persians also managed to get across the ditch, but Belisarius did for them what he had just done for their countrymen on his other flank, sending in his mounted flank guards and reserves, who, together with a counterattack by the troops that had driven back, sent them to the rear. As the Persian troops fell back, they uncovered the left flank of their own center, which was still engaged in an archery duel with the Roman center. Seeing his opportunity, Belisarius took his reserve and his mounted archers (many of them Huns), and drove into the exposed flank of the Persian center, crossing the ditch at the points known to his men. This broke the Persian army, which was harried from the field with great slaughter.

Belisarius' deception at Daras was a tactical one; that is, it gave him an advantage on the battlefield. By digging the ditch across his front, he maximized his resources. While his men could easily cross the ditch, the Persians had to find the passages, a process that cost them men and momentum. It was a simple trick, but it worked, with the help of some desperate fighting.

The Imagination Is Mightier Than the Sword: Belisarius' Bloodless Victory

The best victories are those that are won with the least bloodshed, and the greatest of all are those won without any bloodshed whatsoever. While rare, these almost always depend on some form of deception. A vital ingredient has often been the high reputation of the commander. The Byzantines (Eastern Romans) had several generals of this caliber over the centuries. Indeed, the Byzantines in general were feared by their enemies because of the generally high caliber of their generals and troops. They knew what their reputation was, and used it to the hilt. In the following account, the greatest of East Roman generals used displays, demonstrations, lies, and a formidable reputation to scare away a much larger enemy army. That Belisarius was a skilled trickster, as well as a superb tactician, can be seen in his victory at Charchemish.

In A.D. 542, the Persian king Khosrow once again crossed into Roman territory, with an army estimated at "a little short of two hundred thousand." This figure should be taken with several grains of salt, but in any case, the Persian host was large, certainly too large for the local Roman forces to deal with, depleted as they were by drafts to other areas over the previous few years. Faced with the loss of some of his more profitable real estate, the Roman emperor Justinian I, surnamed "The Great," called upon his most talented and most put-upon commander, Belisarius, a man who had never lost a battle, even against sometimes extraordinary odds. Belisarius headed eastward with his personal regiment, a hand-picked and carefully trained band of about six thousand heavy cavalrymen, probably the finest troops in the world at the time. To these men Belisarius was able to add only about seven thousand more, and those not anywhere as good as his own troops. So if Syria and Palestine were to be defended, it would have to be by some very clever generalship indeed, regardless of the size of the Persian host.

Aware of the arrival of Belisarius (and his reputation) on the scene, Khosrow decided to parley—a ploy to gain time and information. Belisarius agreed, issuing a safe conduct to the Persian emissary. The Persian negotiator and his entourage arrived at Belisarius' camp, near the old city of Charchemish in Syria, on the appointed day. When he arrived, the Persian noted that the troops were occupied in various innocuous activities. Some men were holding athletic contests, others were riding or hunting or engaged in physical exercises. All the men were only lightly armed, with a dagger or a bow or a light ax, and all wore simple white tunics. He noted further that there were men of many different races among the Roman soldiers, and upon questioning them, learned that they were all in the personal service of Belisarius, having enlisted after their native kings had been defeated by their present commander. Suddenly, a trumpet sounded, and the men began to fall in, so that in an instant they had formed into squadrons, and then, upon a second trumpet call, deployed on either side of the road leading to the Roman camp. A man clad much like all the others came forward, courteously introduced himself as an officer, and escorted the emissary to headquarters.

At the end of the road, on a little hill, a tent was pitched, and there the Persian (decked out most luxuriously) was introduced to an ordinary-looking man dressed precisely as were all the others. As the Persian was invited to sit, a trumpet sounded, and the troops dispersed, returning quickly to their former activities.

The Persian's host introduced himself as Belisarius, and apologized for not having had time to prepare a proper welcome, in full uniform with the pomp appropriate to an ambassador from the Great King. When the emissary attempted to impress Belisarius with the size of the Persian host, the latter bluntly informed him that Khosrow had five days' truce to send his army back across the frontier, and brought the interview to a conclusion. There were again various trumpet blasts as the troops re-formed to escort the ambassador out of camp, and then returned yet again to their former activities.

The Persian ambassador reported to his king all that he had seen and heard, stressing that the troops had displayed extraordinary skill and enormous confidence in themselves and their commander, so much so, in fact, that they must represent the vanguard of a great host. Khosrow was inclined to agree, particularly since his reconnaissance patrols had brought word of the movement of additional Roman armies, all apparently converging on Charchemish. As a result, he shortly issued orders for the Persian host to retire. In fact, had Khosrow pressed the issue, even a general as resourceful as Belisarius would not have been able to do much to prevent him from overrunning Syria.

Belisarius' display had been a pure sham. The only troops he had allowed the Persian emissary to see were men of his household cavalry. The troop movements that had been reported to Khosrow were made by a detached body of his household troops, plus some of the best of the available local forces. Ultimately, the Persians had been defeated by Belisarius' reputation and the impression made by his troops. It was Belisarius' greatest victory, and one achieved without a single man dying.

The Kahena of Barbary Routs the Arabs

In the last decade of the seventh century, Arab armies carrying the banner of Islam began an invasion of the Mahgreb, the western portion of North Africa, encompassing the areas which are today Tunisia, Algeria, and Morocco. Roman Carthage, one of the great cities of the ancient world, held out until 698, but in general, the sons of the Prophet proved virtually invincible. However, in the Aurasius Mountains (the modern Aurès) of what the ancients called Numidia, there dwelt a number of Berber tribes, some Christian, some pagan, and some even Jewish, which had never been wholly subdued by the Romans. One of the Jewish tribes was led by a remarkable woman, the Kahena Dahiyan. Patiently welding together a coalition of the Berber tribes and the settled Romanized peoples of the local cities, the Kahena was able to hold off the Arab onslaught for about a decade. As able a commander as she was a diplomat, she made extensive use of deceptions, ruses, guerrilla tactics, and night attacks, taking advantage of the detailed knowledge of the terrain that her warriors possessed. Nor was she afraid to engage in proper battles, though she liked to stack the deck as much as possible. In about A.D. 700, the Kahena inflicted a grievous defeat on a large Arab army at the city of Bagia.

An Arab army estimated at some forty thousand men under one Hassan ibn-Numan had advanced westward on Bagia, held by a Romano-Byzantine garrison and the Kahena's sturdy Berbers. While the Arabs prepared to besiege Bagia, the Kahena took her Berbers out of the city. By a circuitous route through some rugged and rocky ground familiar to her tribesmen, she led her army around into the rear of the Arab forces. She timed her arrival so that she would arrive toward evening, and immediately offered battle. Hassan drew the bulk of his forces from the siege lines about the city to contend with this new threat to his rear. However, by the time his troops had finished getting into line of battle, darkness had fallen. That night, both armies rested on the field. But the Kahena's Berbers got a lot more rest than did the Arabs, who stood to their arms and horses lest the wily queen attempt a night attack.

At dawn, the Kahena unleashed her army in a ferocious assault on the tired Arabs. As the two hosts clashed, the garrison of Bagia sortied, breaking through the thin cordon of Arab troops encircling the city to fall on the rear of the Arab main body. The two attacks, front and rear, caused the Arabs to break. Although Hassan was able to rally a good many troops, the defeat was so decisive that he was ultimately forced to fall back all the way to Gabes on the Tunisian coast, many hundreds of miles distant.

The Kahena had taken advantage of her reputation as an unconventional commander and her intimate knowledge of the terrain to inflict a crushing defeat on a superior foe. She continued in this fashion for several more years. However, although perhaps unmatched as a commander, she was ultimately outclassed as a diplomat by Hassan ibn-Numan. Unable to contend with the Kahena on the battlefield, the clever Arab worked behind the scenes to destroy her. Through bribery, careful politicking, and outright lies, he set Christian against Jew, both against pagans, Roman against Berber, farmer against nomad, and so forth, so that by about A.D. 705, the Kahena no longer headed a powerful coalition. Defeat and death soon followed, and the Arab hosts resumed their march of conquest. So the trickery that won for the Kahena victories on the battlefield was no match for that which won the war for Hassan.

Duke William Runs Away: Hastings, 1066

In 1066, King Edward the Confessor of England died heirless. Harold Godwinson, Edward's brother-in-law, assumed the throne. But Harold's claim was by no means undisputed. King Harald Hardraade of Norway and Duke William of Normandy both had claims of their own. Such a dispute could only be settled by force of arms. Harold Godwinson disposed of Harald Hardraade in a tough battle at Stamford Bridge, in Yorkshire. Then the young Anglo-Saxon king marched south with his army to meet William's expected invasion. By the time Harold reached southern England, William had landed on the southeastern coast of England and was encamped not far from Hastings. Soon afterward, on October 14, 1066, the two armies met on Senlac Hill, about eight miles from Hastings.

Harold had about 7,000 Englishmen, armed mostly with swords, spears, and axes, but with 1,000 archers. He posted his troops, many of whom were militiamen, in a more or less solid mass, along the crest of Senlac Hill. His 2,000 housecarls (personal guards) held the center of a front which was perhaps 800 meters wide. With his flanks well covered by natural obstacles, the only way that he could be attacked would be frontally. Duke William had about 10,000 men (3,500 men-at-arms or knights, 3,500 spearmen, and 3,000 archers), mostly professional warriors drawn from all over Europe by the promise of loot. He deployed on a front of 800–1,000 meters, in three divisions. The flank divisions consisted of about 1,000 each of men-at-arms, spearmen, and archers, while the center was rather stronger, with 1,500 each of men-at-arms and spearmen, plus 1,000 archers. Each division was posted in three lines: archers in front, spearmen behind them, and men-at-arms to the rear. William himself began the battle with the central division.

William opened the fight with a heavy archery barrage, which greatly annoyed the English, who could not respond with equal firepower. Englishmen were melee fighters, experts particularly with the ax. They protected themselves behind a "shield-wall," a defensive position in which some men knelt with their shields to form a solid wall. Their axmen could fight from behind this wall and even advance from the shield-wall to wreck havoc on the enemy as opportunity presented itself. So when William sent in his spearmen, they were roughly handled. With his spearmen falling back, William committed his men-at-arms. The onset of the heavily armored horsemen seems to have partially broken the shield-wall, but for the most part it held, and the attackers were soon streaming to rear, those on the Norman left in a virtual rout. Seeing the enemy fleeing, some of the English (probably militiamen from the shire levies) broke ranks to pursue. At this, William hurled the men-at-arms of his center division into their flank, killing many of the unprotected militiamen and chasing the rest of them back up Senlac Hill. As the English resumed their lines, William renewed the attack, only to be repulsed once more.

Having twice attacked, and twice been repulsed, William decided it was time to try something different. So he attacked again. Yet again his men-at-arms were repulsed, falling back in some disorder. But this time they did so by design. And having thus three times routed the enemy, many of the English thought the battle already over and broke ranks to pursue. As they streamed down after his fleeing cavalry, William let them come for a bit, thus enticing more men to join the wild attack. Then he hit them in the flank, cutting many to pieces. Although the surviving English were able to re-form a line on Senlac Hill, the odds were now greatly against them. William launched a series of coordinated attacks, alternating showers of arrows with assaults by men-at-arms. These wore down the English. Finally, King Harold fell, traditionally to an arrow in the eye. With that, the remnants of the English army disintegrated and the battle was won. While it has been argued that there was no "feigned retreat" at Hastings, William was too capable a commander to keep doing the same thing over and over again. The feigned retreat was a common tactic in cavalry fighting, but wholly unknown in England, where cavalry was rare. So it seems reasonable that William's first attack was designed to feel out the enemy, and the second to set the foe up for the third, which would be a feint.

Alexius Comnenus Slays a Usurper: Calavryta, A.D. 1079.

For many centuries, the Byzantine Empire was preserved from all comers by a highly effective military system developed in the seventh century, a system which, of course, included the extensive use of deception in warfare. However, in 1071, the empire suffered a massive defeat by the Seljuk Turks at Manzikert, near Lake Van in Armenia. A combination of excessive enthusiasm, bad luck, and not a little treachery resulted in the loss of the flower of the imperial army, and the capture of the emperor himself. As a result, the Turks overran large portions of imperial territory, while various magnates and generals frittered away what was left of the old army in a bout of civil wars over who would ultimately wear the purple slippers of the emperor. Even during these internal disputes, the Byzantines displayed a remarkable penchant for trickery.

In 1079, the reigning emperor, Nicephorus III Botaniates (an able, if aged, general who had usurped the throne in 1078), sent his best subordinate, Alexius Comnenus, to deal with Nicephorus Bryennius, who was contending for the imperial dignity in the European portions of the empire. Comnenus' army met Bryennius' at a place called Calavryta. Bryennius' army was rather small, certainly no more than twenty thousand men. But Comnenus' was even smaller, apparently only about fifteen thousand. Moreover, only about half of his army consisted of Byzantine troops, the balance being Frankish and even Turkish mercenaries. So Comnenus decided it wisest to stand on the defensive.

Comnenus deployed on ground very suitable for the defense. It had a narrow front, where the road crossed the slope of a hill and there was some cover for his flanks. He posted his main body, perhaps eight thousand men between Byzantines and Franks. He placed some Byzantine infantry on the right, in a "refused" position, that is with their front somewhat farther from the enemy than that of the main body. Beyond these troops he posted on his far right a contingent of Turkish horse archers. The balance of his army, probably five thousand men constituting his left wing, he concealed in a hollow off to the left of the road, from which they could not be seen by an enemy advancing up the road from the plain in front of his position. These troops were actually posted in advance of his main body. So in effect, just as he had "refused" his right, Comnenus had "advanced" his left, but in a concealed position.

Since he was on the offensive, Bryennius deployed his army in three parallel columns: the right of five thousand men, the left of about three thousand, and the center of perhaps ten thousand. In addition, he assigned several thousand barbarian Patzinak horse archers to move parallel to the army, about a quarter of a mile to its left, with orders to turn Comnenus' right.

The battle opened when Bryennius' army, advancing up the road, came abreast of the hollow in which Comnenus had deployed his left wing in ambush. These Comnenian troops suddenly attacked, taking Bryennius' right column in the flank. This was the signal for Comnenus' main body to attack, and it struck Bryennius' center column in the front. Some hard fighting resulted, and both of Comnenus' attacks were beaten off, with portions of his main body streaming to the rear while other portions were surrounded by Bryennius' victorious troops. Meanwhile, to make matters worse, Bryennius' Patzinaks (nomadic Asian warriors) had attacked the infantry on Alexius' refused right wing, routing it and pursuing it off the field. At this point, the battle looked like it was going badly for Comnenus. However, Comnenus was a resourceful fellow.

During the melee in the center, Comnenus had become engaged with Bryennius' personal escort and had managed to capture the usurper's spare mount, with all of the regalia appropriate to a claimant to the imperial dignity (e.g., "purple housings and a gold frontlet, and . . . the two swords of state . . ."). Cutting his way out of the melee with his own escort and the prize steed, Alexius retired down the reverse slope of the hill for a breather. There he rallied his men, encouraging them by shouting that he had just slain Bryennius, as they could plainly see from the wonderful trophies he had captured. In short order, Alexius amassed a considerable number of freshly heartened men. At the same time, the Turks whom he had posted on his far right and who had so far been unengaged, came up to support him. Placing some of his rallied troops under cover on either side of the road, Alexius advanced up the road once more with the bulk of his men—the Turks acting as mounted skirmishers.

Alexius attacked at a moment when Bryennius' army was in considerable confusion. The usurper's Patzinak mercenaries had returned, only to begin plundering his own camp, while other men were looting the dead, and still others contending with the remnants of

Alexius' center, which had just surrendered. Bryennius himself was basking in the glow of his victory, receiving the homage of some of the surrendered Frankish chieftains. Despite this, Bryennius reacted swiftly to Alexius' renewed attack, ordering an immediate counterattack.

As soon as Bryennius' men came forward, Alexius began to pull back. This continued for some minutes, until Bryennius' men were abreast of the two groups that Alexius had concealed on the sides of the road, on the reverse slope. At that, Alexius halted his retirement, turning to attack Bryennius' troops in the front, even as his flankers emerged from cover to take them in the flanks. The result was a rout, as, beset from front and flanks, Bryennius' army melted away.

In the Battle of Calavryta, Alexius Comnenus demonstrated remarkable skills as a commander. From the beginning, he thought in terms of deceiving the enemy. First there was the concealed left wing, a trick that might well have worked had not Bryennius' troops fought as well as they did. Then there was the business of using the usurper's captured horse to help rally his own troops. And finally, he went on to win the battle using more or less the very technique he had attempted at the start: the concealment of troops in ambush. Altogether a sterling demonstration of the effective use of deception in battle, by a master of the art.

Alexius Comnenus' military skills were matched by his diplomatic ones. On the death of Nicephorus III, he was "called to the purple," reigning for many years as Alexius I. Eminently clever, he arranged for his daughter Anna (who wrote an excellent biography of her father) to marry the son of Nicephorus Bryennius (who wrote a biography of *his* father), thereby cementing the usurper's family and supporters to his cause.

Those Slippery Infidels: Trickery During the Holy Wars

At the end of the tenth century, Western Christendom undertook a Crusade, a war to recapture Jerusalem from the "infidel" Moslems who had held it for more than four centuries. The result was a long series of wars (arguably, not ending until 1399), during which the

Europeans (called the "Franks" by the Moslems) were only temporarily able to secure control of the Holy Land. Aside from the casual massacre in which both sides indulged, not to mention the forced conversions and the desecration of holy places, these wars were characterized by clashes between armies of very different natures. The Europeans relied heavily on their knights, who could be quite effective when used in coordination with spear- and crossbow-armed infantrymen, operating from stoutly fortified castles, while the Moslems relied more on light cavalry, especially horse archers, plus some infantry, supplemented by an impressive arsenal of deceptions, ruses, and tricks.

The most obvious trick the Moslems employed was the feigned flight, enticing the heavier Christian knights to pursue beyond the safety of their infantry supports. They used this to great effect a number of times, mostly against Crusaders newly arrived in the Holy Land. On other occasions, the Moslems would fix the attention of the Crusaders to their front, and then attack them in the flank or rear with hitherto concealed forces. And, of course, ambushes were popular. But there were many even more devious tricks, as can be seen from the following sampler.

Deception played an important role right from the start of the Crusades. Even before the Crusades were formally under way, enthusiastic bands of common people began making their way to the Holy Land in the so-called People's Crusade. This inchoate, ill-trained, ill-equipped, and ill-led horde reached Asia Minor in the late summer of 1096. By October, the Crusaders were on the march south from Nicomedia toward Nicaea, when they met Kilij Arslan, the sultan of Roum (i.e., "Rome," referring to the former Byzantine control over his domains, an impressive swath of the heart of present-day Turkey). The Crusaders were in two separate contingents, and the sultan managed to defeat the leading one handily. Then, before word of the defeat had reached the second contingent, he sent some agents into the Crusader camp at Civtate to spread word that the men of the first column had actually taken the rich city of Nicaea and were even then plundering it to their hearts' content. Kilij Arslan intended that the Crusaders would rush pell-mell to Nicaea in order to share in the spoils, thereby falling into an ambush which he had laid. As luck would have it, a straggler from the first Crusader detachment made it back to the Christian camp even as the sultan's agents were attempting to trick the Crusaders into an advance. This development did not, as it turned out, negate Kilij Arslan's plans, for when the

Crusaders heard that their comrades had been defeated, they rushed out in great haste and some disorder to relieve them, thereby falling into his ambush anyway, to be more or less obliterated in a morning attack on October 21, 1096. The victory didn't serve Kilij Arslan's cause very much, as he went down to defeat at the hands of the real Crusaders, in the great battle of Dorylaeum, a few miles farther to the south, the following July, mostly through a combination of poor reconnaissance and bad luck. But from it, one can at least get a fair notion of the practice of deception at the time.

Shortly after the Crusader defeat in the Battle of Harran (May 1104), one Moslem commander hit upon an excellent way to get into Christian strongholds with a minimum of fuss. Sokman ibn-Urtuk, Atabeg of Kayfa, put about three thousand men into European clothing, gave them Western weapons as well, and rode forth at their head with captured Christian banners flying. In this way, he was able to seize several Christian-held fortresses, by the simple expedient of riding up to them and being admitted.

During the great siege of Acre (Akko in modern Israel) by Richard the Lionhearted in 1189–1191, Saladin Ayubi, the great sultan of Damascus, hit upon a clever ruse by which to resupply the desperate defenders. In June of 1190, he arranged for a very large ship to be loaded with foodstuffs at Beirut. But let the Islamic historian Baha al-Din continue the tale, in a loose translation of the original:

> The ship was manned with Moslems dressed like Franks, with shaven beards, while crosses were affixed to the masts and pigs were displayed on the deck. They set sail for Acre. As they approached the city, blockading Christian vessels challenged them, saying "You are heading for Acre."
>
> The crewmen asked, with some astonishment, "Haven't you taken it yet?"
>
> The Franks, thinking they were dealing with their coreligionists, replied, "No, not yet," to which the Moslem crewmen responded, "Well, then we will anchor near the camp," and, having spotted a Frankish vessel approaching offshore, added, "but there is another ship not far behind us, and you had better warn them lest they sail into the city." The Crusader warships immediately made off to warn the approaching vessel, whereupon the Moslem seamen clapped on all sail and headed into the port of Acre, to be greeted with cries of joy.

And thus was Acre succored, if only symbolically, the amount of food brought into the city being relatively small. But it was not until July 11, 1191, that the Crusaders took the town, a siege prolonged by desperate measures and clever ruses which enabled the garrison to eke out its meager food supplies.

Stupor Mundi Goes into Winter Quarters: Cortenuova, 1237

The Holy Roman Emperor Frederick II von Hohenstaufen (reigned 1212–1250), was a man of enormous intellectual, political, and military skills, so much so that he was nicknamed *Stupor Mundi* (Wonder of the World). He used these skills repeatedly in his protracted struggle with the papacy over primacy in Europe.

In 1237, Frederick was at war with the Lombard League, an alliance of north Italian city-states which supported the papacy. He had managed to capture Mantua and several other cities, and was anxious to secure Brescia before the onset of winter. Unfortunately, he could not invest Brescia because a substantial Lombard army was in the field against him. Under normal circumstances, he would have no trouble disposing of the Lombards, since his army was better than theirs, but if he attempted a siege, he would have to divide his forces, thereby giving the enemy a chance to defeat him in detail. On the other hand, if he marched out boldly to give battle, they would merely move off, aware of his tactical superiority. So he decided that a little ruse was in order.

The season being well advanced into autumn, he began making arrangements as if to go into winter quarters. Early in November, he dismissed his unreliable Italian allies, ordering them to return to their home cities and prepare for operations in the spring. Then, on November 23, he broke up his camp and began putting his infantry and baggage train on the road to Cremona, where he had ordered supplies to be accumulated.

The leaders of the Lombard army soon became aware of these developments and decided to take up winter quarters as well, the Brescians returning to their home city, while the balance of the army was to retire to Milan after keeping an eye on Frederick's retirement to Cremona. To do this, the Lombards moved by a parallel route. While most of his army trod the road from Brescia to Cremona, Frederick halted nearly ten thousand of his finest troops (German men-at-arms, Saracen archers, and the best of his Italian spearmen) about halfway. From there, he sent out patrols to ascertain the location of the Lombard army. Sure enough, the Lombards were not far off, with their attention focused on the portion of his army that was still on the march for Cremona. Nor were they keeping very good order, either, since the war was over for the year anyway. The Lombards had posted no scouts, and their army was stretched along some several miles of road, with their cavalry in front, their infantry in the middle, and their baggage train in the rear.

This gave him his chance. On November 27, Frederick undertook a long forced march cross-country. Timing the movements of the two armies perfectly, Frederick was able to strike the Lombards in the center of their line of march, cutting them in half. A hard fight followed, for while the Lombard cavalry fled, their infantry made a stand around the village of Cortenuova, fighting desperately until nightfall, when they were able to slip away in the dark. The result of Frederick's deceptive retirement on Cremona might have been more spectacular were it not for the treachery of a turncoat who, a few minutes before Frederick attacked the Lombard column, gave the alarm. Despite this, he managed to inflict very heavy casualties on the Lombards, killing many and capturing eleven hundred of their knights.

Getting Snookered by a Guelf

During the Middle Ages, there were a lot of little wars, "private" affairs between various petty lords, fought over who owned what, who owed feudal allegiance to whom, and so forth. In many of these, some occasionally remarkable ruses were effected.

Take, for example, the case of Itri, a small town perched on a mountaintop at the northernmost edge of the kingdom of Naples, some eighty-five miles south of Rome, where it dominates a series of defiles through which the Via Appia winds on its way south. During the fourteenth century, the town was involved in a dispute as to whether the lord of Itri was, or was not, subject to the feudal authority of the duke of Gaeta, a seaport about a dozen miles to the southwest. During the early Middle Ages, Itri had originally been a dependency of the dukes of Gaeta. But that was many years before, and the authority of Gaeta had gradually diminished, until the count of Itri could argue that he was subject to no overlord other than the king of Naples. In the early fourteenth century, this worthy was one Robert of Anjou, widely known as Robert the Good, for his gentle nature and support of the arts. (Aside from being king of Anjou, a large spread in southeastern France, as well as in Naples, Robert was also a friend of the great poet and humanist Petrarch and patron of the great writer and scholar Boccaccio.)

In 1338, the current duke of Gaeta, who happened to be a Ghibelline, or opponent of the temporal power of the popes, decided to assert his authority over Itri. The count of Itri promptly proclaimed himself a Guelf, or supporter of the temporal authority of the pope, which by chance happened also to be the view of good King Robert. With the king's help, the count of Itri managed to win the war, and by a treaty concluded in 1340, was actually paid an indemnity of two hundred ounces of gold. However, the victory settled the matter only temporarily.

Good King Robert died in 1343, and the throne of Naples went to his granddaughter, Giovanna, who was only fifteen years old. She was a bright, noble young woman, but was a poor ruler and had terrible morals. When she ascended the throne, she was already married—to the first of four or five husbands, most of whom met untimely deaths (Giovanna's responsibility in some of these has often been suspected, but never established). In any case, her reign was a disaster, and she lacked any fixed policies. This encouraged the duke of Gaeta to try again to secure the domination of Itri. Gaining the queen's ear, he secured her approval for an attempt to reassert his authority. In 1346, after Count Niccolò of Itri rejected his demands, the duke declared war.

Realizing that his military resources were slender indeed, Count Niccolò, decided that the best way to defeat the enemy would be through the use of deception. As the duke of Gaeta's army advanced on Itri, Count Niccolò issued arms to every man, woman, and child in the town. He then ordered the townsfolk to take to the roofs and upper floors of the houses and there to maintain complete silence. Then he closed all the gates to Itri but one, near which he concealed his most trusted soldiers.

Since the consequences of Count Niccolò's ruse are known, it is pretty clear that the duke of Gaeta was extremely confident. Perhaps he took the unguarded, open gate as a sign of submission, or perhaps he thought the town abandoned by its defenders. In any case, he certainly does not seem to have thought that his troops were walking into a trap when they entered Itri. The results were predictable. As the enemy began to push up the steep, narrow streets, Count Niccolò's select men closed the gate, whereupon the townsfolk emerged from hiding to massacre the lot.

Count Niccolò's ambush of the duke of Gaeta at Itri in 1346 brings to mind Chu-ko Liang's psychological ambush of the Ssu-ma at Hsi eleven hundred years earlier. The principal difference is that at Itri, there really was a trap, but the intended victim suspected it not, while at Hsi, there was no physical trap, but the intended victim thought there was.

Cesare Borgia Throws a Peace Conference

Cesare Borgia was perhaps the ultimate Renaissance prince, a man of culture and learning, a patron of the arts, a capable soldier and general, and an accomplished practitioner of *realpolitik* in the dog-eat-dog world of sixteenth-century Italy. He spent years attempting to cement the miscellaneous petty lordships, tyrannies, and republics of central Italy into a unified state. This was no easy task, even for one having a pope, Alexander VI, for a father, for Italy was not only a mass of petty states, but also a battleground between France and Spain.

In mid-1502, Cesare had, by swift marching and not a little deception, managed to secure Urbino, one of the more important and richer of the small principalities of the Romagna, the area of central Italy just east of the Apennines. He took it from an erstwhile ally, Guidobaldo da Montefeltro, a matter which caused no end of embarrassment to his sister Lucrezia, who was related by marriage to the Montefeltri. Using Urbino as a base, Cesare moved quickly against several other local states, seizing the republic of San Marino, the principalities of Camerino (managing in the process to knock off the local tyrant and three of his four sons), Forli, Imola, and a few others, all of which he promptly strengthened with additional fortifications, some of them designed by his chief engineer, a chap named Leonardo da Vinci. These successes struck terror into some of the other petty tyrants of the Romagna.

Well established in the Romagna, the next item on Cesare's menu was Bologna, the richest and most important city in the Romagna, and an essential acquisition if Cesare was to cobble together a viable kingdom for himself. Bologna was held by the Bentivogli family, themselves a notably tough and wily clan. Cesare laid careful plans to seize the city, securing a financial commitment from France's Louis XII in exchange for a promise of future aid against the Spanish, who held Naples (the fact that the Borgias were of Spanish origins was of little importance), not to mention the able assistance of his Holy Father, who accused the Bentivogli of misgovernment of what was technically a papal town. However, Cesare's preparations were interrupted by a mutiny by his own commanders.

As was the custom of the day, Cesare's army was composed primarily of mercenary companies, each commanded by an independent military contractor, the famous *condottieri*. Most of Cesare's condottieri were themselves lords of small territories in the Romagna. After ably assisting Cesare in despoiling their neighbors of their estates, several of his Romagnian condottieri began to wonder if they were next on the list. On October 7, five of them held a secret conference. They were as fine a bunch of Renaissance men as could be found anywhere—soldiers, patrons, schemers, and so forth: Vitellozzo Vitelli, lord of Citta del Castello (a leading military innovator and noted practitioner of the arts of treachery, who suffered from occasional lapses in judgment due to severe pains from various venereal disorders to which he was subject); Gian Paolo Bagnoli, lord of Perugia (from a family notorious for its bloody internecine quarrels, the most noted member of which is famous for having been murdered after being caught cheating on her husband, a romance immortalized by Dante himself); Oliverotto da Fermo (who, having been taken in as an orphan and raised by his mother's brother, repaid the favor by murdering the fellow and assuming his principality); and two members of the noble Orsini clan, Paolo, lord of Polombara, and Francesco, the duke di Gravina (both of whom harbored old grudges against their current master, but, after all, in the words of Don Vito Corleone, "It was business.").

Joining them were representatives of the Bentivogli, the Montefeltri, and other opponents of the Borgias, all noble men of equal merit. The meeting was full of talk of "liberty" and "honor," as the plotters decided that Cesare (better known as Duke Valentino) must not only be stopped, but destroyed. They pledged to raise an army of six hundred lances (i.e., about eighteen hundred mounted troops) and eight thousand on foot. Then, raising a rebellion in the Romagna, they would undertake an offensive against Cesare, one column advancing from Bologna southward and the other from Perugia eastward, so as to "liberate" Urbino and catch Cesare, who was at the time based at Imola, between two armies. Well pleased with themselves, the conspirators went their separate ways to put their plans into effect. Within days, Vitellozzo, Oliverotto, and Bagnoli were in the field. As a result, Cesare quickly lost control of Urbino, San Marino, and Camerino to the conspirators. But he never lost heart.

Although Cesare is reported by some accounts to have taken fright at these events, his actions reveal more a certain disappointment and indignation, and a lot of quick thinking. In quick succession, Cesare (1) sent an "emissary" to negotiate with the conspirators, separately if possible; (2) offered the Florentines an alliance, the lord of Siena being numbered among both his and their enemies (as their ambassador, the Florentines sent Niccolò Machiavelli!); (3) undertook a letter-writing campaign to string his opponents along with promises of possible negotiation, which they would take as weakness; and (4) began collecting a great number of troops, until he had some fifteen thousand, and a large train of artillery. His maneuvers were disheartening to his opponents, none of whom liked each other any more than they cared for him. And gradually, several of them sent Cesare confidential emissaries assuring him of their love for him, begging his pardon for any offense they might have given, and asking only that they might prove their loyalty by rendering him some service. On December 2, Cesare cracked the conspiracy, concluding a secret treaty with the Bentivogli of Bologna, who promised to support him secretly, with men and money, in exchange for being left to enjoy Bologna in peace.

Meanwhile, Cesare took the offensive and began recapturing places that had only weeks before been taken from him. By late December, the conspirators were in desperate straits. Cesare cornered them at Senigallia, on the Adriatic coast. On December 29, he informed them that if they abandoned the city and surrendered to him, he would be in a forgiving mood. The Orsini, Vitellozzi, and Oliverotti agreed to a reconciliation; Bagnoli, perhaps because his family's homicidal quarrels had honed his instincts for survival, was elsewhere at the time. On the last day of the year, Cesare greeted his four errant subordinates in a lavish ceremony. As their troops paraded outside the walls, the four condottieri accompanied Cesare as he reviewed their men on horseback and then rode into the city. He chatted amiably as they rode, pointed out the houses which he assigned for their use, and discussed various nonessential matters. Then he turned into the entry courtyard of the house which he had requisitioned for himself. Dismounting, he casually invited them to stay awhile. As they began to dismount, he took a few steps up a small stairway, turned, and looked upward. At this signal, the four were seized by Cesare's men. He executed the first two, Vitellozzo and Oliverotto, that very night, and saved the Orsini for later. Meanwhile, he dispersed their troops, recruiting some into his own service.

Cesare Borgia's extraordinary response to the conspiracy of the condottieri is one of the most successful deceptions in history, a wonderful demonstration of how one's enemies can be destroyed through political maneuver rather than military force. Although the most readily available account, Machiavelli's wonderfully titled short work, *Description of the Methods Used by the Duke Valentino When Murdering Vitellozzo Vitelli, Oliverotto da Fermo, the Signor Pagolo, and the Duke di Gravina,* is inaccurate in several instances, it clearly demonstrates the resourcefulness and creativity of Cesare's vengeance. Unfortunately, history was not kind to the duke, for his fortunes declined rapidly after his father died, and he himself later fell in a minor skirmish in a half-forgotten dynastic war in an obscure corner of the Pyrennes.

El Gran Capitán Breaks the Rules: The Garigliano, 1503

The best deceptions are carried off by commanders who are willing to break the rules and take some chances. A good case in point is Gonzalo de Córdoba's successful operations along the Garigliano River in 1503.

In the 1490s, there began a titanic struggle between France and Spain for primacy in Europe, a struggle which was to last for more than fifty years. Italy, adjacent to France on the northwest and controlled by Aragon, one of the Spanish kingdoms, on the south, was the principal theater of this war, and a seemingly interminable series of campaigns took place. One of the most spectacular battles was that of the Garigliano, a considerable river emptying into the sea about 100 miles south of Rome. An unusually long and arduous campaign that had begun in the summer of 1502 dragged on, with a break for "winter quarters," into the following year. By the onset of winter 1503, the Spanish, under Gonzalo de Córdoba, popularly known as *el Gran Capitán* (The Great Captain) for his superb generalship, had succeeded in driving the French northward from the city of Naples. Exhausted, both armies settled into winter quarters on opposite sides of the Garigliano.

The French army of eighteen thousand troops (fourteen thousand infantry and four thousand cavalry with some artillery) was a polyglot mercenary force of French, Swiss, German, and Italian troops under Ludovico de Saluzzo. The better to provide for their provisioning and quartering through the winter, Saluzzo had dispersed his army over a wide area to the north and northwest of the Garigliano, posting small detachments at all bridges and fords across the swollen river, and in addition at virtually all crossroads, villages, and towns (including a commune called Itri, from whence hail the ancestors of one of the present authors).

Not so the Spanish. Córdoba, a seasoned veteran of wars against the Moors and the French, had some 16,000 troops (7,000 Spanish infantry, 3,000 cavalry, with some artillery, plus 4,000 Italians on foot and 1,500 on horse).

Although he too made provision for his men to endure the winter, he kept his troops well in hand, for he planned to undertake an unprecedented winter offensive. Gonzalo concentrated virtually all of his forces in entrenched positions on a very narrow front just south of the river line, with most of them at the lower end of the river.

Shortly before Christmas, Gonzalo declared a special holiday to last for several days. Learning of this, Saluzzo did likewise. But Gonzalo reneged. On December 27, he began putting his troops on the march. And on December 29, Córdoba attacked. Throwing pontoon bridges across the Garigliano at unguarded sites upstream from his original position, his troops soon penetrated the French rear, whereupon his forces at the lower end of the river made a frontal attack. The enemy rapidly disintegrated. In fact, there was no "Battle of the Garigliano"; it was more like a series of desperate rearguard actions by the French, who incurred severe losses: between three and four thousand killed, many thousands more prisoner, and the loss of their entire artillery. By January 1, the fortress city of Gaeta, some eighteen miles northwest of the Garigliano, fell to Córdoba's columns virtually without a shot, effectively driving the French from Neapolitan soil.

Gonzalo de Córdoba's success at the Garigliano was rooted in his willingness to see things from his enemy's point of view. In that era, and for the most part even today, armies did not undertake major operations during winter. So Córdoba went into winter quarters, thereby strengthening Saluzzo's assumption that things would be quiet until the spring. But Córdoba didn't just attack in the winter; he rapidly shifted his forces and attacked precisely where Saluzzo was least prepared to deal with him.

Although it is unlikely that he had ever heard of Gonzalo de Córdoba or the Battle of Garigliano, this was almost precisely the same maneuver that George Washington would pull some 275 years later at a place called Trenton.

The Ages of Reason and Science—and Deception

Ruses and Trickery in Early Modern Times (1648–1900)

The Age of Reason (the latter seventeenth and entire eighteenth centuries) and what might be termed the Age of Science (the nineteenth) were eras of great progress in all aspects of human endeavor. For this reason, these years may also be called the Age of Enlightenment. Obviously, there was a lot of original mental activity going on in this period, as witness the vast number of names the great thinkers conjured up for the era in which they flourished.

In military affairs, these centuries saw the foundations laid for modern warfare. What marks the beginning of this period is the end of the Thirty Years War, a conflict noted for its savagery and lack of finesse. That war was a religious struggle, the last major spiritual combat to ravage Europe. And ravage it did, causing such trauma that the acceptable style of warfare went through a complete metamorphosis. After 1648, warfare became less physical and more mental. There was still a lot of violence, but it was much more organized and well thought out violence. The big change was that armies were much more disciplined, and "living off the land" was generally avoided. Supplies for armies came from stockpiles in fortresses (known as "magazines"). Thus the wars tended to consist of a lot of marching and besieging in order to cut the other army off from its magazines. Battles were rare, if bloody, while sieges were common. Since warfare had now turned into a grand game of chess, the commanders had more opportunity to use deception, and gain more from its use. This was warfare in the Age of Reason, and it endured until the appearance of the conscript armies of revolutionary France in the 1790s, which changed the rules considerably. And then along came one Napoleon Bonaparte, who changed them even more, beginning the trend toward "total war," which saw its fullest development in the twentieth century.

But if it was Napoleon who truly ushered in the age of modern warfare in a technical sense, in terms of deception, he was merely carrying on a tradition that had begun in the dim past. Although Napoleon demonstrated how deception could be regularly employed on a vast, even strategic, scale, the entire era was full of successful deceptions. While many dwell on this period for the rapid technological changes, the flowering of deception has tended to be ignored.

Napoleon was an excellent teacher. Most of his ablest pupils were his many enemies, and they learned the art of deception well enough to eventually overthrow the master.

There was a general period of peace after Napoleon's final defeat in 1815, but the "little wars" through the balance of the nineteenth century (some of which were by no means so little) saw an increasing use of deceptions, and notably so in the case of those that took advantage of the role that technology had in war.

Montecuccoli Puts One Over on Turenne: The Rhineland, 1672

In 1672, Louis XIV invaded the Netherlands, intent on adding a significant portion of the region to France, bringing on a war with both the Dutch Republic and the Austrian Hapsburgs, who between them owned the real estate in question. In order to keep Count Raimundo di Montecuccoli (1609–1680) and his Austrian forces as far from the Netherlands as possible, Louis sent his finest commander, the great Henri Turenne (1611–1675), *Maréchal-Général* of France, to Mainz. Using the French-held fortress city on the Middle Rhine as a base, Turenne would be able to interfere in any Austrian attempt to bring troops up from southern Germany. Through the summer and autumn of 1672, Turenne (with twenty thousand men) and Montecuccoli (twenty-five thousand) played a confusing game of feint and counterfeint, neither being able to get the jump on the other, until the two armies went into winter quarters early in 1673. This was a net defeat for Montecuccoli, who was supposed to bring his army to the support of the Dutch under William of Orange, then campaigning on the lower Rhine.

For the campaign of 1673, Louis renewed his orders for Turenne (reinforced to about twenty-three thousand men) to prevent Montecuccoli from joining the Dutch, and added that he was also to defend Alsace, and to do both while not offending the numerous neutral petty German states. Montecuccoli's orders remained as before, to support the Dutch Army, then preparing to invest Bonn, in the Rhineland. Obviously, if he attempted to move directly to join the Dutch, Turenne would soon get wind of it and begin the interminable ballet of forces that had prevented him from fulfilling his mission the previous year.

However, he, by chance, had become aware of the complex character of Turenne's instructions. This gave Montecuccoli the opportunity to engage in a little deception. He would move his army to the support of the Dutch by tricking Turenne into thinking that he was going to invade Alsace.

In July, Montecuccoli marched northwest from Nuremberg, where he had wintered, with about twenty-five thousand men. To prevent his crossing the Rhine, Turenne advanced from Mainz into Germany. Mindful of his instructions not to offend local German potentates, Turenne had to negotiate to secure a bridge across the Main River at Aschaffenburg. This gave him a very slender line of communications, and complicated his supply situation, a matter not helped by the fact that the local populace in Germany was unfriendly, making it difficult to purchase supplies. As a result, Turenne was anxious to engage Montecuccoli at the earliest opportunity, so that he could withdraw to a securer area as soon as possible.

It was Montecuccoli who offered battle first, at a place called Windsheim, and Turenne promptly began deploying his forces as well. However, Montecuccoli's preparations were a ruse, with only part of his army actually forming up for battle. He quickly curtailed his deployment, gave Turenne the slip, and headed for Marktbreit, where there was a bridge over the Main, controlled by Austrian forces. Reacting quickly, Turenne followed, and by forced marches was able to arrive at Marktbreit first. There, he positioned his army so as to cover the approaches to the town. In doing so, he apparently foxed Montecuccoli, since, although the latter controlled the bridge, any attempt to cross it would expose his army to attack on unfavorable terms. Nevertheless, Montecuccoli deliberately tarried in the area for about a week, creating the impression that he was preparing to cross the Main in the face of Turenne's army. Having thus lulled his opponent into believing a major clash was imminent, Montecuccoli took advantage of the fact that Austrian forces controlled all but one of the bridges over the Main, that one being Turenne's at Aschaffenburg.

By some quick marching, Montecuccoli withdrew from the vicinity of Markbreit and began marching as if to sever Turenne's line of communications at Aschaffenburg. This caused Turenne great difficulty. If Montecuccoli seized Aschaffenburg, Turenne's lines of communication to the Rhine and France would be cut, and, worse, Alsace would lie open to invasion. On the other hand, if Turenne chose to march to the relief of Aschaffenburg and Montecuccoli's threat turned out to have been a feint, then the way would be open for the Austrians to march down the Rhine and join the Dutch.

Confused as to Montecuccoli's intentions and worried about his supplies, Turenne withdrew to the Tauber, there to await developments. This gave Montecuccoli the break he needed. Quickly crossing the Main at one of the many bridgeheads controlled by friendly forces, he marched to the support of the Dutch before Bonn, which fell on November 12, 1673.

Montecuccoli's successful outmaneuvering of Turenne was a brilliant feat of deception, the more so because Turenne was himself an unorthodox warrior. Indeed, the fruits of Montecuccoli's triumph would be minimized when Turenne undertook an unprecedented winter campaign early in 1674.

The Duke of Marlborough Makes a Little Excursion: Germany, 1704

The first member of the Churchill clan to make good was one John Churchill, an English country gentleman who is more generally known by the title which a grateful nation bestowed upon him, the Duke of Marlborough. One of the most capable and innovative commanders of the Age of Enlightenment, Marlborough by no means had an easy task of it, as he commanded the allied British and Dutch armies during the War of the Spanish Succession (1701–1714). The war was a complicated affair (far too complicated to be summarized here), which in essence saw a French bid to dominate Europe through acquisition of Spain and her possessions, including the area now called Belgium. As a result, France saw herself confronted by a coalition of European powers, including England, the Netherlands, and the Austrian Hapsburgs.

Early in the war, the French had undertaken a series of campaigns to secure control of southern Germany, in order to prevent cooperation between the Austrians and their western Allies. By 1703, they had attained a great degree of success, having secured an alliance with Bavaria. Their plans for 1704 envisioned twin offensives, one against the Netherlands, to pin the Anglo-Dutch forces in place, and one against Austria, with the hope of knocking the Hapsburgs out of the war. To counter this, Marlborough, in command of the Anglo-Dutch forces, and Prince Eugene of Savoy, another brilliant officer and the principal Austrian commander, planned to hold the line in the Netherlands, while concentrating an enormous army in the Danube Valley, in order to throw the French out of south Germany. Marlborough's part in the plan was the most difficult. Essentially he had three problems:

1. To convince his Dutch allies that it was in their interest to let Marlborough campaign in Germany, leaving the Dutch frontier "unguarded";
2. To convince the French that Marlborough was not planning on a campaign in Germany;
3. To move an army farther than any was supposed to be able to move in a single campaign.

The third problem was the least difficult. Although it was uncommon for armies to make marches of hundreds of miles in a single "campaigning season," Marlborough believed it could be done in this instance, because for a good portion of its march his army would be advancing parallel to the Rhine, and could thus be supplied by water, which permitted him to leave behind the customary elaborate supply trains of the day. It was his allies and his foes who represented the biggest obstacles to his proposed movement, rather than logistics. And to deal with these twin problems, Marlborough resorted to an elaborate deception.

First he convinced the Dutch that a short advance up the Rhine would put him in position to undertake an offensive into France via the Moselle River. This seemed reasonable to the sturdy burghers of the Dutch Republic meeting in The Hague, and so they gave their blessing to the undertaking (particularly since Marlborough explained how he planned to leave a "substantial" force to protect the Netherlands from the south).

Now came the problem of convincing the French. This took more doing. Fortunately, Marlborough had an elaborate network of spies in France, some of them in high places. This provided him with excellent "feedback," information that permitted him to correct or alter his plans as they unfolded. So Marlborough's army marched.

The army moved with some speed, considering the times. Marlborough was shortly abreast of the Moselle. At this point, of course, the Dutch and the French both expected him to advance upriver. But Marlborough demurred. He held off inquiries from the nervous Dutch by saying conditions at the Moselle were not right for an offensive in that direction, since the French seemed to be onto the plan. He then explained his alternative plan, to cross to the west bank of the Rhine farther upstream at Philippsburg, thereby outflanking the French forces poised to meet him on the Moselle. This calmed the nervous burghers. Of course, the French got wind of the scheme (they had spies everywhere, too), particularly since Marlborough ordered munitions stockpiled at Philippsburg and a bridge started. So the army kept marching.

Soon the army was at Heidelberg, not very far from Philippsburg. There, Marlborough had a lot of new supplies issued (including a new pair of shoes for each man). At this point, the army ought to have turned west. Instead, Marlborough marched eastward, deeper into Germany. The Dutch howled, but it was too late for them to do anything. Marlborough had his army (including ten thousand troops paid for with Dutch money) where he wanted it. Meanwhile, the French struggled to reposition the troops that they had posted to Alsace in anticipation of his Philippsburg operation, but to no avail, for Marlborough was actually between France and Germany. And on August 13, 1704, Marlborough, in cooperation with Prince Eugene, dealt the French and their Bavarian allies a devastating blow at Blenheim, a small village on the Danube in southern Bavaria.

Marlborough's feat in bringing forty thousand men two hundred and fifty miles from the Netherlands to Bavaria in five weeks, with the loss of only twelve hundred men to straggling or disease, has always been regarded as remarkable for the age—a logistical wonder. Yet, his most important achievement in the campaign was the great finesse with which he deceived everyone as to his intentions. Everyone, that is, except Prince Eugene.

Dan Morgan Does a Cannae: The Cowpens, January 17, 1781

One of the most remarkable and complete victories ever attained by American arms took place in an almost pristine wilderness, by a minuscule army under the command of a rustic frontiersman named Daniel Morgan.

The War for Independence in the South was characterized by an almost unbroken series of reverses for the American cause, which nevertheless led ultimately to the liberation of more and more territory from British control. The one significant victory in this campaign was at a place called the Cowpens in the western reaches of South Carolina.

The American force in the area was led by Daniel Morgan, a frontier farmer with a fair modicum of military experience reaching back to the French and Indian War. Morgan's force was modest, under five hundred "Continentals," troops regularly raised and trained by Congress; some five hundred unreliable militiamen; and about a hundred light cavalry. It was being pursued by a British force of about the same strength, composed almost entirely of "Tories," Americans loyal to the Crown, and commanded by Banastare Tarleton, one of the most ferocious and hated Tories in the colonies. Tarleton's men were better trained and better equipped than Morgan's, and fresher to boot.

Finding himself unable to retire anymore, Morgan decided to fight. Taking advantage of the ground and every wile that he possessed, he deployed his troops very carefully. He selected a modest rise as his main line of resistance, and then posted his men in three lines and a reserve. About 450 yards to the front of the rise he posted 150 militiamen, men from the Carolinas and Georgia. Reminding them that they had often boasted of their prowess with the rifle, Morgan ordered them to let the enemy get within 50 yards, whereupon they were to fire two rounds and retire on the second line. About 150 yards behind these men would be a second line, of 300 militiamen. They were ordered to await the retirement of the men of the first line, holding their fire until the enemy was within 50 yards, when they were to fire twice and retire to the left and rear of the army. Now, since militiamen were notoriously inclined to run rather than fight, Morgan's orders to fire and flee made a virtue out of a vice. Some 150 yards behind these troops, 150 yards short of the crest of the rise, Morgan posted his best men, his 450 or so Continentals. In reserve, Morgan posted his light cavalry and some mounted militiamen, for a total of 125–150 men. Morgan deployed some time before the enemy could come up. This enabled him to rest and feed the troops, no small thing before a battle. He also deployed with a river not too far in his rear, making retreat difficult, and this served as an encouragement to the fainthearted (who knew they could not retreat easily).

Arriving in the vicinity of the Cowpens after an arduous four-hour march, Tarleton deployed his men in a much simpler formation, with about 800 infantry (including 200 recruits) and two light cannon on the front line, covered by 50 light cavalry on each flank, with another 200 light horse in reserve. Tarleton gave his men some rest, but he roused them before dawn and had them ready to advance while it was still morning twilight. Although a small engagement, a lot of interesting things went on at the Cowpens, but the basic course of the battle can be simply told.

After some preliminary skirmishing between British cavalry patrols and American riflemen, Tarleton's infantry advanced against Morgan's line of skirmishers at 7:00 A.M. The American skirmishers promptly fired their two rounds and fell back on the second line, where the balance of Morgan's militiamen waited. These militiamen in turn also fired two rounds and departed rather precipitously for the left rear of the army. These volleys temporarily halted the British, but, despite losses, they pressed on and at 7:15, ran into Morgan's Continentals. A heavy firefight now ensued, as the British regulars met their American counterparts. Outnumbered, the Americans gave ground, slowly falling back toward the crest of the rise, folding back ("refusing") their flanks when the British light cavalry became rambunctious. By 7:30, the British were pressing the Americans hard.

However, in the rear, Morgan's officers were rallying the militiamen, who had fallen back from the first and second lines. By 7:45, as the Continentals were being pressed down the reverse slope of the rise, which marked his main line of resistance, Morgan was ready. From his left rear emerged about 120 mounted troops, while from his right rear emerged the 450 or so rallied militiamen. These two attacks took Tarleton's troops in the flanks as the Continentals returned to the attack against their center. Although Tarleton and his men fought bravely, they were pocketed as neatly as had been the Romans by Hannibal at Cannae, some two millennia earlier. More neatly, indeed, for about a third of the Romans escaped, while of Tarleton's army, only his reserve cavalry and a few other mounted men (including Tarleton) managed to get away.

The Cowpens remains one of the finest examples of using all of one's assets, even those likely to be regarded as debits, to their maximum. It is also the neatest victory in American military history. Not bad for a band of uneducated farmers and frontiersmen led by a commander with no formal military training.

George Washington Attacks New York: 1781

Warfare in the American colonies depended greatly on deception. And the British never quite got used to it. For the Americans, the extensive use of deception seemed to come naturally. Perhaps it was the legacy of frontier warfare, but the use of deception also sprang, as it often does, from sheer desperation.

George Washington spent much of the American Revolution within about a hundred miles of New York City. This was because the Big Apple (which wasn't very big in those days) was the principal British base in the colonies from the time they captured it in mid-1776 until the very end of the Revolutionary War in late 1783. As Washington's army—rarely more than 8,500 men—marched and countermarched across New Jersey and southern New York State, it posed a constant threat to the British position in New York City, thereby pinning down 12,000–17,000 men, a sizable chunk of the king's troops in America. Meanwhile, a lot of fighting was going on elsewhere, notably in the South.

In the spring of 1781, frustrated at being unable to run Nathanael Greene's tiny patriot army to ground in Georgia and the Carolinas, the British commander in the South decided to launch a major campaign in the South. In May, Cornwallis had invaded Virginia, hoping to strike at the heart of the insurgent-held territories. American forces under the youthful marquis de Lafayette promptly led him on a merry chase all over the eastern parts of the state. The result was that by August 4, Lafayette had succeeded in running Cornwallis and 8,000 British troops to earth at Yorktown on Chesapeake Bay. Normally, a British army in fortified positions with its back to the sea was virtually invincible. But this was not to be in 1781. At about the same time that Cornwallis found himself cornered at Yorktown, Washington learned that the French comte de Grasse was sailing from Santo Domingo with a powerful fleet and 3,000 troops to support the American cause. After conferring with the comte de Rochambeau, the commander of the 5,500-strong French expeditionary force based in Rhode Island, Washington resolved upon a daring undertaking.

For several years, Washington had made no secret of the fact that he would like to undertake an offensive against New York. In fact, only a year or two earlier, Washington had concentrated a considerable force, upward of 12,000 men (Continental regulars and militiamen), in anticipation of such an attempt. Upon closer examination of the British defenses, however, Washington had thought better of the attempt, and sent many of the troops home. Even with Rochambeau's aid, the place seemed too strong. Washington now proposed to use his cherished offensive against New York to mask an even more daring operation.

Soon troops began to move; some 4,000 Frenchmen from Rhode Island marched westward toward New York, while Washington drilled his Continentals harder than ever and ordered the militia to turn out in great numbers, equipped for some weeks' service, and promptly put them into the lines around the city. A number of small skirmishes and raids were undertaken to grab isolated British posts around New York, and there was some fighting in what is now the Bronx and along the New Jersey side of the Hudson. Orders were given to collect supplies and boats and to expand the number of army bakeries. All seemed in readiness for a major assault on the city. And on August

21, leaving behind perhaps 3,000 regulars plus some thousands of militiamen, Washington and Rochambeau began a rapid march southward with 7,000 troops, while a French naval squadron sailed from Rhode Island with Rochambeau's heavy artillery. As Washington's army marched south, the British in New York prepared to meet a major offensive, one which never came. For the three Franco-American forces (the army and the two fleets) rendezvoused in Chesapeake Bay, and by September 18, Washington disembarked in Virginia and promptly went to Lafayette's aid before Yorktown. A siege ensued, which, aided by the successful outcome of a fleet engagement between the French and the British, resulted in Cornwallis' surrender with 7,000 men on October 19, 1781. With that, the War of the American Revolution was effectively over.

Washington's cover plan for the Yorktown operation is a classic example of what a proper deception plan ought to include. Not only was an assault on New York City a plausible operation, indeed one in which Washington had several times expressed an interest, but it provided a convenient mask for the concentration of Rochambeau's army on his own. The skirmishes and raids tended to alert the British to the possibility of such an attack, as did the gathering of supplies, the building of additional bread ovens in the principal American camps, and the collection of boats. Perhaps the finishing touch was when Washington called out the militia, for the militia was notoriously unwilling to serve for very long, and the move suggested that action was imminent. Altogether a neat, tidy, and highly successful deception.

The Battle of Arcola: Achieving Victory with Twenty-nine Men

In 1796, a young general named Napoleone Buonaparte was given command of the French army in Italy. Although it was an ill-supplied, demoralized force, the young officer (who shortly Frenchified his name to the more familiar "Napoléon Bonaparte," the accent mark usually dropped in English) led his troops on an offensive to seize "the most fertile plains of the world. The rich provinces and great cities" of Lombardy and Venetia, for the greatness of the French

Republic and their own "honor, glory, and wealth." And so it was, as Bonaparte soon demonstrated himself to be a commander of remarkable skill and great resourcefulness (see "Bonaparte's Favorite Trick," page 104). One of his most impressive victories was at Arcola, a small town about a dozen miles southeast of Verona, in a bloody fight that lasted three days (November 15–17, 1796).

As part of a maneuver to get behind an Austrian army of twenty-four thousand men preparing to invest Verona, Bonaparte had advanced his army of nineteen thousand along the south side of the Adige River. Having crossed the Adige, portions of the army were to cross the small Alpone River at the bridge at Arcola, which would put them on the flank of the enemy's army. Cleared of details (some of which are quite spectacular), the battle amounted to an effort by the French to seize the river crossing at Arcola from the Austrians, who held it in strength. A tough fight ensued, for the only practical way to assault the town was across the bridge itself, the Austrians having secured control of the whole eastern side of the Alpone.

The French repeatedly tried forcing the bridge, all to no avail. At one point, Napoleon himself led the French charge, with the colors in his hands. Nor were efforts to entice the Austrians into a counterattack successful. But as the fighting went on at the bridge, Bonaparte was planning a more creative response to the problem presented by the Austrians holding the line of the Alpone in strength.

On November 16, a small force managed to cross the Alpone by boats just above its junction with the Adige. Rapidly establishing a defensive perimeter, they provided cover for engineers to begin building a pontoon bridge across the little river. Although this bridgehead was sorely beset by Austrian forces, it held. Early on November 17, Bonaparte renewed his attack across the bridge at Arcola, using General André Masséna's division of 8,000–8,500 men, while sending General Pierre Augereau's division of 9,000–9,500 men across the pontoon bridge, roughly a mile to the south. As this meant pulling Masséna out of a movement to the northwest at Belfiore di Porcile, where he had been lightly engaged with the other half of the Austrian Army of 13,000 men under the count of Hohenzollern-Sigmarinen, Bonaparte posted a demibrigade of 1,000–1,200 men to maintain Masséna's old position to the northwest, thereby creating the impression that the place was held in strength.

At Arcola, the Austrian commander Baron Mitrovsky, who had 8,000 men under command along the Alpone, held Masséna's attack, and then made a furious counterattack of his own, sending the bulk of his troops across the disputed bridge, to drive back the French. Meanwhile, he dispatched a column of 2,500 men to contain Augereau. Catching Augereau before he could emerge from the swampy ground along the Alpone, the Austrian column was able to block the French from advancing. At this point, Bonaparte could easily have lost the battle. Thinking quickly, he dispatched a troop of guides, his personal bodyguard, and four trumpeters, a total of 29 men, to sweep around the enemy's rear. At the same time, he ordered up a detachment of 800 men of the garrison of Legnano, a small town a few miles to the southeast.

Riding first southward a half mile to the town of Albaredo, the small mounted detachment then turned east and northeast for two miles, then west, so that by 3:00 P.M., they were less than a half mile in the rear of the Austrians blocking Augereau's advance. At this point they attacked, with a great sounding of trumpets, pounding of hooves, and firing of pistols. Shortly afterward, the detached battalion from Legnano came up to strengthen Augereau's right rear. Amid the smoke and confusion of the battle to his front, Baron Mitrovsky believed that the French had managed a massive turning of his left flank. At the same time, word came that his counterattack at Arcola had faltered. Masséna, having executed an ambush which threw back his troops in great disorder (leaving 3,000 of them prisoners), was once again storming the bridge. It appeared to Mitrovsky that the game was up. He ordered a retreat.

Arcola was a hard-fought battle. Including the losses incurred in Masséna's fighting to the northwest of the town on the first two days, the French suffered about 4,600 casualties, the Austrians 6,000, most from Mitrovsky's division. But if the battle was a testimonial to courage and devotion of the troops on both sides (charging across a bridge is not likely to enhance one's actuarial profile), the battle itself was won because of a series of deceptions: the pinning of Mitrovsky's attention to the bridge and of Hohenzollern-Sigmarinen's to Belfiore, to divert attention (and troops) from events elsewhere; Masséna's ambush defense of Mitrovsky's counterattack at Arcola; the 29 mounted guides and trumpeters attacking from the rear; and the troops from Legnano coming up to "reinforce" Augereau—a total of five tricks all running simultaneously. Not a bad afternoon's work.

Bonaparte in Britain: The French Invasion Scare of 1805

With a couple of brief intermissions, England and France were at war for most of the years from 1792 to 1815. It was the classic struggle between the alligator and the lion, each formidable in his own environment, but relatively weak in the other's. Despite Britain's powerful maritime resources, by 1804, it was beginning to become apparent that Napoleon was making strenuous efforts to invade England.

Napoleon appears actually to have contemplated an invasion of Britain from sometime in 1803, shortly after hostilities were renewed following a short peace. He began issuing orders for the accumulation of a great number of vessels at Boulogne and other Channel ports. In addition, he created the "Camp of Boulogne," an elaborate cantonment and training base, and began concentrating an army there, which was soon ominously named *L'armée d'Angleterre* (The Army of England). In 1804, he revised his plans in anticipation of an undertaking against England the following year.

The final version of the proposal envisioned the French Navy's Toulon squadron (eleven battleships) and a third of the Rochefort squadron (five battleships) making a sortie to the West Indies to draw the Royal Navy away from the Channel. This done, the Brest squadron of 22 battleships and the balance of the Rochefort squadron of 10 battleships would combine to lift an army corps of about 14,000 men and land it in Ireland, as a diversion. The fleet would then sail around Scotland, to arrive in the Channel from the north, and, picking up the Dutch Navy's 12 battleships at the Texel, escort the movement of the bulk of the army (four corps totaling 112,00 men) to England, in a series of short trips over three days. Held in immediate reserve in France would be a further pair of army corps totaling 50,000 men.

Napoleon's subordinates made exhausting efforts to accumulate the hundreds of light craft that would be necessary to effect the landings. In addition to an already intensive training program (actually the only one they would ever get as an army throughout the long years of war to come), the troops practiced boarding and debarking from landing barges. Meanwhile, Napoleon collected all available literature on Britain—geographic, political, economic, historical, and cultural as well as military—and interviewed numerous persons who had recently been there, including not a few Frenchmen who had been held there as prisoners of war until the peace of Amiens in 1802.

Napoleon's proposed invasion of England elicited an immediate response from the British. In 1804, there were relatively few troops in the British Isles, perhaps 80,000 regulars. The militia was immediately called to duty, adding 50,000, and a new body, the Supplementary Militia was activated, for another 26,000, making a total of 156,000 troops available, albeit that only half of them were properly trained, a third more partially trained, and the balance wholly untrained; but given time, both numbers and training would improve. At the same time, the country's maritime defenses were revamped, provisions being made for the repair of older fortifications and the erection of new ones. And, of course, the Royal Navy put everything it had into commission. So by the spring of 1805, the British were at least half ready to meet Napoleon. Then interesting things began to occur.

It was in March of 1805 that the plan began to be put into effect. On March 30, the French Toulon squadron of eleven battleships, under Admiral Pierre de Villeneuve, evaded Admiral Lord Nelson's blockading British squadron of twelve battleships, and disappeared into the Mediterranean, while Nelson swept the seas in search of it. Meanwhile, Napoleon pressed his preparations for the invasion, so that on August 8, it was determined that he had available, on paper, 2,343 vessels of all types, capable of lifting 167,590 men and 9,149 horses.

Unfortunately, he didn't have command of the Channel, for by early August, Nelson, having scoured the seas in vain for Villeneuve, had brought his squadron into the English Channel, temporarily raising British strength in battleships in home waters to about fifty. (Villeneuve, having led Nelson on a merry chase across the Atlantic, was anchored at El Ferrol, in northwestern Spain, preparing to sail for Cádiz, there to join the Spanish fleet, to effect a concentration of thirty-five ships of the line. Nelson, with twenty-seven battleships, would annihilate this fleet at the Battle of Trafalgar, October 21, 1805.)

Now, while Nelson had been chasing Villeneuve across the Atlantic and back, and Napoleon had been gathering his invasion fleet and training his army at Boulogne and other coastal areas, Britain had concluded the so-called Third Coalition, bringing Austria and Russia into the war against France. This was not precisely unanticipated by Napoleon, as it was only a matter of time before these two powers would throw in with Britain.

These two new enemies would bring enormous forces against him: Austria had nearly 190,000 men in Italy and southern Germany, and Russia would soon be coming up behind them with 100,000 more within a few weeks. Napoleon began making some surprising moves with impressive speed. On August 23, even before Austria had made a hostile move, he quietly issued orders for the army to march eastward. The next day, he instructed his cavalry to impose a tight screen along the middle Rhine. They took up this task with alacrity, for Napoleon's cavalry had been cantoned not with *L'armée d'Angleterre,* at the Camp of Boulogne on the coast, but in Alsace, 250 miles to the east, curiously near that very Rhine River which they were now being told to screen. On August 25, three of Napoleon's most valued officers, Murat, Bertrand, and Savary, crossed incognito into southern Germany to commence reconnoitering the lay of the land. And two days later, the army, which was shortly afterward renamed *Grande Armée*, began leaving its camps along the coast. By September 25, the *Grande Armée* was on the Rhine, over 165,000 strong, while at the old Camp of Boulogne there remained only some 30,000 second-line troops, with orders to close up shop. Over the next few weeks, Napoleon won his great enveloping victory at Ulm (see "Bonaparte's Favorite Trick," page 104) and began moving eastward to confront the advancing Russians. And the invasion of Britain? It was never officially canceled, and indeed Napoleon would issue inquiries and proposals several times over the next few years, but they never went anywhere.

In fact, if not from the first, then certainly from late 1804, the idea of an invasion of England appears to have been a ruse, a carefully planned deception designed to mask the preparation of an enormous army for a clash with Britain's probable continental allies, Austria and Russia, who by that time were clearly aligning themselves against Napoleon.

There were, in fact, too many flaws in Napoleon's preparations for the plan to have been a completely serious one. To effect a successful landing in England, Napoleon needed local command of the seas for weeks, during all of which time he would require favorable weather. This would enable him to get his army ashore and keep it supplied until a sufficiently large lodgment was secured to permit the troops to live off the land. But while his army was a good one, in England, it would have to operate under a number of handicaps, the most serious of which would be a desperate shortage of horses. At best, there would be transport available for only nine thousand horses, a fifth of the normal allocation for an army the size of which Napoleon planned to land. So it would be an army lacking transport, lacking cavalry for reconnaissance, and lacking any real mobility for its artillery. And that was a "best-case" situation. So it seems very likely that the whole plan was a ruse; indeed, the speed with which the *Grande Armée* abandoned its preparations for the invasion and set out on the march eastward, and the interesting deployment of the French cavalry during this period, both strongly confirm this conclusion.

In fact, given the Royal Navy's traditional sensitivity about the security of the Channel, Napoleon would never have made it to England. It's worth recalling that as soon as Nelson determined that he had no idea where the French fleet was, he immediately sailed for the Channel, Britain's most vulnerable point. In the words of Admiral Lord Barham, First Sea Lord, "I do not say they cannot come, I only say they cannot come by sea." But of course, it was perhaps not so much for the benefit of Britain that the whole invasion scheme was laid on, as for that of Austria and Russia, who were to become its primary victims. So Napoleon's invasion of Britain was one of the most successful deceptions ever attempted.

Bonaparte's Favorite Trick: The *Manoeuvre sur le Derrière*

For a commander with so spectacular a reputation, it is surprising how few tricks Napoleon actually had up his sleeve. He liked one of them so much that by one count, he used it over two dozen times, in virtually every case to great success. This was the *manoeuvre sur le derrière*, or "movement about the rear." It might be termed Napoleon's "strategy of superiority," for its proper execution required that he have more manpower than his foes, as well as lots of maneuvering room.

Put simply, the *manoeuvre sur le derrière* was a vast turning movement in the face of the enemy. While one or two army corps were detached to pin the attention of the enemy to his front, Napoleon would take the bulk of his army on a swift, wide march around one of the enemy's strategic flanks, behind a thick screen of cavalry, optimally with some substantial geographic feature providing a "curtain of maneuver." As he advanced toward the enemy's rear, he would thrust a corps or two and some cavalry in the direction from which the enemy might expect reinforcement, to prevent any such from interfering in the main event. He would then fall upon the enemy from the rear, having severed his lines of communication and retreat. In the ensuing battle, he had high hopes of annihilating the enemy.

Napoleon built his reputation on the *manoeuvre sur le derrière*, using it at least three times in his very first campaign, in Italy in 1796. This maneuver made possible the smashing strategic victories of Marengo in 1800, Ulm in 1805, Jena and Auerstadt in 1806, Friedland in 1807, and Wagram in 1809. Perhaps the best way to illustrate the *manoeuvre sur le derrière* is to have a look at the campaign against Prussia in 1806, arguably the most perfect campaign in Napoleon's career. In this campaign, virtually everything went according to plan, leading to the complete defeat of the much-vaunted Prussian Army in but eight days, and the total annihilation of the Prussian state in the following weeks.

To begin with, Napoleon's approach to organization must be understood. He divided his army into a number of separate commands called *corps d'armée* (literally "bodies of army"), each composed of several divisions of infantry, some cavalry, and some artillery, all under the command of a seasoned officer. Each corps was capable of holding its own for a day or two in a defensive battle against a considerably superior enemy force. No two army corps were identical in composition, their size often ranging from as few as ten thousand men to as many as fifty thousand, thus making it difficult for the enemy to ascertain the overall strength of the French Army. When moving, the army corps were disposed in what Napoleon called the *bataillon carré* (literally, "battalion square"), a sort of lozenge formation, with the corps advancing along parallel roads about a day's march apart. The idea was that the corps at the leading point of the lozenge was the advanced guard, those in the rear the reserve, and those on the lateral points the flank guards. Should the enemy be detected on, say, the right flank, the right flank corps would turn in his direction, thereby becoming the advanced guard, while the old advanced guard would become the left flank guard, the old reserve the new right flank guard, and the old left flank guard the new reserve. The *bataillon carré* advanced behind a thick cavalry screen to prevent the enemy from ascertaining Napoleon's movements, while enabling him to feel out theirs.

The campaign of 1806 began on October 6, by which time Napoleon had established his base of operations at Würzberg, in northern Bavaria. Concentrated in an area only 95 miles across by 75 miles deep, he had an army of some 180,000 men (**1** on the map, represented by the thick lines), at a time when the Prussians and their Saxon allies (the thin lines) had not yet completed their concentration (although they had begun mobilizing before the French), having some 90,000 men north of the Thuringian Forest (**2**), an additional 40,000 southeastward of that, in Saxony, and some more men deep in the rear, along with large Russian armies which had barely begun moving. The Prussians expected that Napoleon would move directly upon them from his base area. Instead, forming a *bataillon carré* of three major groups (40,000 men on the left flank, 50,000 on the right flank, and a 70,000-strong central body, partially committed as the advanced guard and partially as the reserve), he pushed his forces northeastward behind his cavalry along three roughly parallel routes (**3**), using the Thuringian Forest as a "curtain of maneuver," to mask and screen the operation. The movement was conducted with almost total immunity. The first actual contact with the enemy did not take place until October 10, when a Prussian division under Crown Prince Louis Ferdinand was soundly thrashed at Saalfeld, the prince losing his life in the process.

This incident served to alert the Prussians to the French maneuver and they attempted to shift front, fall back, and reconcentrate at the same time, since it was clear that Napoleon's movement was going to put him between their two forces (one north of the Thuringian Forest and the other in Saxony). But such a regroupment proved impossible, as Napoleon's army began turning northward and then westward (**4**), further separating the Prussian forces while threatening their communications with their depots in eastern Prussia. On October 13, Marshal Jean Lannes, one of the finest advanced guardsmen in the French Army, encountered the main Prussian Army in the vicinity of Jena. He immediately offered battle, thereby pinning the enemy, and sent couriers spurring back to report to Napoleon, who was with the French main body less than a day's march behind. The next day saw the simultaneous overwhelming French victories of Jena and Auerstedt.

The broad outlines of the *manoeuvre sur le derrière* are quite easy to understand. However, there was great risk in this strategy. Only

The Strategy of the Indirect Approach as illustrated by the Campaign Against Prussia in 1806

Situation and General Outline of Movements October 6 through October 13.

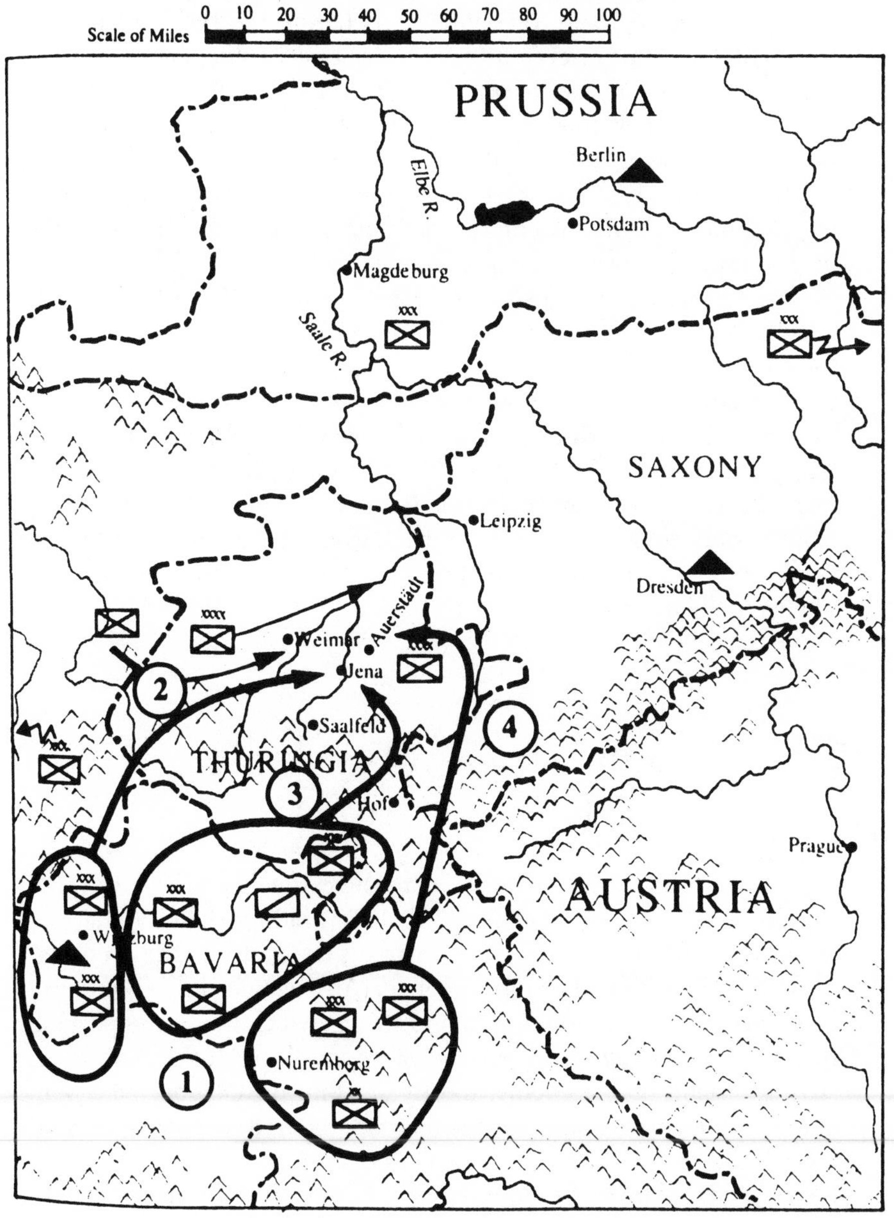

Reprinted from *Napoleon at War*, edited by Albert A. Nofi (New York: Hippocrene Books, 1984). Copyright©, Albert A. Nofi, 1984, used with permission.

bold execution, swift movement, and aggressive use of the pinning forces and the cavalry, could make it work. If the enemy gained any notion of what was afoot, as in 1807 when an order captured by the Russians revealed Napoleon's intentions before the Battle of Eylau, he might slip away, or even attack the relatively vulnerable marching columns. In his early campaigns, Napoleon had the advantage of having an army which moved much faster than those of his enemies, and was employing an innovative technique. What is difficult to understand is that he kept using the technique, and that it kept working. Napoleon didn't patent the *manoeuvre sur le derrière,* and it's been used a number of times in warfare since his day. The duke of Wellington used it in Spain in 1813 with considerable effect, and it was used twice during the Chancellorsville Campaign, in 1863, during the Civil War, once by Union General Joseph Hooker to get the drop on Confederate General Robert E. Lee, and then, during the same campaign, by Lee to discombobulate Hooker. Even Operation Cobra, George S. Patton's breakout from the Normandy beachhead in 1944, was something of a *manoeuvre sur le derrière.* With all of these examples, one would think that the trick would be getting a little obvious. But apparently, it doesn't hurt if one's opponents are intellectually limited.

Joachim Murat Tells a Little White Lie

Napoleon's initial strategic objective in the campaign of 1805 was the capture of Vienna, on the theory that by holding their capital, he would make it impossible for the Austrians to resist for long. In order to do this, first he had to dispose of the Austrian armies in southern Germany, nearly a hundred thousand strong, which was accomplished by a spectacular *manoeuvre sur le derrière,* resulting in the capture of sixty thousand of the enemy, mostly at Ulm, on October 20. Even before Ulm surrendered, Napoleon had ordered elements of the *Grande Armée* eastward, to pursue the remnants of the Austrian forces. With Marshal Joachim Murat's cavalry in the lead, and Marshal Jean Lannes' hard-marching infantry close behind, the *Grande Armée* headed eastward down the Danube, the surviving Austrian forces fleeing before them. Even Austria's allies were unnerved, causing Russian General Mikhail Illarionovich Kutuzov's army to fall back as the French advanced. By the evening of November 12, Murat and Lannes were at Spitz, near Vienna, where there was a bridge across the Danube, having covered a "crow flies" distance of about four hundred miles in only twenty-six days, during which several important skirmishes were fought.

Reconnoitering the bridge at Spitz, Murat and Lannes saw that Austrian troops held the other end in strength, and were making preparations to blow it up. Thinking fast, they issued some hasty orders and then, accompanied by their staffs, boldly strode onto the bridge. As they approached the Austrian troops—some of whom were already preparing to light the powder trains that would send the bridge sky high, while others were raising muskets and pistols to shoot down the elaborately uniformed officers—Murat shouted, "Armistice! Armistice!"

The cry took the Austrians by surprise, and they hesitated. Murat approached them, explaining that an armistice had been concluded a few hours earlier, an assertion in which he was strongly seconded by Lannes and their several aides. The Austrian officer in command said that he had received no word of such.

"How is it that you have heard nothing about it?" asked Murat, incredulously. Giving his word as a Marshal of France, he went on to say, "Peace is being negotiated; lead us to your general." The officer agreed, and summoned General Prince Auerfperg. Meanwhile, the two French marshals swapped pleasantries with the bridge guards, Murat reportedly sitting or leaning on an Austrian gun.

As the French officers and Austrian troops awaited Auerfperg, a detachment of French grenadiers was quietly approaching the foot of the bridge. When they were close enough, the column suddenly charged across, overrunning the defenders with scarcely a shot being fired. Within minutes, the French held the key to Vienna. All for a little white lie.

This sort of trick is an ancient one in warfare. General Kutuzov himself used precisely this ruse to seize a bridge later in the same campaign. And it has even been used in the supposedly less trusting twentieth century, as can be seen in two examples drawn from World War II.

In June of 1940, German Major Georg Michael, of the 22nd Cavalry Regiment, was leading a patrol of six mounted men and a motorcycle dispatch rider, scouting ahead of the German Army as it advanced into France. The patrol entered the hamlet of Fremontierla-Petit and encountered a handful of French troops, who promptly surrendered. Within minutes, a French staff car drove up, and Michael's men promptly captured it as well, bagging several officers, including the commander of a battalion of Senegalese infantry. Meanwhile, one of Michael's men, whom he had posted outside the village, reported a large enemy force approaching. This presented a problem, as Michael had just reported to his regiment that the village was undefended. Promptly taking his dispatch rider's motorcycle, he drove off to investigate, with the dispatch rider following behind in the French staff car.

Not far beyond the village, Michael encountered some French Senegalese troops, who began firing. Taking shelter, he called for a parley. A white French officer agreed to talk, and approached him, asking how he came to have the battalion commander's vehicle. Michael replied that the commander had been captured, and that he had been sent to accept the surrender of the unit, which was even now under the guns of an entire artillery battalion. The French officer refused to surrender, but Michael kept talking. "Do you really want useless bloodshed?" he asked, observing that an entire German cavalry division was practically on top of them at that very moment. He then offered honorable terms, including the right of officers to retain their side arms. There was a brief and heated discussion among the French officers, but shortly, the seniormost agreed to surrender. So with a little lie, Major Michael and his dispatch rider captured five hundred fine infantrymen.

It happened again in newly liberated Cherbourg on June 28, 1944. U.S. Coast Guard Commander Quentin R. Walsh was leading a party of armed coastguardsmen on a reconnaissance to determine the extent to which the Germans had damaged the port facilities before surrendering the city. A German sailor whom they captured informed them that about fifty American soldiers were being held prisoner in one of the harbor forts, guarded by some German troops who refused to surrender. With another officer, Walsh approached the fort and called for a parley. A German officer escorted the pair into the fort, where they confronted the German commander. When the German officer flatly refused to surrender, Walsh informed him that he had eight hundred men outside, ready to take the fort by storm if necessary. Hearing this, the German commander suddenly lost interest in dying for his *Führer*, and surrendered. He must have felt pretty stupid when he discovered that Commander Walsh actually had only seven armed men waiting outside.

So if you're a soldier, don't believe what your mother told you about telling lies. They sometimes come in quite handy.

Deceiving the French Stomach

When Napoleon marched on Moscow in 1812, he did so with the mightiest army yet raised to that time. Over half a million troops, armed with muskets and cannon, moved toward the Russian capital of Moscow in early 1812. The Russians resisted in several battles, losing to the French every time. But the Russians were exercising a clever deception. The French found that most food and supplies in their path were being destroyed. By delaying the French with harassing attacks and all those lost battles, the Russians did not have to abandon Moscow until just before the ferocious Russian winter set in. The French fell for it, thinking to the last that the Russians would surrender once Moscow fell. But the Russians had no reason to surrender. Their country was vast, and the Russian government, and most of the population, simply abandoned Moscow. The French marched in and found an empty city. Moscow was particularly empty of food. Shortly, Napoleon found himself in such desperate straits that it seemed prudent to abandon Moscow in the middle of winter, and march toward food supplies that had been accumulated farther west. This was what the Russians were ready, and waiting, for. Russian raiders sniped at the retreating French, picking Napoleon's army to pieces. Few French got out of Russia alive.

This was not as deliberate a deception as the Russians would later declare. Yet, over the centuries, the Russians had learned to fall back in the face of a superior foe, and to destroy food and shelter in the enemy's path. Russia is a large country, and sparsely populated. Through the nineteenth century, armies tended to live off food supplies of the local population. This meant that if there were no local people, or the locals had destroyed their food supplies, the troops would go hungry and would eventually either starve or fall back. Naturally, this tactic was rough on civilians, but the Russians are a tough bunch and practiced this tactic into the twentieth century. Napoleon, who preached the study of past campaigns, was remiss in ignoring these time-honored Russian tactics. For their part, the Russians were simply doing what they generally did when invaded. Many of the senior Russian commanders knew that if the French kept on coming, winter and starvation would become Russian allies before long, and the French would be in dire straits. Such was the case. Russia was saved and Napoleon's army was lost.

The Ruse That Was Better Than the Plan: Fort Niagara, 1814

In early 1814, Major General Jacob J. Brown, a very fine officer, was the senior American commander on the northern front, between New York and British-held Canada. On March 9, while at Sackett's Harbor, at the eastern end of Lake Ontario, Brown received a curious communication from Secretary of War John Armstrong. The secretary wanted to capture the British naval base at Kingston, on the northeastern corner of the lake, not far from where it empties into the Saint Lawrence River. In his message, Armstrong explained that it was believed that the British intended to capture Detroit, nearly four hundred miles to the west, as the crow flies. To do this, he explained, it seemed probable that the British would have to weaken their garrison at Kingston. Therefore, he suggested that, if circumstances were right, Brown might undertake a surprise attack on Kingston, which would unhinge the entire British position from there westward, delivering a stunning strategic blow to their war effort.

Armstrong proceeded to outline the conditions which he believed had to prevail in order for Brown to undertake such a venture: The British garrison at Kingston had to be reduced; weather and road conditions had to be suitable for the venture; and the U.S. Navy's squadron on Lake Ontario had to agree to cooperate in the undertaking. Having said this, Armstrong added that if Brown was willing to essay an operation against Kingston, he should "use the enclosed letter to mask your object." The letter in question was an order for Brown to dispatch Brigadier General Winfield Scott, one of the ablest young officers in the army, to take a strong column to attack Fort Niagara, a British-held post on American soil, about one hundred eighty miles to the west, adding that if Brown felt that the American hold on Sackett's Harbor was sufficiently secure, he might take most of the balance of his forces and proceed there himself. Armstrong presumably intended for Brown to somehow arrange for this second set of instructions to fall into British hands, thereby convincing them to weaken Kingston in order to strengthen Fort Niagara.

Considering the weather and the strategic situation on the eastern end of Lake Ontario at this time, Brown did not feel that an offensive against Kingston was a reasonable undertaking. But looking at Armstrong's bogus set of orders, he decided that they actually embodied a better plan. After all, the British were expecting him to attack Kingston; therefore, presumably they would be less prepared to meet him at Fort Niagara. So Brown sent a brigade of infantry marching westward. And on April 7, while at Batavia, New York, only about twenty-five miles from Fort Niagara, an interesting thing happened. General Scott, who had, in fact, not been with Brown at Sackett's Harbor, showed up with his brigade of infantry and his artillery train, in obedience to separate orders from Armstrong issued some weeks earlier. And the very next day, Brown received a letter from Armstrong dated March 20, in which the secretary gave him a mild scolding for "misunderstanding" his earlier instructions, but then went on to approve the Fort Niagara operation, saying, "Good consequences are sometimes the result of mistakes."

As a result of Brown's movement to the western end of New York, he was able, that summer, to conduct a moderately successful offensive into Canada, capturing several forts, albeit not Fort Niagara, and almost securing complete control of the Niagara River, until the British victory in a bloody five-hour clash at Lundy's Lane (July 25, 1814), a battle that the Americans almost won. Although forced to retreat, Brown was able to maintain a substantial hold on the Canadian side of the Niagara until winter, when he was forced to withdraw to Buffalo.

It is to this day unclear precisely what Secretary Armstrong had in mind with his two sets of orders to Brown. If the Fort Niagara undertaking was a ruse, it was one that seemed a more reasonable undertaking than the officially sanctioned movement against Kingston. But if it was a ruse, why were Scott's orders framed in such a way that they complemented Brown's movements? And why did Armstrong seem so resigned to the consequences? So it may well be that the ruse was really the plan.

Santa Anna, Trickster Supreme

Several times president of Mexico (through usurpation or election), usually with disastrous results for the republic, "villain" of the Texas

War for Independence (1835–1836) and the Mexican-American War (1846–1848), and arguably, one of the slimiest characters ever to disgrace the Western Hemisphere, Antonio López de Santa Anna Perez de Lebron (1794–1876), to give him his full name, was a dictator, traitor, drug abuser, lecher, thief, and all-around rotten guy. But Santa Anna was also an organizer and commander of some skill, and something of a master at deception. There is a tale told about him that is interesting not because it is true, but because his reputation made it believable.

In February of 1836, Santa Anna was leading an army of six thousand men into Texas for the purpose of extirpating the Texas Revolution. His initial objective was the Alamo of San Antonio, a fortified mission held by a brave band of rebels. On the evening of February 22, many members of the garrison were treated to a Washington's Birthday Ball by the good citizens of San Antonio. According to legend, Santa Anna, whose army was only a day's march away, appeared in disguise at the ball. He supposedly engaged in conversation with many of the guests, and took a couple of turns around the dance floor with some of the ladies, thoroughly enjoying himself, while gathering some valuable intelligence. This is all very dubious. Santa Anna was personally acquainted with several of the people at the *fandango*, so it is highly unlikely that any disguise would have been quite good enough to fool everyone. And, in fact, had Santa Anna actually been at the ball, probably he would have been able to capture the Alamo single-handedly that very night, rather than have to invest it for nearly two weeks and then take it by storm, since all but ten men of the garrison were at the party, and, according to one tradition, most of them were beastly drunk by midnight.

So the tale is not true. But it developed because Santa Anna did have a penchant for deceptions and ruses, and owed not a little of his success as a politician and commander to them. For example, he had on one occasion disguised himself as a woman to reconnoiter enemy positions. Another time, when he entered a town during a battle and suddenly realized that the place was still occupied by the enemy, he eluded capture by boldly striding about proclaiming that his men had just taken the place, until he was in a position to make his getaway.

In 1838, during a short war between Mexico and France, a French raiding party attacked Santa Anna's quarters. The general (who was, as usual, passing his time with a young woman of easy virtue) fled, clutching his clothes to his naked body. By chance, he ran right into some of the French raiders, who called out, "Where is General Santa Anna?" Not pausing for an instant, Santa Anna turned, said "There!" and pointed at the quarters occupied by another general, one of his rivals for political power. He subsequently lost a leg largely by accident (to a French cannon shot) during this same "war," and managed to paint the unfortunate incident in such heroic terms that he was almost immediately elevated to the presidency again. But it was his capture of Vera Cruz from the Spanish in 1822 that was something of a masterpiece.

Santa Anna had begun his military career as an officer in the Spanish Army, helping to surpress the Mexican War for Independence (1811–1822). But in 1821, along with a number of other disaffected officers, he had thrown in with the revolutionaries. As a result, October of 1822 found him besieging Vera Cruz, the country's principal port, on the Gulf of Mexico. The place was well fortified. After attempting to storm the city, Santa Anna decided that treachery was likely to be more profitable. Since the place was not formally under siege, it was possible to get messengers in and out with relative ease. By this method, Santa Anna was able to bribe a citizen to open one of the gates, through which he slipped a large body of troops under cover of a convenient rainstorm. After some confused fighting in the darkened streets and plazas, the city was in Santa Anna's hands by dawn of October 6, the Spanish having withdrawn to the offshore fortress of San Juan de Ulúa. From this bastion, over the next few days, the Spanish launched repeated harassing raids on the Mexicans. As long as they held the fortress, Vera Cruz would be neither secure nor usable as a port. But Santa Anna had no way to take the fortress, lacking a fleet. So he decided that a little more trickery was in order. If he could not defeat the Spanish in San Juan de Ulúa, he would entice them to commit themselves to a major operation ashore, where he would be able to destroy them.

Secretly contacting the commander of the Spanish garrison, Santa Anna offered to surrender Vera Cruz upon payment of a substantial bribe. After careful negotiations, the Spaniard agreed. The bribe was paid, and Spanish troops began secretly occupying portions of Vera Cruz on October 26, only to be jumped by Santa Anna's men, who had been carefully concealed. After some desperate fighting, the Spanish were again able to retire on the fortress, so Santa Anna's trap ultimately failed. But it did inflict a severe reverse on the enemy, once again reminding us that even the cleverest deceptions can sometimes be frustrated by hard fighting.

"The Little Napoleon" Meets "Prince John": Yorktown, 1862

By March of 1862, the American Civil War had already lasted nearly a year. Yet, aside from Bull Run in Virginia, the previous July, and Wilson's Creek in Missouri, that August, there had been no battles of consequence, and the war seemed to be going nowhere. But Major General George Brinton McClellan, commanding general of the Union Army of the Potomac, nicknamed "The Little Napoleon," for his stature and his supposed skills, had a secret plan to win the war. Rather than take his 150,000-strong army from Washington and attempt to bull his way the hundred miles to Richmond, the Rebel capital, past an "enormous" Confederate army, he would use the Union's command of the seas to lift the army around the Rebel flank, depositing it on the lightly defended Virginia coast. From there, he would be closer to Richmond than would Confederate General Joseph E. Johnston's army in northern Virginia. A quick march, a little fighting, and Richmond would fall, and with it, the Rebellion.

McClellan's army began to take ship on March 17, to begin landing at Union-held Fort Monroe, on the Virginia coast, on March 23. By April 2, McClellan had some fifty thousand men on the so-called Virginia Peninsula, the area between the James and York rivers, only about sixty miles from Richmond.

Two days later, the Army of the Potomac began advancing up the peninsula, with historic Yorktown as its first objective. The leading elements of the army arrived before Yorktown on April 5. The town was held by Confederate Major General John Magruder, nicknamed "Prince John," for his lordly manner, and only about ten thousand men. Yet, Yorktown did not fall until exactly one month later.

There were a number of things that impeded McClellan's advance: His maps turned out to be wildly inaccurate; a river turned up in an inconvenient place; and the rain turned the roads to mud. In addition, that "enormous" Confederate army in northern Virginia was giving Lincoln pause, so he decided to hold back a portion of the reinforcements scheduled to support McClellan. And then there was a strange fellow named Thomas J. Jackson, nicknamed "Stonewall," who was raising so much hell with Union forces in the Shenandoah

Valley that Lincoln felt it necessary to take still more of McClellan's reinforcements and send them thither. But the principal reason was that "Prince John" Magruder knew precisely what was needed, not to defeat McClellan, but to deceive him into inactivity.

Actually, when McClellan's first troops had debarked at Fort Monroe, Magruder had only about three thousand men under command. But the newly landed Yankees remained inactive, enabling him to bring up thousands more. He promptly set these men to work building a thin defensive line across the seven-mile-wide peninsula, partially covered by the Warwick River. The defenses were essentially a facade, but were made to look as convincing as possible. To strengthen the illusion, Magruder liberally sprinkled the works with "Quaker guns," logs trimmed and painted to look like cannon. As a result, when McClellan finally made contact with the "Yorktown lines," they looked formidable indeed. McClellan attempted a few probes (April 5–16) to test the strength of the defenses, but by holding most of his troops behind the lines, Magruder was always able to move them to the post of greatest danger, where they took care to reveal themselves, as if in great numbers.

This convinced McClellan that the position was well fortified and defended by perhaps a hundred thousand men. Against such a position it would be suicidal to attempt a frontal assault. Since he was unable to use his sealift to turn the Yorktown lines—the York River approaches to the town being well covered by some very real Confederate batteries at Gloucester Point, and the James River side covered by the presence of the Rebel ironclad *Virginia*—McClellan decided that there was nothing to do but to undertake a formal siege, despite the fact that he had not attempted a really vigorous probe. So the Army of the Potomac dug a first parallel, emplaced siege mortars, and began evacuating trenches to advance the lines. All of which took time. This gave Joe Johnston the chance to bring his army, not very "enormous" at forty thousand men, down from northern Virginia. Johnston could have attempted to hold the Yorktown lines, what with fifty thousand men under command. But it was clear that McClellan was about to begin a very heavy bombardment. So on May 3, the Confederates abandoned their defenses at Yorktown. McClellan did not learn of this for two days. And he never admitted that he had been foxed, literally omitting all mention of the deception in several written accounts of the Yorktown operation.

A clever deception, indeed. However, what is particularly interesting about Magruder's deception at Yorktown is not that it succeeded, nor even that there were ample precedents for such ruses in military history. Rather, this was the third time in several months that McClellan had been fooled by some hastily erected earthworks and some painted logs! In the autumn of 1861, the Confederates had for a while ostentatiously held a place called Munson's Hill, overlooking Alexandria, Virginia, thereby effectively bottling up McClellan close to Washington, using a handful of troops and some "Quaker guns," only abandoning the place when Union patrolling became more intensive. Then, through most of the winter of 1861–1862, Joe Johnston's army had been concentrated at Centreville, about twenty miles southwest of Washington. But, anticipating a Union offensive, Johnston had evacuated his fortified camp at Centreville on March 9. McClellan soon learned of this, and on March 11, advanced to "capture" the place, only to discover that in many instances what appeared to be well-entrenched positions liberally garnished with artillery were weakly supported piles of earth festooned with "Quaker guns."

As Civil War historian Richard L. DiNardo noted, "The most successful deceptions are those which convince the enemy of what he already believes." And at both Centreville and Yorktown, McClellan was so fearful, that he needed little convincing.

David Porter Fools the Rebels

David Dixon Porter, foster brother to David Glasgow Farragut, was one of the ablest American naval officers during the Civil War. Despite a dour, tough appearance, Porter was possessed of a mischievous sense of humor, which made him one of Abraham Lincoln's favorite officers. He was also a bold sea dog with a fertile imagination. These traits all combined to serve him in good stead when, while commanding the federal flotilla on the Mississippi in February of 1863, a potentially disastrous crisis suddenly arose. Porter saved the day by means of a clever deception. But let Porter explain it himself, from his amusing and lively *Incidents and Anecdotes of the Civil War* (New York, 1885).

> Colonel Charles Ellet, Jr., a young man of twenty-two, commanded the *Queen of the West,* a ram improvised from a river steamboat.
>
> I ordered young Ellet to pass the batteries of Vicksburg at night, proceed to the mouth of the Red River, intercept the supplies for

Vicksburg and Port Hudson, and capture everything he could overtake.

I don't know whether it was from love of glory or from want of judgment, but, instead of taking advantage of the darkness to run the batteries, Ellet chose early daylight, got well hammered as he passed the forts, and nearly defeated the object of the expedition. Not being accustomed to strict discipline, Ellet did not realize the necessity of carrying out his orders to the letter.

After Colonel Ellet reached [the] Red River he captured several steamers loaded with provisions for Port Hudson, and having on board a number of Confederate officers; and hearing that other steamers were on their way down [the] Red River, his youthful ardor led him to go up that stream.

He arrived at Fort De Russy, and there, by the treachery of his pilot, was run on shore near the batteries. The enemy opened fire on the *Queen of the West,* killing and wounding numbers of the crew and cutting the steam-pipe. The vessel was now helpless, and Ellet and all his officers and men who were able jumped overboard and drifted down the river to a point where one of their prizes lay, got on board of her, and made their escape.

In the meantime, I had prepared the ironclad *Indianola* and sent her down to assist the *Queen of the West.* The *Indianola* passed the [Vicksburg] batteries at night with little damage, and met Colonel Ellet and his men coming up in their prize steamer *New Era.*

The *Indianola,* with two coal-barges in tow, continued down until she reached the mouth of the Red River, then turned back and proceeded by river again until near the plantation of Mr. Joseph Davis, the brother of the Confederate President.

At daylight the next morning, after the *Queen of the West* had been abandoned, the Confederates took possession and soon repaired damages.

The Confederate ram *Webb* joined the *Queen of the West* from Alexandria, and the two vessels, well manned and armed, proceeded in search of the *Indianola,* came up with her at Davis's plantation, rammed her, and she ran into shoal water and sank, February 24, 1863.

We heard of the disaster a few hours after, and all my calculations for stopping the enemy's supplies were for the time frustrated; but I took a philosophical view of the matter as one of the episodes of the war. However, it was necessary to try and prevent the rebels from raising the *Indianola,* and, as I was not ready to go down river [with my squadron] myself, as it would interfere with an important military movement [Grant's offensive against Vicksburg], I hit upon a cheap expedient, which worked very well.

I set the whole squadron at work and made a raft of logs, three hundred feet long, with sides to it, two huge wheel-houses and a formidable log casemate, from the port-holes of which appeared sundry wooden guns. Two old boats hung from wooden davits fitted to the "ironclad," and on her wheel-houses was painted the following: "Deluded Rebels, Cave In!" An American flag was hoisted aft, and a banner emblazoned with the skull and crossbones ornamented the bow.

When this craft was completed, she resembled at a little distance the ram *Lafayette,* which had just arrived from St. Louis.

The mock ram was furnished with a big iron pot inside each smoke-stack, in which was tar and oakum to raise a black smoke, and at midnight she was towed down close to the water-batteries of Vicksburg and sent adrift.

It did not take the Vicksburg sentinels long to discover the formidable monster that was making its way down the river. The batteries opened on her with vigor, and continued the fire until she had passed beyond the range of their guns.

The Vicksburgers had greatly exulted over the capture of the *Queen of the West* and the *Indianola.* The local press teemed with accounts of the daring of the captors, and flattered themselves that, with the *Indianola* and *Queen of the West* in their possession, they would be able to drive the Union navy out of the Mississippi. What was their astonishment to see this huge ironclad pass the batteries, apparently unharmed, and not even taking the trouble to fire back!

Some of our soldiers had gone down to the point below Vicksburg to see the fun, and just before reaching Warrenton the mock monitor caught the eddy and turned toward the bank where these men were gathered.

The soldiers spent several hours in trying to shove the dummy off into the stream, when daylight overtook them in the midst of their work, and the *Queen of the West,* with the Confederate flag flying, was seen coming up the river and stopping at Warrenton. . . . In the meanwhile the military authorities in Vicksburg had sent couriers down to Joe Davis' plantation to inform the people on board the *Webb* that a monster ironclad had passed the batteries and would soon be upon them. The crew of the *Webb* were busy in trying to remove the guns from the [*Indianola*], and, when they heard the news, determined to blow her up.

Just after the *Queen of the West* made the Warrenton landing, the soldiers succeeded in towing the mock ironclad into the stream, and she drifted rapidly down upon the rebel prize, whose crew never stopped to deliberate, but cut their fasts and proceeded down the river. Their steam was low, and for a time the mock ironclad drifted almost as fast as the *Queen of the West*; but at length the latter left her formidable pursuer far behind.

The *Queen of the West* arrived at the point where the *Indianola* was sunk just as the people on board the *Webb* were preparing to blow her up, bringing the news that the "great ironclad" was close behind. So the *Webb* cast off and, with her consort, made all speed down the river.

The *Webb* had been so greatly injured in ramming the *Indianola* that she had to go to Shreveport for repairs, and the *Queen of the West* was shortly after recaptured and destroyed. . . . The *Indianola* remained embedded in the mud until after the fall of Vicksburg [July 4, 1863], when we raised her.

The Vicksburg people were furious at the trick we played on them, and the newspapers reviled their military authorities for not being able to distinguish an old raft from a monster ironclad!

Porter's ruse with the mock ironclad has all the characteristics of a classic deception. It was simple, it played on the enemy's fears, and things could not be made worse in the event of failure.

"Old Rosey" Foxes Braxton Bragg: The Tullahoma-Chattanooga Operation, 1863

William Starke Rosecrans was one of the most successful generals in blue during the Civil War, a man who, in the words of one biographer, reached to the very "edge of glory," before his career was shattered in a few bloody hours along Chickamauga Creek in mid-September of 1863. But the story of how Rosecrans got to Chickamauga Creek is actually more interesting than the fatal battle itself, for in a series of operations beginning in late June and continuing for nearly three months, Rosecrans and his sixty-five-thousand-strong Army of the Cumberland succeeded in driving back Confederate General Braxton Bragg's forty-five-thousand-strong Army of Tennessee over a hundred miles, culminating in the capture of the vital railroad junction at Chattanooga, all virtually without firing a shot or incurring any casualties, in one of the most masterful demonstrations of maneuver and deception in the Civil War.

In mid-June, Rosecrans' army was concentrated around Murfreesboro, in central Tennessee, about fifteen miles north of Bragg's army, which was deployed to prevent his advance along the Western and Atlantic Railroad, the only line linking central Tennessee with Chattanooga and points south. The two armies had been in roughly the same positions since the end of the bloody battle of Stone's River on New Year's Day. On June 26, Rosecrans undertook an offensive. With his right wing (most of his cavalry and a small army corps), Rosecrans feinted at Bragg's left. As this drew Bragg's attention westward, Rosecrans threw the bulk of his army (three army corps and some cavalry) around Bragg's right, with Colonel John T. Wilder's "Lightning Brigade" leading the way. Now, Wilder's brigade was unique in the army, for although it consisted of only a couple of thousand mounted infantrymen, they were all armed with the Spencer breech-loading repeating rifle, making them the first sizable military unit in history to be so equipped, and incidentally giving them an extraordinary firepower. As a result, Wilder's brigade had been in the thick of the fighting since it received its new weapons, always proving a formidable opponent.

This maneuver forced Bragg to fall back, lest Rosecrans sever his line of supply, the railway to Chattanooga. So by June 30, Bragg was concentrated about Tullahoma, some fifteen miles behind his position of four days earlier, while Rosecrans was concentrated at Manchester, a few miles to the northeast of the Rebel commander's army.

After only a day or two to rest, Rosecrans once again moved out. And once again, he feinted against the Rebel left, drawing Bragg's forces in that direction, and then threw the bulk of his army around the Rebels' right, with Wilder's brigade in the lead. So once more Bragg fell back, this time all the way to Chattanooga, which he reached on July 4 (the very day that Vicksburg fell to Ulysses S. Grant, and Robert E. Lee began his painful retreat from Gettysburg). In nine days, Rosecrans had thrown Bragg back about a hundred miles, at a cost of only five hundred and sixty casualties.

After Bragg retired on Chattanooga and the line of the Tennessee River, Rosecrans dispersed his army somewhat. To permit it to rest and replenish supplies, he spread it over an eighty-mile front, and kept it about twenty miles behind the front line, along the Tennessee, which was but lightly held. A meticulous planner and extraordinarily concerned about his supplies, Rosecrans tarried in this position for more than a month, resisting all efforts by the president and Chief of Staff to pressure him into advancing. Rosecrans would advance when he was ready (what would have happened if a more energetic commander than Bragg had been his foe is a matter of conjecture). And Rosecrans was ready on August 16. Once more he intended to turn Bragg's position. Having held his army back from the line of the Tennessee River, he was well positioned to do this, since Bragg would be unable to ascertain his initial movements.

Now Bragg was perfectly aware that Rosecrans was getting ready to move (there were spies everywhere). But Rosecrans could try to turn his position by crossing the Tennessee above or below Chattanooga. There were a number of fords suitable for the movement of an army, and Bragg could not cover them all. So Bragg paid careful attention to his intelligence reports.

As part of his preliminary movements, Rosecrans quietly sent Wilder's mounted brigade and two infantry brigades, a weak force indeed, to a position on the Tennessee just upstream from Chattanooga. Bragg's spies soon apprised him of this development. Concluding that Rosecrans was preparing to cross the Tennessee not far upstream from Chattanooga, he positioned one corps in the city itself, and his other just to its right, covering the most likely crossings. Sure enough, skirmishing began along the river upstream from Chattanooga. Meanwhile, the balance of Rosecrans' army marched quickly to the south and began crossing the Tennessee some sixty miles downstream from Chattanooga, at Chaperton's Ferry on August 20, and other sites over the next few days. Rosecrans had foxed Bragg prettily. But the game was not yet up.

It would take several days for the entire Army of the Cumberland to cross the Tennessee. Moreover, the terrain on the southeastern side of the Tennessee, between the river and the Western and Atlantic Railroad, was extremely difficult to traverse. Rosecrans would have to keep Bragg's attention elsewhere for a bit longer. So while his cavalry imposed tight security on his expanding bridgehead, Rosecrans continued his deception operation before and upstream from Chattanooga. Not only did demonstrations and skirmishes intensify, but the city itself was shelled from across the river, further fixing Bragg's attention in the wrong direction.

Bragg did not receive convincing proof that Rosecrans was across the Tennessee in force below Chattanooga until September 8, eighteen days after the Army of the Cumberland had begun to cross. By then, the entire army was across the river and advancing eastward in three columns. At that, Bragg abandoned Chattanooga, to shift his forces and confront the advancing Union Army. Elements of Rosecrans' army entered the city the next day.

In the Tullahoma-Chattanooga operation, Rosecrans outthought Bragg at every turn, conducting three successful turning operations, each preceded by a feint to the opposite flank. The fact that Wilder's brigade had led in the first two such operations was the bait on which Bragg was hooked for the third. In the end, it was not Rosecrans who fooled Bragg, but Bragg who fooled Bragg.

Nathan Bedford Forrest Puts One Over on the Yankees

One of the better amateur soldiers of the Civil War, Nathan Bedford Forrest had a reputation for his aggressive, unorthodox approach to warfare. One of his most interesting tricks occurred in late September of 1864, when he was leading forty-five hundred mounted troops on a raid from northern Alabama against Union lines of communication through Tennessee, in an effort to forestall the fall of Atlanta to Union General William Tecumseh Sherman. As always on a raid, time was of the essence. Unfortunately, there were some fifteen hundred Union troops well fortified at Athens, Alabama, a post which would require time to reduce by conventional methods, time which Forrest did not have, particularly as there was a column of Union reinforcements on the march for the post even as his own troops approached it.

Sending off a detachment to keep the Union reinforcements entertained, Forrest surrounded the Union post during the night of September 22. At 7:00 A.M. on September 23, he began to shell the fort. Shortly afterward, Forrest requested that the Union commander, Colonel Wallace Campbell, surrender. Campbell rejected the demand, whereupon Forrest suggested that the two meet, in the interest of discussing ways to avert unnecessary bloodshed. Campbell agreed, and much to his surprise, Forrest promptly took him on a tour of the investing forces. What he saw greatly impressed the Union officer, for he could see that there were over eight thousand Rebel troops, amply supplied with artillery. Deeply impressed by what he saw, Campbell rode back to his troops and ordered them to surrender in the face of such greatly superior opposition, and some fifteen hundred Union troops were soon in Forrest's hands.

Meanwhile, the Union reinforcement column had been driving back Forrest's screening force, and he was now able to turn his attention to that, which he promptly bagged as well, since it consisted of only about seven hundred men. So altogether Forrest had captured over two thousand Union troops, plus an important military post, with virtually no loss to himself. But where had he gotten all those troops which so impressed Wallace Campbell?

Forrest had pulled a simple trick. He divided his command into several contingents. As Campbell inspected one Confederate position, troops whom he had just visited packed up and moved quietly, but quickly, to another position. In this way, Forrest managed to make his four thousand men appear to be twice as many. Campbell may seem to have been somewhat dim-witted, since arguably, he ought to have smelled a rat from the moment Forrest suggested that he inspect the investing troops, but there was, in fact, precedent for just that. Only two years earlier, in September of 1862, Union Colonel John T. Wilder, an amateur soldier who later became one of the Union's best cavalrymen, had found himself similarly invested, and requested that the Confederates allow him to inspect their lines in order to assist him in deciding on an honorable and humane course of action, and had surrendered when satisfied that further resistance would only waste lives. But Wilder's opponent was Confederate General Simon Bolivar Buckner, an honorable man. And things were generally a lot more honorable in 1862, than in 1864.

Forrest's little trick was by no means a new one, but one with a long and honored history. Its last notable use in American military history prior to the Civil War occurred during the War of 1812, when the great Indian leader Tecumseh marched his three hundred warriors several times around Fort Detroit to help convince the garrison that resistance was impossible.

In our own century, the Soviets used much the same ruse to create the impression that they had a huge fleet of intercontinental bombers at a May Day parade in the mid-1950s, flying the same couple of dozen new bombers (all they had of that model) round and round so that an almost continuous stream of awesome new aircraft seemed to by flying over Lenin's tomb. It's actually a simple trick, playing as it does on the enemy's fears. However, it's also a dangerous one, since if it fails, the enemy is likely to be better informed about one's strength than had hitherto been the case.

Crazy Horse at Lodge Trail Ridge

Deceptions and ruses were a commonplace of warfare on the Great Plains, and few Native Americans were as adept at their use as was Tashunca-Uitco, more popularly known as Crazy Horse, an Oglala Sioux war chief. One of his most masterful feats occurred in late 1866 near Fort Phil Kearny, not far from present-day Banner, in north-central Wyoming.

The Civil War had barely interrupted the westward movement of pioneers, and its end saw an increase in the flood of settlers to the west. In mid-1866, negotiations were begun between the United States government and the Oglala Sioux for the use of the Bozeman Trail into Montana. Even before the negotiations had reached a conclusion, the government ordered the army to begin establishing posts along the trail. As a result, Colonel Henry B. Carrington led a mixed force of infantry and cavalry out from Fort Laramie to set up three new posts: Forts Reno and Phil Kearny in Wyoming, and C. F. Smith in southern Montana. Since they had not yet granted permission for the government to do this, the Oglala Sioux understandably took offense and began harassing the posts. Fort Phil Kearny, where Carrington had his headquarters, received a lot of this attention.

In the garrison at Fort Phil Kearny were several officers with distinguished combat records from the Civil War, most notably, Captain William J. Fetterman, of the 27th Infantry. A veteran of the Army of Tennessee, Fetterman had marched with Sherman through Georgia and won a brevet (honorary) promotion to lieutenant colonel for gallantry. But Fetterman despised his commander, for Carrington was an older man who had seen no service in the Civil War and precious little before it. Nor did he think the Indians particularly difficult opponents, once asserting that with eighty men he could take on the whole Sioux nation.

On December 21, 1866, a small party left Fort Phil Kearny to cut firewood in a small forest nearby. Just as the woodcutters had begun their work, they came under harassing attack from a small Indian war party. As this could be seen and heard from the fort, Carrington called for volunteers for a relief party. Fetterman demanded to be allowed to lead the relief column. Carrington acceded, but gave him strict instructions that under no circumstances was he to pursue the enemy beyond Lodge Trail Ridge, a prominent rise just visible from the fort. Soon Fetterman, Captain Frederick H. Brown—another distinguished war veteran with a dim opinion of both his commanding officer and the Indians—and eighty-one other men, including a number of civilians (thus exceeding Fetterman's magical "eighty men" by two!), were riding out to the rescue of the woodcutters.

The relief column had not even reached the woodcutting party when the attacking Sioux began riding off. Putting spurs to horse, Fetterman gave chase, following the fleeing Indians right over Lodge Trail Ridge, and beyond it, out of sight. Even as the woodcutting party made its way back to Fort Phil Kearny, the garrison could hear the sounds of a fire fight taking place beyond the ridge. Carrington put together another, much larger, relief column, but even before it could set out, the sounds of firing grew less. Cautiously advancing across the ridge, the relief party soon found Fetterman and his entire command, dead.

When Fetterman's party crossed the ridge, it was jumped by between twelve hundred and two thousand additional Sioux. Crazy Horse, who was in command, had planned and executed a perfect trap. No one actually knows exactly what happened that day on the other side of Lodge Pole Ridge, as none of Fetterman's men survived, and none of Crazy Horse's ever told the tale to anyone who bothered to pass it on. But it was the worst disaster the U.S. Army suffered at the hands of the Plains Indians until George Armstrong Custer encountered that same Crazy Horse along the Little Big Horn River almost exactly a decade later.

Crazy Horse's successful ambush of the Fetterman party points up another important element in effecting a successful deception: It doesn't hurt if your enemy thinks he's smarter, or tougher, than you are.

Taking Your Eye Off the Water

The American Indian way of war largely took the form of raiding. This had long been the customary form of entertainment and a means of discovering which men were the best leaders. The raids were conducted not just to steal horses, women, and goods from neighbors, but also to demonstrate who was the bravest and boldest warrior. There was a fair amount of danger involved, as the victims often caught the raiders, and fatalities sometimes resulted. But death was not the goal of raiding, and a host of deceptions were developed to enable the raiders to sneak into their neighbors' territory, grab the goodies, and get away without getting caught.

When the first European settlers, the Spanish, began to appear in the American West, the Indians greeted the newcomers in the traditional fashion, with raids. These newcomers were seen as a bunch of strange ducks, possessing lots of nifty gadgets (metal implements, cloth, trinkets, etc.) that the Indians had never seen before and were most eager to possess. Moreover, the odd-looking intruders appeared to know little about the art of raiding, or protecting themselves from same.

The settlers were from a different culture, that of Europe, where raiding had largely gone out of fashion some centuries earlier. They did not take the Indian raids in the proper spirit, and several centuries of warfare ensued. Unlike the Indians back east, who had also practiced raiding, but on foot, western Indians were largely mounted by the eighteenth century, and were superb horsemen. When the U.S. cavalry proceeded to chase after the Indian raiders, they found themselves at the short end of many deceptions. The most lethal one was the "water hole race." The American West was relatively dry, and water was essential to keep man and horse going. During dry periods (which were frequent) or in areas with few water sources (which was common), it was important to get your horses and troops to the next water hole before both were rendered ineffective from dehydration. The Indians were well aware of this, and played an often fatal game with the palefaces.

It worked like this. If a body of cavalry, or mounted travelers, were crossing "hostile" territory, they were careful to keep an eye out for Indian war parties. Through nearly all the Indian "wars," the Americans were more heavily armed than the Indians, and used this firepower to force the Indians to keep their distance. The Indians used this to their advantage. Once the Indians had spotted a party of Americans, they would come up on all sides. But the Indians would keep out of rifle range, and subtly "nudge" the Americans away from the next water hole. Attempts would be made to get the Americans to stop and fire on the Indians, usually to no effect. What the Indians were doing was wearing down the Americans and their horses as well as forcing them to use up their ammunition. Finally, some of the Indians would be waiting at the water hole when the worn-out and thirsty Americans finally arrived. Although the Indians often did not have firearms, they did have other weapons and these could be used to good effect in defending the water.

Eventually, the desperate Americans would be forced to risk all and attack the Indians defending the water. The Americans usually lost. Americans with some experience in the ways of the Indians would be more concerned about getting to the water first, or moving off to another water source. But the Indians had most of the advantages. In addition to being accustomed to living out on the plains, the Indians could get by on less and move faster than their more "civilized" opponents. The Indians' ponies fed on grass, rather than grain, and were generally able to get by on less water. The Americans had to get from one source of water and horse feed or perish. The Indians knew this and knew how to harry the Americans to death. For this reason, the sight of some Indians in the distance brought dread to those experienced in the clever ways of Indian warfare.

Total Deception: Nelson A. Miles Fools Everyone

Sometimes, you have to fool everyone, including your superiors, in order to deceive your opponent successfully.

In April of 1898, the United States went to war with Spain. In rather short order, some spectacular victories were gained in the Philippines and Cuba. But if Manila Bay and San Juan Hill and Santiago garnered most of the headlines, there was another operation which, although essentially a sideshow, demonstrated the professional skills and wiles of American military leaders far better than did those more famous actions: the brief campaign in Puerto Rico from July 25 to August 13, 1898.

Actually, before the war broke out, no one had given much thought to capturing the "Pearl of the Antilles"—Cuba and the Philippines were far more splendid prizes. But Puerto Rico was there, and it did belong to Spain, and what with a war going on, it seemed like a reasonable move. Besides, the Commanding General of the Army, Nelson A. Miles, a tough Civil War veteran and Indian fighter, wanted to get into action somewhere (President William McKinley and Secretary of War Russel A. Alger didn't like him very much), and Puerto Rico was the only place left. So Miles planned an invasion of Puerto Rico.

His plan was simple. He would land a considerable army (upward of thirty thousand men) at Cape Fajardo or Cape San Juan, on the east coast of the island, and move directly on the fortified city of San Juan, only a few dozen miles to the northwest, thereby striking at the heart of Spanish power in Puerto Rico. In retrospect, Miles' plan looks a little foolhardy. San Juan was a pretty solidly fortified place, and about half of the fifteen thousand troops on the island were concentrated within a short distance of the city. So Miles was planning on attacking the enemy's strength. Or was he?

The first contingent of Miles' expedition put to sea from Santiago in Cuba on July 21, and stayed at sea for the next four days. This put Miles totally out of communications with Washington, where his plan was apparently public knowledge. For most of the four days, Miles' little fleet steamed slowly eastward, passing north of Hispaniola, and "obviously" heading for his objective, the east coast of Puerto Rico. Then, on the night of July 24, when the little squadron was only a few dozen miles to the northwest of San Juan, Miles ordered a change of course westward and an increase in speed. And on the morning of July 25, as a cruiser bombarded distant Cape San Juan, the first of Miles' troops began disembarking at Guanica, a small port on the southwestern coast of the island. The landing caught the Spanish completely by surprise, and there was virtually no resistance. The Spanish were never able to recover from Miles' little trick. The best of their troops were concentrated to meet a threat to San Juan, a threat all the more real for the fact that Miles' initial plan seemed to confirm the importance of the place, and so they were unable to prevent the Americans from overrunning most of the island, offering only feeble resistance.

Given the odds, and the strategic situation, the United States would have seized Puerto Rico even without Miles' deceptive movements. As it was, by the time the United States and Spain concluded an armistice, Miles' troops had occupied about 75 percent of Puerto Rico, at a cost of fifty men, killed, wounded, or missing, as the Spanish kept falling back to cover San Juan.

The Spanish weren't the only people whom Miles fooled. In fact, he fooled everyone, including the president, who had no idea where the elusive general was for about a week. So Miles fulfilled the classic Chinese definition of deception: "Make a noise in the east, and attack in the west"—in this case, quite literally.

4

Deception Comes of Age

The Early Twentieth Century (1900–1939)

The onset of the "industrial warfare" of the twentieth century enhanced the importance of deception. Radically novel weapons like the airplane, the submarine, and telecommunications altered war from a two-dimensional to a three-dimensional undertaking, and in a sense, even a four-dimensional one (time became even more critical). The new technologies meant that new tricks were needed and were possible. Nor were the older ones neglected, for new applications were found for many of them.

The object of the trickery, of course, remained the same: to gain a march on the enemy. A lot of the new tricks were really logical developments from the old ones, using (or abusing) new technology.

Perhaps the most important "new" development in deception was the greater emphasis on camouflage. This actually began late in the nineteenth century, with the introduction of more subdued uniform colors, such as Britain's khaki (meaning "dusty" in Hindi, which was appropriate, since they found the color useful while fighting the sharpshooting warriors of India's dusty northwest frontier). The Germans adopted *feldgrau* (field gray, gray-green actually) uniforms. Most countries followed suit, with the notable exception of France, which continued to sport red trousers, blue coats, and white kepis, with white gloves for officers! All this was a result of the increasing range and lethality of infantry weapons. World War I (1914–1918) greatly accelerated this trend. The horrendous butcher's bill at the front in 1914–1915 convinced even the French to give up their *pantalons rouges* (red pants), while trench warfare soon found the troops improvising all sorts of camouflage, if only to enhance their chances of survival.

In addition, the use of airplanes for reconnaissance led to the introduction of camouflage even for rear-area installations. Camouflage was the principal form of deception used during World War I, but there were some important strategic deceptions as well in that war, despite the generally dismal record of the strategists.

World War I was not the only source of deceptions during the first generation of the "violent century." There were many political and diplomatic deceptions among the Great Powers. Some of these were modern variants on hoary old chestnuts, while a few were new, invented to meet the needs of a changing technology and an evolving international situation. It is a period of particular importance for the student of deception, for a number of the ruses employed are likely to be with us for some time to come.

The Great White Fleet, 1907–1908

The sight was a magnificent one, that December 16, 1907, as the pride of the U.S. Navy, sixteen new battleships and their escorts, all painted a sparkling white, stood majestically out from Hampton Roads onto the broad bosom of the Atlantic, on the first leg of a world cruise, crewed by the finest officers and men that America had to offer. It was a wonderfully theatrical gesture, thought up by President Theodore Roosevelt, to demonstrate to the world the goodwill and technological abilities of the great republic.

The voyage was a smashing success as the fleet steamed its way down the east coast of South America, rounded Cape Horn, then proceeded up the western side of the Americas, and thence across the broad Pacific to the mysterious East, visiting Japan, the Antipodes, China, and India, before passing into the Mediterranean, where it lent a hand succoring the victims of the Messina earthquake of 1908, and then dispersed to pay courtesy calls to all the nations of Europe, before reuniting for the trans-Atlantic voyage home. Everywhere the fleet went it was given a tumultuous welcome, as cities from Rio to New York vied with one another to entertain the youthful representatives of the great republic, and be impressed by the products of American science and technology. And then the "Great White Fleet" (the ships were painted white to lower their internal temperature in the tropics) passed into memory.

Roosevelt's idea that the "Great White Fleet" was a demonstration of the greatness and goodwill of the United States was accepted by most people. More perceptive observers saw in it Teddy's "Big Stick," a subtle reminder to the rest of the world that America was a military power to be reckoned with. Only a handful of people knew the real purpose of the "Great White Fleet," a couple in Washington and a few more in Tokyo. But they were all that mattered. In fact, the ostensible reason for the voyage was an elaborate deception, masking a brutally clear threat bluntly aimed at Japan.

Tensions between the United States and Japan had been quietly rising for many years. Both were new to the ranks of the Great Powers, and both had conflicting interests in the Pacific. In 1898, the United States had acquired Hawaii and the Philippines, real estate upon which Japan had its eye. The following year, U.S. Secretary of State John Hay had proclaimed the "Open Door" with regard to China, arguing that all powers have equal access to the vast markets of the dying Celestial Empire, much to Japan's annoyance, since it wanted a piece (or perhaps all) of that pie. And then there was American racism, blatant, often violent, discrimination against the Japanese in California. Although American public opinion had sided with the Japanese in their war with Russia in 1904–1905, and U.S. President Roosevelt had brokered what most of the world thought was a good peace between the two, this had done little to reduce tensions, for many ultranationalist Japanese believed Roosevelt had deliberate robbed *Dai Nihon* (Japan) of its legitimate spoils.

Japan's nationalists and militarists were riding high in the first decade of the century. They had raised their nation from a technologically backward, militarily weak, minor feudal country to the ranks of the Great Powers in little more than a generation, in the process defeating both China (1894–1895) and Russia (1904–1905), thus demonstrating to the world the skill, quality, and character of Japan's new armed forces. In short, by 1907, the possibility of a war between the United States and Japan was by no means remote. Which is what prompted Roosevelt to send the fleet around the world.

Of course, the president did not do this merely to show the Japanese how many ships we had. They were well aware of the fact that we had a larger navy. Sheer numbers were not as much of a deterrent to war as it might seem. The Japanese had, after all, just recently defeated Russia, which also had a larger navy. A Japanese-American war would be fought in the western Pacific, while virtually the entire U.S. Navy was in the Atlantic. It was this that gave Japan's militarists their hope for victory, should it come to a fight. In order for the U.S. Navy to come to grips with the Imperial Navy, it would have to make the long voyage from the Atlantic Coast to the western Pacific. The Russians had essayed precisely the same undertaking, sending a battle fleet equal to that which Japan possessed from the Baltic around Africa, across the Indian Ocean, and into the China Sea.

By the time the Russian fleet arrived off Japan, after a voyage of seven months (October 1904–May 1905), it was in no condition to fight, and was literally blown to pieces. With their more experienced navy, Japan's militarists thought they would be able to work the same routine on the United States if it came to a fight. And that was why Teddy Roosevelt sent the "Great White Fleet" around the world, to impress upon the Japanese that the U.S. fleet would arrive in Japanese waters in fighting trim, making the outcome of a Japanese-American war likely to be much different from that of the Russo-Japanese War.

The enormous subtlety of Roosevelt's threat to Japan constitutes one of the most successful deceptions in history. The threat was veiled so effectively that the only people who were aware of it were those who counted: the president and the war party in Japan. The "Great White Fleet" has important lessons for the present. In our increasingly chaotic world, it may be necessary to resort to equally subtle maneuvers to attain strategic goals, maneuvers which must needs be concealed not only from the enemy, but also from the public. The most successful deceptions are those which work without anyone knowing about them. And the object is to win, not necessarily to win by fighting.

"Where Is Count Luckner?" Naval Deception in World War I

Commerce raiding is an ancient naval tradition—sending ships out to intercept the enemy's maritime trade, in order to disrupt his economy. Always an attractive notion, particularly to a power weak on the sea, commerce raiding underwent a number of changes in the twentieth century, due to technological innovations. In the "good old days" of wooden ships and iron men, commerce raiders could operate independently, literally for years, plundering stores and even munitions from their prizes. But the advent of steam power (which required frequent coaling), the development of rapid-fire weapons, and the invention of radio made the old-style commerce raider increasingly obsolete. The radio, in particular, made it much more difficult for raiders to operate freely, because now, each attack on a cargo ship was likely to result in a radio message that would let the enemy know almost instantaneously the location of the raider. For ages, the raiders had often been armed merchantmen. A little faster than their prey, and much more heavily armed, this was sufficient to allow the raider to roam the sealanes looking for victims. Now, in the new century, faster enemy warships equipped with radio made the raiders' world a lot smaller.

At the outset of World War I, Germany essayed some commerce raiding with her warships, particularly those on distant stations, but, despite some spectacular successes, these were quickly hunted down and sunk by the Royal Navy. So the Imperial High Seas Fleet decided to try a couple of new ideas: the submarine and the disguised merchant raider.

Both of these had their advantages and disadvantages. Submarines were wonderfully "stealthy" and capable of slipping through the Royal Navy's blockade of German-controlled ports. But they could not, in accordance with the existing rules of war, rescue the passengers and crews of their victims. Nor could they, once again in accordance with traditional rules of war, give "fair warning" to their victims, since they were highly susceptible to even the slightest damage; and, moreover, their victims might then be in a position to broadcast their positions by radio. Eventually, the Germans decided that the only way to use submarines effectively was to scrap the "rules of war" and undertake unrestricted submarine warfare; in effect, sinking everything in sight without warning or assistance. This ultimately created difficulties with the United States, and in the end, led to American participation in the war.

The other route, the disguised armed merchant cruiser, was rather effective, albeit by no means as effective as the submarine, although politically less explosive. In essence, the disguised armed merchant cruiser was an ordinary merchant ship converted into a warship through the installation of concealed guns and even torpedo tubes, then provided with a large supply of paint, canvas, and timber (to enable the crews to alter the appearance of their vessels); given false papers (to get them by suspicious British blockade officers) and false colors (which had to be hauled down before shots could be fired). In one case, such a ship even carried a small seaplane (to enable the ship to hunt for prey more effectively). These disguised cruisers were then sent to sea after enemy merchantmen. Much less vulnerable than submarines, much faster, and much more innocent-looking, merchant cruisers still could operate within the existing rules of war. Germany commissioned eight commerce raiders during World War I (and even more during World War II). The four raiders that managed to get to sea (of the balance, one ran aground and was lost before her first voyage and three were intercepted during their breakout attempts or soon afterward) accounted for eighty-seven ships, for a total of about a half-million tons of shipping; not a bad record.

Several of these raiders had adventurous cruises. Once, in order to more convincingly fool British blockade officers (who would come aboard under cover of a light cruiser or two), a captain dressed a junior officer up in woman's clothing and introduced the man as his "wife" (it was not unusual for merchant marine captains to take their wives along in those days). Another cruiser was lost in a duel with two British auxiliary cruisers, armed merchantmen like herself, accounting for one of the enemy in the process. But undoubtedly the greatest adventures were those of the raider *Seeadler*, commanded by Count Felix von Luckner.

Seeadler (sea eagle) accounted for only sixteen British and Allied ships, and those small ones (her total score was 30,100 gross registered tons, or about 1,900 GRT per victim), hardly in the same league with the raider *Mowe*, which sank forty ships (182,800 GRT, or 4,600 GRT per ship). But Luckner was unique in other respects, for *Seeadler* was an old, full-rigged, three-masted bark, wearing nearly 28,000 square feet of canvas when under full sail. Although she did have an auxiliary diesel engine which could propel her along at a respectable nine knots, *Seeadler* did most of her work under sail. Armed with only two 105-millimeter (4.1-inch) guns and a couple of machine guns, the ship had a crew of only sixty-four. She ranged the seas from December 21, 1916, when, under Count Felix von Luckner, a seasoned old salt and publicity hound of extraordinarily blue blood, until August 2, 1917, when she was swamped by a tidal wave that struck her while anchored off Mopelia in the Society Islands, near Tahiti.

Since when under sail *Seeadler* was unable to overtake steamers, Luckner's usual victims were sailing ships, still in that era an important factor in maritime commerce. His normal method of intercepting a ship was to employ a simple ruse. Under cover of neutral colors, such as Norwegian or Swedish, he would use flag signals to indicate distress or a desire to communicate. The unsuspecting vessel would usually come about, allowing Luckner to close. As she did, Luckner's men got into their naval uniforms and manned their concealed weapons. Once the victim was within range, Luckner would replace his neutral colors with the Imperial naval battle flag (thereby maintaining the niceties prescribed in the rules of war), and unmask his guns. That was usually enough to convince the helpless merchantman to surrender, but if her skipper decided to make a run for it, betting on his engines, Luckner would put a round or two across his bow, which always had a salutary effect.

The ship would then be boarded by an officer and armed party of bluejackets. After plundering her cargo of anything useful (such as five hundred cases of cognac and twenty-three hundred of champagne) and removing the passengers and crew to special quarters on *Seeadler*, the ship would be sunk, usually by scuttling, but occasionally by gunfire. Luckner's happiest hunting was in the relatively narrow portion of the Atlantic between Africa and South America, where from late January to mid-March of 1917, he accounted for eight sailing ships and a large steamer. At that point, with his prisoners beginning to overflow the available quarters, and fearing that those waters were likely to become too hot for him quite soon, Luckner decided that a change of scene was in order. Putting his prisoners on a newly captured sailing ship (his ninth in slightly less than two months), he sent them to internment in neutral Brazil, and then headed for the Pacific Ocean, rounding Cape Horn in the Southern Hemisphere's autumn.

To his misfortune, the Pacific did not provide anywhere near as good hunting as had the Atlantic. By June of 1917, *Seeadler* had added only four vessels to her score. In need of some maintenance and running short of supplies (some crewmen were showing signs of scurvy and beriberi), Luckner put in at Mopelia, an uninhabited island in the Society group, in late July. A few days later, a tidal wave swept the ship up on a reef—a total loss. Although Luckner took a small group of men twenty-three hundred miles to Fiji in a lifeboat, hoping to hijack a new ship to resume raiding, he was captured. *Seeadler's* career was over.

Considering the return on their investment, Germany's merchant commerce raiders were a rather profitable venture. At an individual cost of conversion not much different from the price of a submarine, the raiders, even including the four that never sank a single ship, racked up an average number of sinkings exceeding that of the average submarine, and caused the Royal Navy considerable headaches, as it tried to spread its available light cruisers over the ocean in an effort to intercept them.

Footnote 1: *Auxiliary Merchant Cruisers.* Like many powers before World War I, Germany believed that large, fast passenger liners could be converted into commerce raiders by the simple expedient of mounting a few guns. This proved a disastrous experiment, the ships in question being much too large (for example the liner *Cap Trafalgar* was 18,700 GRT) and consuming enormous amounts of coal (350 tons a day was not unusual). Although in the opening days of the war three of the six converted liners (all were armed at sea by "prepositioned" supply ships) did fairly well (twenty-nine ships, about 105,000 GRT), they all were rather quickly run down and sunk by British cruisers or ran out of coal and were forced to intern themselves in American ports. It was the failure of the armed merchant cruisers that prompted the Imperial Navy to try merchant commerce raiders, which were much more successful.

Footnote 2: *Armed Merchant Raiders in World War II.* Encouraged by the success of her merchant raiders in World War I, the Germans sent ten to sea during the later war. These used many of the same ruses that the World War I raiders used, and met with some success. The raiders accounted for ninety-eight vessels, including one Australian light cruiser, for about 590,000 GRT, or nearly 6,000 GRT per victim, an average score considerably lower than that of the World War I ships, even including those which had not made it to sea (59,000 GRT per World War II raider, as against 125,000 for the World War I ships). For the record, it should be noted that both Italy and Japan also sent some armed merchant raiders to sea during the Second World War, meeting with very little success.

Gallipoli, the Most Masterful Retreat

The Anglo-French attempt to throw Turkey out of World War I by a massive amphibious landing on the Gallipoli Peninsula, to be followed by a quick advance through Thrace and the capture of Constantinople, was one of the most spectacular failures in military history. Almost everything that could go wrong in an amphibious operation did, and the campaign was one of the most remarkably mismanaged in a war that saw numerous examples of incompetent generalship. But there was one spectacularly successful chapter to the whole sorry tale of Allied failure at Gallipoli, and that was the remarkable success with which the British and French were able to abandon the operation, by means of a well-planned withdrawal, greatly aided by an elaborate deception plan.

The initial landings, in April of 1915, had left the British (most of whose troops were actually Australians or New Zealanders) and the French (who only contributed a modest contingent to the operation) clinging to two narrow beachheads separated by several miles. An attempt to break out from the beachhead in August, by means of an "end run" amphibious landing, only managed to enlarge the more northerly beachhead ("ANZAC," for "Australian–New Zealand Army Corps," and later, "ANZAC-Sulva") somewhat, while no gains at all had occurred at the other lodgment area (Cape Hellas, on the tip of the peninsula). By the end of 1915, the Allies had committed over 400,000 ground troops to the operation, of whom over 250,000 had become casualties, including some 50,000 dead. Already dissatisfied with the enormously wasteful operation, one which he had opposed from the first in any case, late in 1915, Lord Kitchener convinced the British and French governments that the best thing to do was to abandon the entire enterprise.

Of course, this would not be an easy undertaking. There were a dozen divisions on the beaches, over 150,000 men, who were closely invested by about 100,000 Turkish troops. In some places, the trenches were within five yards of each other, and they were never more than about three hundred yards apart. Moreover, the Turks held virtually all of the higher ground and could see almost everything going on in the rear of the Allied forces. In a reverse of what had hitherto been extremely sloppy staff work, the evacuations were meticulously planned, and careful attention was paid to ensuring that the enemy be kept in the dark.

The northern beachhead, ANZAC-Sulva, was to be evacuated first, beginning on December 10, 1915. During daylight hours, the normal leisurely movement of barges carrying reinforcements and supplies to the beachhead continued. But at night, those same barges took off troops, animals, and equipment under blackout conditions, and with considerably greater speed. By the end of the first week, about forty thousand men remained at ANZAC-Sulva, and by dawn on December 19, there were only twenty thousand in position to fight. As the number of troops at the front thinned out, those remaining were assigned to maintain the same level of casual sniping as was customary when there was no "Big Push" going on. Each man was assigned several firing positions and would move from one to the other and let off a shot or two, while fewer and fewer artillery pieces maintained the same leisurely harassing fire against enemy positions. Meanwhile, tentage and storage facilities were kept up, to further the illusion that all was normal. The last British and ANZAC troops were taken off the beaches at 5:00 A.M. on December 20. The last men to leave the trenches activated various simple mechanisms that maintained the incessant sniping, devices such as water-filled counterweights which slowly dripped away their contents until they activated a rifle's trigger. As a result, the Turks did not find out for several hours that the Allies had left.

This, of course, put the Turks on the alert, which made the evacuation from Cape Hellas all the more risky. It began shortly after New Year's, following several days during which "reinforcements" were openly landed, creating the illusion that the troops pulled out of ANZAC-Sulva were merely being transferred to Cape Hellas. The French pulled out first, with the British moving in to cover their lines. All of the same ruses were used to convince the Turks that the trenches were being held in strength. Of course, the Turks were being more careful. On January 7, 1916, they essayed a determined probe of the front lines, only to suffer a bloody reverse by some very determined rearguards supported by gunfire from the Royal Navy. This convinced the Turks that the Allies were at Cape Hellas to stay. And by dawn of January 9, the last British soldier left the beach. Once again, it was some time before the Turks realized what had happened.

The *Stosstruppen*

World War I (1914–1918) was not a conflict noted for extensive use of deception. For the most part, it was just straight-ahead firepower and massive casualties among the infantry. But toward the end of the war, many senior commanders began to wake up, and often the solutions they introduced depended heavily on deception.

One of the more amazing, and effective, forms of deception was embodied in the German *Stosstruppen* ("storm" or "assault" troops) tactics, usually called "infiltration tactics" in English treatments. These new tactics emphasized the classic deception of doing what the enemy doesn't anticipate and being where the enemy doesn't expect you.

In the first three years of the war, tactics had fallen into a wearily repetitious pattern. Masses of troops moved forward to be mown down by machine guns and artillery. The defenders had the deception advantage, as their positions were dug in and they had even deeper bunkers in which to survive the attackers' extensive artillery bombardment (often lasting days, and sometimes even weeks, thereby totally losing the element of surprise). But all the attackers knew was that when they finally came out of their trenches, there were always some defenders left shooting at them with machine guns. Since the defenders were in trenches and fortifications themselves, and these were often cleverly hidden positions, the attacker was constantly deceived as to where all this lethal firepower was actually coming from.

The Germans developed a better way to attack. First of all, instead of masses of troops marching across no-man's-land, they trained small groups to operate like a patrol. Effective patrolling techniques had developed during the war, if only because the generals didn't bother themselves with minor details like that and thus left the troops to use their judgment. While a major attack was hard to miss, and was often announced by several days of artillery fire, a patrol would slip out into no-man's-land at night, enter the enemy positions, grab a prisoner or two, and scurry back to friendly lines. Patrols depended on deception, sneaking out at night when they could not be seen, and generally staying out of sight as much as possible. Actually, a patrol to snatch a prisoner was often the culmination of a series of patrols. Night after night, the patrols would go farther into no-man's-land to map the area and determine the easiest route into enemy positions. Thus the final assault would be well rehearsed, and a complete surprise for the enemy.

The Germans adopted these patrol tactics on a large scale for their Stosstruppen. But that was not all there was to it. Like their enemies, the Germans had built up a large artillery force. Yet, unlike past attacks, the artillery would not pound enemy positions for days before an attack. Instead, the Germans developed the "hurricane" barrage. This was a fifteen- to thirty-minute shelling of key enemy positions. The objective was to fire quickly and put as many shells on the enemy position in as short a time as possible. Experience had shown that most of the damage to the target was done in the first hour or so of the barrage. After that, the enemy was either underground in his bunker, or dead where caught in the open. The hurricane barrage also hid the advance of the Stosstruppen groups (usually less than thirty men each), who were ready to leap into the enemy's positions shortly after the German barrage lifted (stopped).

The artillery was also used to stop the enemy artillery from putting defensive fire in front of its infantry fortifications. Additional shelling was done on roads, or open areas, over which enemy reinforcements could approach.

The key element of the Stosstruppen tactics was that the troops did whatever they could to keep advancing. Once they broke through enemy lines, the Stosstruppen kept going. Bypassed enemy troops were left for the second or third wave of Stosstruppen to take care of. The first groups of Stosstruppen only attacked enemy fortifications if they were in the way. If the Stosstruppen could go around the enemy, they did so. Unlike the traditional World War I offensive, the Stosstruppen attacks did not put most of the troops in the first wave. The first groups were "combat patrols" that would fight their way through the enemy and then indicate (usually by a flare signal) that they were through. That group would then be reinforced immediately. Because the ground had not been torn up by days of shelling, the Stosstruppen were able to move quickly, and drag small carts containing ammunition and heavy weapons (machine guns and mortars) behind them. A Stosstruppen attack quickly broke through the lines of an enemy that did not know what they were up against. This was a form of warfare that was called "blitzkrieg" in World War II. The only difference was that the blitzkrieg used tanks to make and exploit the breakthrough.

Once the Stosstruppen were past the enemy front lines, they could easily attack artillery and supply units. This usually caused a panic among the enemy, with many of the frightened troops fleeing.

The Germans developed and tested Stosstruppen tactics in Russia during 1917. Late that same year, Russia sued for peace. So the Germans moved many of their Russian front troops to the west and, on March 21, 1918, launched a Stosstruppen attack against several British armies in northern France. The Allies had ignored Stosstruppen operations in Russia and were generally unprepared for the attack. The German deception was complete, and the offensive almost succeeded. The Germans attacked at dawn when fog still covered the battlefield, thus making it more difficult for the Allied troops to see what was coming.

But during the first week of April, after the Germans penetrated forty miles into enemy territory, the Allies managed to stop the German offensive. It wasn't just the desperate Allied response that stopped the Stosstruppen, but the Germans' own lack of mobility. Once the Germans had gone ten miles past the enemy lines, they were out of range of their own artillery. The several hundred yards of no-man's-land was difficult to get wheeled vehicles (artillery and supplies) across. Moreover, the Germans had to build roads across their own and the British frontline trenches. This took time, while the Allies were able to use their own rear-area road network.

Most important, the Allied railroad network was still there, and thousands of reinforcement troops could be brought in to meet the advancing Germans. Several of the divisions facing the Germans were American, as the United States had entered the war in 1917 and was just beginning to arrive in large numbers in early 1918. The American troops were fresh and eager, and their divisions were larger than those of the other nations. Thus the battle was a near-run thing. Without the Americans, and a little luck, the Allied cause might have been lost. Over four hundred thousand troops from both sides were killed, wounded, or captured in this German "Victory Offensive."

The lessons of the German Stosstruppen tactics were generally lost on the Allies. For twenty-two years later, the Germans used the same tactics, but with tanks and trucks, to successfully conquer France.

While German deceptions played a big role in both the 1918 and 1940 offensives, Allied self-deception was a crucial component.

Ironically, and perhaps revealingly, the American and British term "Stosstruppen tactics" (or "infiltration tactics") is unknown (to this day) to Germans. At the time, the Germans were simply rewriting their book on infantry and artillery tactics. They called the key assault units "Stosstruppen" and the British adopted the term "Stosstruppen tactics." The Allies did recognize that the Stosstruppen were "infiltrating" enemy positions, so the term "infiltration tactics" was also used. To confuse things further, these new tactics were also called "von Hutier tactics," after the German general who first used them on a large scale in Russia. So, in effect, the British and Americans engaged in a little self-deception as to exactly what these new tactics were by giving it an inaccurate title(s). Even during World War II, despite the fact that German infantry operations were generally more efficient than those of their opponents, few Allied commanders fully understood how the Germans were doing it, much less that these German techniques were over twenty years old.

Richard Meinertzhagen Loses His Dispatch Case

Although World War I was not characterized by clever deceptions, a surprisingly simple ruse greatly facilitated the British victory in the Third Battle of Gaza (October 31–November 6, 1917), during the Palestine Campaign.

The British had already made two attempts to break the Turkish hold on Gaza, both abysmal failures. These failures resulted in the promotion of General Sir Edmund Allenby to command the British forces in Palestine. A bright guy, particularly by the standards of World War I, Allenby decided that rather than once more attempting to capture Gaza by frontal assault, he would try to take it from the rear, with a massive cavalry sweep into the Negev Desert. To do this, he had to capture the Turkish outpost at Beersheba. Of course, any preparations for an offensive would soon come to the notice of the enemy, so Allenby put his staff to work developing ways to focus the attention of the Turks and their German military advisors on Gaza.

Now, Allenby had a fondness for clever officers, and on his staff was a fellow named Richard Meinertzhagen, who had already acquired a reputation for undercover operations (during the East African Campaign, he had conducted some remarkably successful reconnaissance while pigmented and garmented as a native). Meinerzthagen came up with an elaborate, but ultimately convincing, ruse.

To begin with, from late September, British radio messages began to contain suggestions that there was to be a reconnaissance in force toward Beersheba, as a diversion to cover a frontal attack on Gaza. Other messages hinted that General Allenby would be away on leave from October 29 to November 4, suggesting that the offensive against Gaza was not scheduled to begin until days after his return. To confirm the impression that Gaza was to be the object of a major offensive in mid or late November, a dispatch case was to be "accidentally" lost by an officer conducting a reconnaissance.

Into this case went a standard British Army staff officer's notebook, between the pages of which were:

1. A twenty-pound note, a great deal of money in 1917 (worth about a thousand dollars in 1995 currency).
2. A letter from the officer's wife announcing the birth of their son (actually written by Meinertzhagen's sister).
3. A private letter in which the officer's correspondent remarked upon the postponement of the offensive and commented on the preparations for naval gunfire support against Gaza.
4. A copy of the purported agenda for a staff conference, which essentially confirmed item 3.
5. A copy of purported instructions from Allenby's headquarters concerning preparations for the offensive.
6. A copy of a telegram from Allenby's headquarters to the commander of the British Desert Mounted Corps, ordering an officer to conduct a reconnaissance toward Beersheba, and specifying what he was to look for.
7. A few miscellaneous notes in a simple code not known to the enemy. (It was hoped that they would "break" this, and thus be susceptible to messages sent in it at a later time.)
8. A short memo critical of the value of Beersheba to the British.

After a couple of false starts, Meinertzhagen finally managed to "deliver" the notebook to the enemy on October 10.

Going out on patrol in the direction of Beersheba, Meinertzhagen "accidentally" got too close to the Turkish lines, and found himself under fire. He promptly turned tail and ran. In the process he "dropped" not only the notebook, but also his field glasses, his water bottle, and his rifle, which thoughtfully had been smeared with horse's blood, to create the impression that his mount had been injured. To strengthen the ruse, Meinertzhagen arranged for himself to be semipublicly reprimanded for losing valuable documents, and had a general order issued to the army concerning the "careless" way in which officers were carrying sensitive items while on patrol. To ensure that the Turks also received this message, a copy of the order was used to wrap a sandwich, which was also "lost" in the very same area that Meinertzhagen had dropped the notebook, by a patrol which was rather obviously looking for something. Finally, the British artillery lent a hand, plastering Gaza heavily, while paying no attention whatsoever to Beersheba.

The ruse (which was known to only four other men besides Meinertzhagen) caused a considerable stir in the Turkish Army, as demonstrated by documents captured during the offensive. The Turks began extending and strengthening their trench lines before Gaza, adding a new division to those already in the lines and bringing two more into reserve close behind the city, while ordering an additional fifty aircraft to be shipped into the area.

The British offensive began on October 31 with a heavy bombardment of Gaza, followed by a two-infantry-division attack supported by the guns of the Royal Navy, further pinning enemy attention to the western end of the front. Then, suddenly and with a minimal preparatory bombardment, four British infantry divisions advanced on Beersheeba from the southwest, while two Australian and New Zealand mounted divisions swung around the eastern side of the city's defenses, and by late afternoon had ridden into the rear of the Turkish position. The capture of Beersheba unhinged the entire Turkish Army, forcing it to fall back, in a retirement that did not end until December 30, by which time the British had taken Jerusalem, some sixty miles beyond Gaza.

Meinertzhagen's ruse at Beersheba was one of the most successful of World War I, and was instrumental in breaking a deadlock that had lasted for more than six months. It was not the only example of what may be termed the "lost order" trick, a particularly ancient one.

The cover plan for the American offensive at Saint-Mihiel in September of 1918 was an offensive through the Belfort Gap, at the southeastern end of the Vosges Mountains in Alsace. One of the many deceptions employed to suggest that this was the American objective involved Colonel A. C. Conger, of General John J. Pershing's staff. Conger had checked into the most expensive room at the best hotel in Belfort, where he held meetings with Major General Omar Bundy and his staff (Bundy commanded VI Corps, which was to spearhead the notional offensive, but was not himself in on the ruse, believing it to be the real thing). Later, back in his room, Conger prepared a one-page typewritten report on the status of the pending operation. He promptly dispatched both the original and a carbon copy to Pershing's headquarters, while "carelessly" tossing the crumpled-up carbon paper into the trash. He then retired to the hotel bar for a few drinks. When he returned to his room, he examined the wastepaper basket and found the carbon missing.

Within a few days, it became clear that the Germans had not only purloined Conger's carbon paper, but were convinced that it, and the other deceptions involved in the cover plan, were the real thing. As a result, they began shifting troops and ammunition into southern Alsace to strengthen their defenses in what had been a "quiet" sector for most of the war. The ruse was only partially successful. It did not divert German attention from Saint-Mihiel, where they were expecting an American attack, but it did convince them that the attack would be on a much broader front. However, the tricks probably helped them decide to undertake a fighting withdrawal, rather than make a decisive stand at Saint-Mihiel, thereby making things a lot easier for the doughboys who went "over the top" on September 12.

This brings out a problem of the "lost order" ruse—that the enemy might not believe it. It is, after all, an extremely old trick. Which reminds us that on September 13, 1862, a Union cavalryman at Frederick, Maryland, found a couple of cigars wrapped up in a piece of paper. Unwrapping the treasure, he noted that the paper was official stationery from the headquarters of Confederate General Robert E. Lee, whose Army of Northern Virginia was just then engaged in an invasion of Maryland. Passed on to Union Major General George B. McClellan, whose Army of the Potomac was looking for its opponents, the document proved to be a copy of Lee's Special Order Number 191, issued on September 9, detailing the movements of his army over the next few days. McClellan's initial reaction, seconded by that of several other senior officers, was that the document was a plant, deliberately lost so that he might be led into a trap. As a result, although McClellan more or less conducted his movements over the next few days in accordance with the Confederate positions as suggested in the order, he did so with great caution, thereby losing the chance to annihilate Lee's army in detail. The "lost order" in this case was genuine, and fear of a ruse caused McClellan to lose the opportunity of a lifetime.

Footnote: *Meinertzhagen Smokes the Enemy.* After the war, Meinertzhagen claimed to have (and was widely credited as having) engaged in a wholly unauthorized special operation designed to further improve the chances of a British success during the pursuit after Beersheeba, as the Turks were attempting to make a stand at Sheria, about a dozen miles to the rear, in early November. According to his memoirs, for some time, he had arranged that an occasional airplane be sent over the Turkish lines in the evening, dropping a few packs of cigarettes wrapped in propaganda leaflets. By the time of the Beersheba operation, this had become a fairly routine procedure. But on the night of November 5, Meinertzhagen arranged to drop several thousand packs of cigarettes, each heavily laced with opium, so that by morning, when the British renewed their offensive, "a high percentage" of Turkish troops were so doped-up as, in Allenby's words, to have "put up a jolly poor fight." If true, this too ranks as a notable ruse of war, and one of the first uses of "nonlethal weapons" in the twentieth century.

The Greatest Deception of the Great War

On April 16, 1917, the French Army undertook its most massive "Big Push" of World War I, the so-called "Nivelle Offensive," which saw 1.2 million troops go "over the top" on a forty-mile front along the Chemin des Dames in Champagne. Since the outbreak of the war in August of 1914, nearly four million Frenchmen had become casualties in an endlessly repeated series of unsuccessful offensives. This time it would be different, for French Commander in Chief Robert Nivelle promised a quick breakthrough and a rapid end to the war. But within hours, it was clear that the highly touted offensive was a failure. By the time the overoptimistic Nivelle called off the assault, 10 percent of the attackers had become casualties—120,000 men (German losses were 21,000)—with negligible gains. And on April 29, the French Army began to mutiny. Within days, some 50 infantry divisions, of 109 in the army, had mutinied. In some cases, the "mutiny" was more like a strike; the men (in many cases tacitly supported by their officers) would defend their positions if attacked, but would not resume the offensive. But in other cases, the troops refused to occupy the trenches, and some even began abandoning the lines. Soon, only a dozen or so elite infantry divisions and the seven cavalry divisions could be considered wholly reliable. At one point, the Minister of War was informed that of the regular infantry divisions, only two might be considered absolutely reliable. For weeks, a major section of the front lay in the hands of the mutineers.

To save France from the twin disasters of revolution and defeat, General Henri Philippe Pétain was given command of the army. While he juggled loyal (or at least not openly mutinous) units into the line from the reserve or quiet sectors, to replace mutinous ones, Pétain worked tirelessly to restore the discipline and effectiveness of the army, personally addressing the men of nearly ninety divisions in just a few weeks, improving the food supply, instituting pay increases, creating new decorations, revamping the officer training system, expanding the medical service, revising the leave system, reorganizing the training establishment, and above all, promising no more offensives without adequate support—by which he meant waiting for more artillery, more tanks, and more Americans. Nor was he afraid to use the court-martial and the firing squad. However, although the number of men executed for the mutinies is still a secret, it is clear that Pétain did not order wholesale decimations of mutineers or call down artillery fire on mutinous units, as has been reported from time to time (these would have been counterproductive acts, likely to intensify rather than end the mutinies, and costly in manpower, to boot).

Surprisingly, the enemy did not notice anything at all, and even the British were unaware of the extent of the disaffection in the French Army. This was the result of an extraordinary effort to conceal the truth.

Almost as soon as the first mutinies broke out, GQG, French general headquarters, firmed up its already tight grip on all information. Not only was news passing from the Zone of the Armies to the Zone of the Interior carefully censored, but GQG even omitted references to the mutinies from official communications with the government. Members of parliament and other officials who exercised their right to tour the front were given carefully controlled access to sectors held by loyal troops—in many cases a thin line indeed. Yet, the front was held, with odd battalions of regular infantry, supplemented by Foreign Legionnaires, elite chasseurs, dismounted cavalrymen, colonial troops, and Senegalese. It wasn't much, but it was enough to convince the Germans, who were occupied with containing the disastrous British offensive at Passchendaele, in Belgium, and who had no plans for an offensive in any case, that the French were merely licking their wounds.

It was not until July that some German intelligence officers began to suspect that something was afoot. When they apprised their superiors of their belief that the French Army was on the verge of a collapse not unlike that which had just overtaken the Russians, the German brass reluctantly decided to test the waters. They shortly essayed a series of violent local attacks, only to find that resistance was fierce. The case of the French 77th Division, which had been one of the most mutinous units in the army just two months earlier, was typical. Holding the line near the Chemin des Dames, on July 19, the division was subjected to an intensive bombardment, followed by an assault by four divisions. The 77th gave ground and then, in a skillful and vigorous nighttime counterattack, threw back its assailants. Several similar probes all met the same fate. Meanwhile, at about the same time, although there was still a lot of work to be done to restore morale fully, Pétain began a series of limited offensives of his own. At that the German high command concluded that the rumors of widespread disaffection and even mutiny were just that, rumors.

So ended the most successful, and most important, deception of World War I. Had the Germans discovered the true state of the French Army in the spring of 1917, they might have won the war easily. Reliable information as to the extent of the mutinies did not begin to come out until several years after the war, and much information about them still remains a secret.

Armageddon, 1918

Soon after Jerusalem fell to General Sir Edmund Allenby's army on December 9, 1917, the Palestine front stabilized about fifteen miles north of the city, as both sides were too tired to continue fighting. There the front stayed for more than nine months due to severe shortages of men and supplies on both sides. For the Turks, these shortages were endemic, the national resources having been virtually exhausted. Allenby's problem was that the Western Front demanded more and more of Britain's resources, leaving him to operate on slender rations. During the long pause, both sides prepared for a final clash, the Turks expecting the British to make a bid for Damascus sooner or later. This, of course, was precisely what Allenby wanted to do. And by the summer of 1918, he had the wherewithal to do it. However, how to go about taking Damascus presented some problems.

Allenby did have about 70,000 men with 550 pieces of artillery against only 45,000 Turks with 430 guns—a comfortable superiority. But by the standards of World War I, this was hardly an overwhelming advantage, for the front was narrow, only about sixty-five miles long, from where the Jordan River flows into the Dead Sea to the Mediterranean coast a dozen miles north of Jaffa. This allowed for little maneuvering room, unlike the situation that had prevailed at Gaza and Beersheeba, the previous year. Moreover, the Turks were well entrenched. With about forty men per hundred meters of front, the Turkish rifles and machine guns would provide a formidable obstacle to the advancing British. A Western Front general would have just thrown masses of men and shells at the enemy in an effort to bull his way through. This probably would have worked, but Allenby was frugal with the lives of his men. He wanted a better way, and he found it.

Allenby evolved a simple plan. A series of diversions and deceptions would draw the attention of the Turks and their German commander (General Otto Liman von Sanders) to the eastern end of the front. This would suggest that Allenby was planning an offensive northward up the Jordan Valley or northeastward, toward Amman. This was a move that would permanently sever Turkish rail links with their forces in Mecca, far to the south, while posing a direct threat to Damascus, just a few-score miles to the north. Then Allenby would attack in great force along the coast, driving the Turks away from the sea, so that his mounted corps could drive northward, spreading out in the Turkish rear to cut off their retreat.

Many supporting operations were developed, each designed to draw the enemy's attention eastward. Several local attacks were made to secure a bridgehead across the Jordan on the main road between Jericho and Amman, while Colonel T. E. Lawrence "of Arabia" and the Arab Army conducted raids and reconnaissance into the area between the front and Damascus. Just behind the newly won bridgehead, engineers and camouflage experts constructed a series of dummy camps, complete with tents, artillery parks, horse lines, and ammunition dumps, suggesting the presence of several divisions. Meanwhile, at the other end of the front, equally elaborate measures were taken to conceal the fact that 35,000 infantrymen (four divisions) and nearly 12,000 cavalrymen (three divisions) with 400 guns were crammed into a space only about ten miles on a side just north of Jaffa. To make sure that the enemy saw only what he was supposed to, the Royal Air Force managed to keep the skies clear of enemy aircraft, save for a few that slipped through on the eastern side of the front. All of this activity convinced Liman von Sanders that Allenby's blow was to come in the east.

In accordance with his understanding of Allenby's intentions, Liman von Sanders reorganized his forces. He concentrated about a third of his forces, including all of his cavalry, east of the Jordan, holding a front of only about fifteen miles, confronting what he believed to be four or five divisions but which actually was only three brigades. He did this despite objections from some of his Turkish subordinates, who had not been convinced by Allenby's machinations.

Allenby opened his offensive at 4:30 A.M. on September 19 with an artillery barrage along the entire front. Then the infantry went in, once again all along the front, with the troops on the coast being supported by gunfire from the Royal Navy. Turkish resistance was fierce, but their lines were thinly held. Within three hours, a breakthrough had been achieved, as the Turks fell back from the coast, curving their flank northward. Into the gap Allenby passed his Desert Mounted Corps, which was soon fanning out in the Turkish rear area, supported by the Royal Air Force. By dawn on September 20, the Turkish Eighth Army had disintegrated, while their Seventh, under Mustafa Kemal, later more famous as Atatürk, fell back eastward. That effectively ended the Battle of Megiddo, for all significant Turkish resistance ceased. Over the next few weeks, Allenby's forces would harry the Turks northward, until an armistice was arranged on October 30, 1918.

The cover plan for Allenby's Megiddo offensive was one of the most successful of World War I. Yet, it could easily have failed. Had Liman von Sanders listened to his Turkish subordinates, he would have disposed his forces differently. The result, at best, would have been a costlier British victory, and at worst, might readily have been a disastrous defeat. But Liman von Sanders was the one who had to be tricked, not his entire army, nor even his entire staff. For an army is an extension of its commander's mind. And at Megiddo, Liman von Sanders, otherwise an extremely capable commander, was tricked most effectively.

"Wild Bill" Donovan Visits Ethiopia, 1935

Late in 1935, President Franklin D. Roosevelt asked William Donovan, a prominent Wall Street attorney and war hero (he had won a Medal of Honor at the head of New York's "Fighting 69th" during World War I), to undertake a little intelligence mission. This was not an unusual request, for in those days, the United States had nothing resembling an intelligence service, and over the years several presidents had made such requests of prominent persons of impeccable reputations and credentials. In fact, Donovan had been doing that sort of thing since early in World War I, when he had been sent by President Woodrow Wilson to gather information in Europe, in the guise of surveying the relief needs for refugees and other persons in German-occupied areas.

Roosevelt's mission for Donovan was simple: to find out what was happening in Ethiopia, which Italy had recently invaded. This was an important mission, since there was a great deal of uncertainty in the international community as to what should be done about this blatant act of aggression. Some British politicians, including Sir Anthony Eden, wanted to make Ethiopia a test case in the control of aggression. But the suggestions for imposing rigid economic sanctions, and even barring Italian vessels from using the Suez Canal, were acts that might easily have precipitated war between Italy and Britain. Roosevelt himself was under some pressure to impose an embargo on oil shipments to Italy. Information about the war was sparse, Ethiopian pronouncements of victories being dismissed as propaganda by the Italians, and vice versa. So off went Donovan.

He arrived in Italy late in December of 1935 and managed to secure an interview with Mussolini, who was impressed by Donovan's decorations. The two were soon swapping soldiers' stories from the "good old days" in the trenches. *Il Duce* gave Donovan permission to visit Ethiopia, ostensibly to look into investment possibilities, and, supposedly, to satisfy an old soldier's curiosity as to the efficiency of Mussolini's new Roman legions.

Donovan was given the red-carpet treatment in Ethiopia. He met with Marshal Pietro Badoglio, the supreme commander, was shown the mountainous stockpiles of munitions, visited the various medical facilities, visited a number of air bases, inspected many troops, and was even given a tour of part of the front. Everywhere he went, Donovan noted that the senior officers with whom he met were uniformly confident that victory was imminent. Needless to say, Donovan was greatly impressed.

On his way home, shortly after New Year's Day, Donovan stopped in Libya, where he was received by the Governor General, the aviation hero Italo Balbo. Much to his surprise, Balbo appeared pessimistic about the war in Ethiopia. However, Donovan seems to have put that down to the fact that Balbo was missing out on the "glory" and other rewards which victory in Ethiopia would bring to the commanders.

Within a day or so, Donovan arrived in London, where he held a clandestine briefing for Anthony Eden. Donovan informed Eden that there did not appear to be anything that could stop the Italians from conquering Ethiopia. Economic sanctions would not have time to take effect before the war would be over, and given the enormous stockpiles of munitions already in Ethiopia, even closing the Suez Canal would probably not seriously impair the war effort. As a result, the British government confined itself to a relatively mild rebuke of the Italians, rather than taking more stringent measures. Then Donovan flew home to inform the president of his findings, in consequence of which, Roosevelt decided against putting an embargo on oil shipments to Italy. As a result, little was done to interfere in the Italian war effort, and on May 9, 1936, Mussolini went out on his balcony at the Palazzo Venezia to proclaim the annexation of Ethiopia.

Not until some time afterward did the truth come out. Donovan had been the victim of an elaborate hoax. Mussolini suspected that Donovan was working for the president, and would probably speak to the British as well. As a result, he ordered Badoglio to lay it on thick. In fact, during the period that Donovan was in Ethiopia, the Ethiopians had undertaken what came to be called their "Christmas offensive," a vigorous assault on the Italians. From December 15 until well into January, the Italian lines were subject to vigorous and often skillful assaults, with the result that in some places, they were thrown back, and in others, stretched almost to the breaking point, while their own attacks were held to minimal gains or even beaten off. So desperate did the fighting become that two days before Christmas,

Mussolini authorized the use of poison gas, which was probably the single most important factor in permitting the Italians to hold their lines. Donovan had been completely taken in by the Italian deception, despite the fact that Balbo's pessimism ought to have tipped him off that all was not right in the new Roman Empire. Had he taken the hint, his advice to Eden and the president might well have been different, with momentous consequences for world peace.

The trick that Mussolini and Badoglio pulled on Donovan is by no means a rare one. It is known as a "Potemkin Village," named after Russian Prince Grigory Aleksandrovich Potemkin, who erected false fronts on impoverished villages to cover up his shortcomings as a civil administrator during the reign of Catherine the Great, in the late eighteenth century. In the mid-1930s, while Germany was rearming, Adolf Hitler resorted to precisely the same ruse on numerous occasions to convince the world of the newfound strength of the Reich, and particularly that of the *Luftwaffe*. Among the foreign dignitaries convinced that Germany had an enormous air force were Italian Air Marshal Italo Balbo, British Air Vice Marshal Christopher Courtney, and French Air Force General Joseph Vuillemin, not to mention the American Charles Lindbergh. The devices used were simple: a tour of various air bases, at which numerous aircraft were displayed, aircraft which were secretly flown from base to base as the tour proceeded, and then hastily painted with new squadron markings and shown off once again. Thoroughly deceived, these men returned to their homelands and preached accommodation and appeasement, with considerable success. And the trick has been played even more recently.

In the late 1980s and early 1990, Iraqi dictator Saddam Hussein began to invite foreign journalists and military analysts to have a look at the Iran-Iraq war from his side of the lines. The visitors were given the full treatment: tours of the front, reviews of the troops, and the like. All of these events were carefully staged, so that only the best was exposed to scrutiny. In fact, many of the troops reviewed were Republican Guards, who, although lavishly equipped, had seen little combat, while all of the "front lines" were carefully constructed dummies considerably in the rear of the real front, which was mostly in swamps. Virtually all of the Western observers came away highly impressed with the character and skill of the Iraqi Army. So when Saddam invaded Kuwait, many of these journalists and military analysts voiced extreme opposition to the use of force to eject him, some verging on the hysterical, with predictions of fifty thousand or more deaths at the hands of the "well-trained, battle-hardened veterans" who comprised "the fourth-largest army in the world." Although other observers had more correctly pointed out the extensive flaws in the Iraqi armed forces, these tended to be ignored. One result of this was the extraordinary amount of firepower which the Coalition forces brought to bear.

For a variety of reasons, it is likely that the "Potemkin Village" ruse will become more common in the "New World Order." Third World armed forces exist primarily for show, designed to impress neighboring powers, and, more important, overawe the local citizenry. Such forces are often elaborately uniformed, can perform intricate drills, and usually have impressive-looking equipment, but are often devoid of serious training. Against a properly trained army, they would quickly fall to pieces; but they can engage in combat against other armies of their ilk, and are very capable of casually massacring their fellow citizens. In a confrontation with a major power, a Third World dictator might try to bank on the impressive appearance of his army; for example, by manipulating the media, most members of which are woefully ignorant of military matters. It is worth recalling that in some circles, the panic over Saddam Hussein's military capabilities had some Americans anticipating hecatombs of dead, while others hysterically predicted nuclear war, and still others sent their children for psychiatric counseling—all of which could easily have impeded the nation's ability to respond to the crisis.

DECEPTION ON A GRAND SCALE

World War II (1939–1945)

World War II saw more deceptions than any other conflict in history. One source lists over fifty major deception plans alone for the Western Allies in the European Theater, thereby omitting dozens of similar cover plans of the Germans and the Russians, not to mention literally hundreds of minor tricks and ruses employed by everyone from theater commanders down to infantrymen. Deception came of age during World War II, where it became firmly enshrined in the arsenal of standard military practices. A more pragmatic reason for the emergence of deception in this war may have had to do with the vast number of people in the military during the conflict. Nearly a hundred million troops were under arms at one time or another during World War II, a greater number than in any previous war. The vast majority were not professional soldiers in 1939, and their presence in the ranks transformed warfare by bringing to the battlefield a lot of fresh ideas and insights. Not being bound by military tradition, these recently armed civilians looked for better ways to avoid getting killed. Deception was one of the more commonly encountered results.

The professional soldiers were also of a different mind than in any previous war. Since the late nineteenth century, there had been an immense growth in the education levels of the industrialized countries. Nearly all the troops were now literate, and many of the officers had university educations. So the peacetime military professionals were better prepared to cope with changes. And changes were desired. World War I had vividly demonstrated the bankruptcy of doing things the traditional way. So the professional soldiers were receptive to new ideas, as were their millions of new soldiers conscripted from the civilian population. While many new ideas (the blitzkrieg, strategic bombing, etc.) and new weapons (tanks, heavy bombers, aircraft carriers, etc.) are recognized as the legacy of World War II, deception was the one growth industry that was not recognized. Yet, the use of deception in World War II was so common and energetic that it appears to have been taken for granted. Such was not the case. Never before in history was deception practiced with such vigor and regularity.

Deception by Press Release

Often the best deceptions are those that develop from unexpected opportunities. One such case was the German He-100 fighter. Designed in the late 1930s as a more modern replacement for the Me-109 (which first flew in 1935), the He-100 did not go into production in 1938 because it was discovered that the Me-109 could be more economically upgraded, and was already a proven fighter aircraft. Besides, the He-100 wasn't all that good in many respects, except on paper. With war imminent, it was no time to introduce a new fighter aircraft.

The He-100 did have some respectable performance stats. In fact, it set the world speed record in early 1939 (463 MPH). All of this caused the Nazi propagandists to come up with the idea of using the unwanted He-100 for a deception. So they renamed it the He-113, painted different squadron markings on the dozen they had, and then flew them around to be photographed. The photos were distributed widely, with the announcement that Germany's "new fighter" was about to enter service. Foreign nations already knew that the Me-109 was a capable and widely used fighter. Now here were the Germans announcing the mass production of an even more capable successor.

To assist in the deception, the three preproduction aircraft were sent to Japan for evaluation (the Japanese were not impressed), and the Soviet Union was to receive the only twelve production aircraft for evaluation. These dozen He-113s were completed in 1939, but the evaluation arrangement was called off.

This minor deception effort caused no end of consternation in foreign nations, particularly after the war got going with the invasion of Poland in late 1939. It was a while before Germany's enemies could convince themselves that, while the Me-109 was pretty lethal, a more capable He-113 wasn't going to replace it. The dozen production models were assigned to the defense of a Heinkel factory throughout the war.

Coming Through the Woods

One of the great deceptions during the early part of World War II was the unexpected advance of German forces through the Ardennes Forest in May of 1940. The Allies had not expected this, as they knew that the Ardennes area was hilly, heavily forested, and covered by few roads. This was hardly the kind of terrain you would drive a panzer army through. The Germans thought so too, but they looked at the situation a bit more systematically. What the Germans found was that, with careful planning, they could quickly move large armored units through the Ardennes. Their movement plan was a masterful work, which was tested using realistic war games. The plan depended on strict discipline and precise observance of a master schedule. These were two things the Germans were good at. The Allies never suspected that the Germans would try such a thing, and the Germans maintained good security regarding their ten-division romp through the woods.

On May 10, the Germans sent seven armored (panzer) and three motorized infantry (panzergrenadier) divisions into the Ardennes.

This was just the first wave, but it was the most important one. When these divisions exited the Ardennes on May 13, they would have to fight their way across the Meuse River in the face of now-alert Allied commanders.

Although the Allied commanders quickly appreciated the gravity of the German deception, there wasn't much they were able to do about it. The Allies thought that, as in 1914, the Germans would come through Belgium, to the north of the Ardennes. It was in Belgium that the Allies placed most of their troops, and the Germans obligingly put on what amounted to a large feint. Once the German panzers were across the Meuse River, they were into the Allied rear area. Within a week, the Allied situation was desperate, and France was soon lost.

The deception the Germans used was a classic one. Appearing to do what the Allies expected of them, the Germans massed troops for an offensive in Belgium, while keeping their motorized divisions in the rear. Once the battle was joined, it was a simple matter for the panzer divisions to take a quick left into the Ardennes instead of proceeding north to the Belgium plain. The Germans used most of the other available deception measures, including false and planted information, camouflage, concealment, and the like. Had the Allies wised up any earlier and sent some units into the Ardennes, the German ploy would not have worked, or would not have worked nearly as well. The Germans took a chance, and it worked.

It's quite likely that the Germans would have won without the Ardennes deception. But had the Germans not employed their dash through the unguarded forest, the Allies would have held out longer and killed a lot more Germans. That said, the march through the Ardennes can be said to have been a success.

Desperate Deceptions After Dunkirk

By the end of June 1940, Great Britain was in desperate straits. The Germans had just conquered France, and defeated the British forces that had been sent across the Channel. The British had gotten most of their troops, but not their equipment, away via a desperate evacuation from Dunkirk. But now Britain faced the prospect of a German invasion. With practically no ground forces, there was only the Royal Navy and Royal Air Force (RAF) to stop the Germans. It was crucial that British air power be preserved as much as possible. If the Germans could obtain air superiority, they could attempt a landing. The Germans had more aircraft, and the initial Nazi priority was the destruction of the RAF and its bases. Deception was widely used to weaken the effectiveness of German bombing attacks and preserve British fighter strength. The principal deception was false airfields and, to a lesser extent, false aircraft factories and aiming points.

The false airfields came in two flavors. The first type was meant to deceive the Germans at night, when many of the bombing raids took place. Over a hundred of these fake "night" (or "Q") sites were built, and they received twice as many bombs as the nearby real airfields they protected. The "Q" sites would not fool anyone during the daytime, because they depended on lights for their deception. An airfield at night would show landing lights around the landing strip and the lights of landing aircraft, or planes moving around on the air base. The "Q" sites had lights rigged (using strung lines, pulleys, and carriages) to emulate, at least from the air, a functioning airfield. When the British knew German bombers were in the area, the real airfield would switch its lights off and the "Q" site would turn its on. Once it was determined that the enemy bombers had seen the "Q" site lights, those lights would be turned off. This was to be expected by the Germans, as an airfield would go to blackout conditions once bombers were known to be in the area. However, once the "Q" site lights were seen by the German bombers, they would usually go for the "Q" site area and drop their bombs. The people running the "Q" site were underground in a bunker, and were some of the few people in Great Britain at the time who actually felt joy when they heard (and felt) the bombs exploding overhead.

The next day, when a German recon aircraft flew over the nearby airfield to check for damage, they were dismayed to find none. The "Q" site, some few miles away, didn't look like an airfield in the daylight. If the German recon plane happened to notice the "Q" site bomb craters, they would think that the bombardiers involved in the previous night's bombing needed some more training.

Whenever the Germans thought they could get away with it, they would make daylight raids. These attacks allowed fighters from both sides to get involved, and bomber losses were always higher. But there was no mistaking an airfield in daylight. Or was there? The British also built sixty "K" site dummy airfields that were expected to stand up to German scrutiny. They didn't. Although the "K" sites were filled with some four hundred convincing-looking dummy aircraft, the Germans soon figured out which fields were real and which were fake. It wasn't that the British hadn't done a good job of creating the fake airfields, but the Germans knew that the original airfields had been there for some time. The British had little reason to build new fields, and the Germans knew this. Any new airfield was scrutinized very closely, and soon uncovered as a deception. The Russians had been more successful creating dummy airfields, but largely because they were always building new fields as the battle line advanced and the Germans were thus easier to fool.

Aside from the airfields and radar facilities (which the Germans had little success damaging), aircraft factories were the other prime target. The British constructed four fake aircraft factories. Two were actually bombed, thus sparing the real factories nearby. The fake facilities might have gotten an extended pounding if the air war over Britain had gone on longer. But during 1941, the Germans moved most of the air power east for the planned invasion of the Soviet Union. The Germans did learn from this experience, and improved on the British deceptions when the Allies began bombing Germany heavily from 1942 on. By 1945, more bombs had dropped on fake German factories than on British ones.

Deceptions were also used to support the worst case—that the Germans would get ashore with troops. In addition to building fortifications near likely landing areas, there was a comprehensive effort to camouflage the real defensive works and create dummy ones. Thus the German aerial photos of British defenses showed only the fakes, which would have left the landing troops in for some rude, and fatal, surprises when they discovered all those haystacks that were actually concrete machine-gun posts.

The Germans could have used their spies in Britain to reveal some of the deceptions. But all (well, nearly all) German spies in Great Britain were quickly captured and most were turned into double agents. But that's another story, and yet another deception.

Little Bits of Business

There were extraordinary numbers of deceptions used during World War II, many of which can be explained in no more than a sentence or two. Forthwith, a sampler of some of the numerous "little" ruses practiced during the war—many of which had important consequences.

- During one of the great kamikaze attacks on the Fifth Fleet as it lay off Okinawa in the spring of 1945, naval historian Samuel Eliot Morison observed two aircraft apparently locked in a deadly dog fight. Suddenly, they broke off the fight, and both dived on American warships. The dog fight had been a clever, and successful, ruse to keep American fighters from interfering with the two kamikazes, and to keep ships' antiaircraft gunners from taking the risk of shooting down a "friendly" airplane.

- Since the Japanese eventually grew to realize that just prior to American amphibious landings "frogmen" would be sent in to reconnoiter the beaches and blow up obstacles, the U.S. Navy began employing frogmen to make bogus reconnaissances of literally scores of beaches on which it had absolutely no intention of landing.

- Admiral Sir Andrew Browne Cunningham, commanding the British Mediterranean Fleet, once left his flagship late at night toting a small overnight case, as if going away for a few days to enjoy the fleshpots of Alexandria. As the city was full of spies, word quickly got to Italian naval headquarters, which, not anticipating any British activity, let down its guard. Meanwhile, Cunningham quietly returned to his flagship, to lead the fleet on a highly successful sortie.

- When U.S. Army Rangers, after a desperate struggle, finally stormed the top of the cliff at Point du Hoc in Normandy on D-Day, June 6, 1944, they were surprised to discover that what appeared to be a battery of six-inch guns was in fact a batch of "Quaker guns"—logs pointed to look like artillery pieces—the Germans not having had time to install the real ones (stored nearby) in time for the invasion.

- Since there was no defense against German V-2 rockets, the British hit upon the idea of reporting V-2 hits that landed in the heart of London as occurring east of the city, to which the Germans responded by adjusting their aiming, so that the missiles began to land harmlessly west of the city.

- To convince the Germans and Italians that the expected Allied landings in Sicily were still some days away, in the late spring of

1943, American General George S. Patton went on "leave," flying from Tunis to Cairo, where he was lavishly entertained at a small dinner party in an elegant restaurant, before turning in late at night in anticipation of a day of sightseeing. Then he was almost immediately flown back to take command of the invasion forces.

- During the Battle of Britain, once the Germans realized that the odd-looking towers on the coast of England were radar installations, they attempted to put them out of service by massive air strikes. Although they were successful in eliminating several stations, the British quickly established dummy radar transmissions from the same sites, which convinced the Germans that it was extremely difficult to knock out radar installations.

- In order to convince the Japanese that the United States was planning an invasion of the Kurile Islands from the Aleutians, American troops shipping out from West Coast ports for the South Pacific were occasionally issued winter underwear and shoulder patches of units stationed in Alaska.

- To facilitate an assault on the Italian-held Siwa Oasis in Libya, the British dropped a large number of dummy parachutists in the vicinity, with special sound-making equipment to create the impression that a considerable force had landed, convincing the defenders that they were greatly outnumbered and that their line of retreat had been severed, whereupon they abandoned their post with minimal resistance.

- During their long retreat from the Alamein Line to the Mareth Line, in late 1942 and early 1943, Field Marshal Erwin Rommel's Italian and German troops used land mines to help slow down the pursuing British. Since they didn't have enough mines, they would often bury junk, old tin cans, spare shell casings, and so forth. British detection gear would indicate the possible presence of mines, and the pursuit would have to be halted until the mines were cleared. Of course, after doing this two or three times, the British would decide that the enemy was playing tricks big time, and ignore the signals. Often, they did so to their misfortune, since it was customary to include a few real mines every so often, just to keep the Brits on their toes. The RAF used much this same trick after discovering that it didn't have enough real mines to plant in the Danube to disrupt German barge traffic. They mixed in a large number of dummy mines with the real ones on the accurate theory that the Germans would have to sweep them all.

- The feint that the U.S. 2nd Marine Division conducted against a beach on the southern coast of Okinawa, in an effort to draw Japanese attention away from the actual landing sites in the spring of 1945, was so convincing that Japanese Lieutenant General Mitsuru Ushijima, commander of the defenses of the island, believed he had beaten off a major amphibious assault.
- One could sometimes convince the enemy that an airfield was so heavily damaged as to be useless, by painting false craters on the runways, a trick that did not work if overlapping reconnaissance photos and stereoscopic optical equipment were used.
- After torpedoing the Canadian corvette *Spikenard*, the skipper of a German submarine had his radio operator imitate her call signal and the "fist" (i.e., the distinct electronic "signature" each operator had because of the slight variations in how the telegraph key was tapped) of her deceased radioman, signaling the convoy that *Spikenard* had lost its bearing. An unsuspecting escort gave the submarine the correct heading, and the convoy suffered several additional losses as a result.
- All World War II combatants routinely placed carelessly camouflaged dummy aircraft on runways, while carefully concealing the real ones, so that enemy air strikes would do as little damage as possible.
- After Louis B. Mayer, the head of MGM, kept pestering the army's camouflage people to "do something" to conceal his Hollywood studio to protect it from possible enemy bombing, they eventually responded with an elaborate scheme that they said would shield the entire installation from view. Only after the war did Mayer learn that the studio had been disguised to resemble a nearby aircraft factory. The nearby aircraft factory was in turn camouflaged to resemble a suburban development, complete with streets, swimming pools, and parks.

U.S.S. *Canopus* at Bataan

Among U.S. Navy ships in Manila Bay when Japan bombed Pearl Harbor was the submarine tender *Canopus*. Her duties were mundane: to supply and repair submarines, which were to do the actual work of war. But by Christmas of 1941, all of her charges were gone, transferred south to Australia, whence they could carry on the war more effectively. Meanwhile, of course, the Japanese had overrun most of the Philippines. As they were closing in on Manila, American and Filipino troops retreated to the Bataan Peninsula for a desperate last stand. On Christmas night, *Canopus* went with them, steaming the few miles across Manila Bay to tie up at the dock at Mariveles, a small port at the southern tip of the rugged peninsula. There, covered with vines and brush, she set up shop as a general repair facility to support the troops at the front.

Despite her jungle camouflage, Japanese aviators managed to spot her, and on December 29, 1941, she was subjected to a serious air attack. Working hard, the ship's crew was able to put out the fires and repair most of the damage, so that within a few days, she was seaworthy again. On January 5, 1942, she was hit again, albeit lightly. And once again, the damage was quickly repaired. However, recognizing that the ship was an easy target, her skipper, Commander Earl L. Sackett, decided that a little deception was in order.

By judicious flooding of some bilge spaces, Sackett caused the ship to acquire what naval historian Samuel Eliot Morison termed a "pathetic list." Some of the ship's most visible injuries (those which would not affect her seaworthiness or operability) were left unrepaired. Her cargo booms were slewed around as if broken or jammed. A few smudge pots were sited here and there on her decks, to give the impression that the ship was slowly smoldering below. No attempt was made to restore most of the ship's camouflage of brush and vines, and, most telling of all, she was allowed to rust. Finally, as much as possible, the ship's company assumed a nocturnal routine. These measures soon gave the ship the look of an abandoned hulk. As a result, although Japanese aircraft striking at installations in and around Mariveles Harbor repeatedly overflew the ship, none of them bothered to attack the obvious derelict.

Now safe from Japanese attack, *Canopus* began to play an important role in the defense of Bataan. While her magazines supplied munitions and spare parts to the defenders, her machine shops kept weapons and equipment in good repair, and converted several motor launches into improvised gunboats ("Mickey Mouse battleships," the troops called them), and her bluejackets turned to with rifles for combat duty from time to time. In this fashion, *Canopus* helped prolong the defense of Bataan until the end. It must have come as a surprise to Japanese observers when, on April 9, 1942, that terrible night of surrender, the "destroyed" *Canopus* suddenly came to life, cast off from the dock, and under her own power backed out of Mariveles Harbor into the dark waters of the South China Sea, where her crew promptly scuttled her.

"Major Martin": The Man Who Never Was

After the Allies expelled the Axis from North Africa in the spring of 1943, the question arose as to "What next?" Although the United States (and the Russians) wanted an immediate landing in northwestern Europe, there were logistical, military, and political obstacles which could not be overcome in time. So they settled on a landing in Sicily, by which they hoped to finally clear the central Mediterranean of Axis interference and, just possibly, knock Italy out of the war. In preparation for the invasion of Sicily, the Allies adopted a number of cover plans, designed to suggest that their next major move would be against Greece or possibly Sardinia. Several tricks were used to get the message across to the enemy, the most clever of which was "Operation Mincemeat," more commonly known as "The Man Who Never Was."

With the cooperation of the deceased's family, British deception specialists procured the body of a man who had just died of pneumonia (which leaves liquid in the lungs, similar to a drowning victim). This man's identity was not revealed until more than thirty years after the war. The corpse was clothed in the uniform of a major of the Royal Marines. In addition to the normal identification that an officer of his rank would carry, "Major Martin" was provided with the common sort of things that accumulate in a man's pockets. For example, his wallet contained some theater stubs, a past-due bill, and a photograph of an attractive young woman in a bathing suit, with "Pam" and a date scrawled across the back (she was secretary to the officer in charge of the operation). There were several keys in his pockets, a reasonable amount of cash, and much else besides. A courier's briefcase was manacled to his wrist. This contained a variety of falsified, medium-sensitive documents, along with a few genuinely secret items of not great value.

But included were two forged documents of particular importance.

The first was a letter from Britain's Chief of the Imperial General Staff to General Sir Harold Alexander, one of the most senior Allied commanders in the Mediterranean. It explained that in preparation for planned landings in the Peloponessus, the southernmost peninsula of Greece, a cover plan was being laid on to suggest that the actual Allied objective was Sicily.

The second letter was a less formal one, from Lord Mountbatten, head of Britian's Combined Operations and an expert in amphibious warfare, to Admiral Sir Andrew Cunningham, senior Allied naval officer in the Mediterranean. It introduced "Major Martin" as a specialist in the use of landing craft, and went on to say, "Let me have him back, please, as soon as the assault is over," concluding with, "He might bring some sardines with him."

The intent of these items was to hint that the next Allied move in the Mediterranean would be against Greece, or, albeit less likely, an invasion of Sardinia. All of this was done within a few days of the man's death, while his body was kept refrigerated.

The "major's" body was then loaded into a special canister, put aboard a submarine, and dropped into the sea a few miles off the southern coast of Spain, where the currents would be certain to carry him ashore shortly. Meanwhile, a routine announcement was made of the loss of several men, including a "Major Martin," in an airplane accident over the Mediterranean. Within a few days, the Spanish government informed the British that the body of one of their officers had washed up on the coast. The British consul took charge and arranged for the man's burial in a local cemetery with full military honors. Meanwhile, of course, the local Spanish authorities had allowed a German agent to examine the body, and Berlin was soon apprised of the findings. Although Hitler was apparently inclined to believe the deception, the principal Axis commanders in the Mediterranean, General Vittorio Ambrosio and Field Marshal Albert Kesselring were not, and managed to bring him around to their view, that the object of the next major Allied offensive in the south would be Sicily. So, despite a very successful book and film about this ruse, "Operation Mincemeat" was a failure.

This brings up one of the less-touted aspects of trickery in warfare: that your enemy may *not* buy your little scheme, for whatever reason, and so, you'd be well advised not to be overly reliant on clever plots. In this case, failure to mislead the enemy with planted information was not particularly critical. In other cases, this was not so.

Kursk: The Grand Deception in the East

The greatest tank battle of World War II, Kursk, was won largely because of deception. In fact, most of the success the Russians had against the Germans during that conflict sprang from a greater appreciation for, and practice of, the Art of Deception.

In early 1943, the situation in Russia had stabilized. The Germans had taken a tremendous beating at Stalingrad, and southern Russia in general, during the previous six months. In the first two months of 1943, the Germans had lost an entire army at Stalingrad, largely because of the Germans' own arrogance regarding Russian capabilities. The late-1942 Russian offensive that had surrounded Stalingrad and led to the destruction of the German Sixth Army was a masterful piece of work. But the Germans thought they had learned from their mistakes as they massed troops for a 1943 summer offensive. The Germans had always walked over the Russians during the warm weather, and believed they could do it in 1943, as they had in 1941 and 1942.

What the Germans failed to appreciate was that the Russians were successful at Stalingrad largely because of their first large-scale use of deception. The Russians had carefully concealed the location, size, and timing of the surprise offensive that cut off the German Sixth Army in Stalingrad. While the Germans recognized that they had been snookered, they did not catch the implication that the Russians would note the success of deception and redouble their efforts to use it on the Germans in the future. The key issue here was the German disdain for Russian military capabilities. German successes since 1939 had blinded them to the fact that the loser tends to get more capable. Defeat is a wonderful incentive, and the Germans were cursed with an attitude that did not include any feelings of having been defeated during the Russian winter offensive in late 1941, or the Stalingrad debacle of late 1942. The Germans felt that those losses were due to German errors, not Russian skill. The Germans were soon to reap the full measure of their presumptuousness during 1943.

The target for the German 1943 offensive was a huge bulge in the Russian line. This bulge contained the city of Kursk, and hundreds of thousands of Russian troops. It also contained a huge deception that the Germans fell for, from which the Russians profited handsomely.

When the chaos of the winter campaign finally died down in March and April of 1943, it was pretty obvious to both sides that something was likely to happen with the Kursk "bulge." Either the Germans would pinch it off, or the Soviets would use it as a jumping-off position for their next winter offensive. While the Germans had been hammered pretty severely the previous winter, they still had never been bested by the Soviets in warm-weather campaigning. But the Germans were aware of the increased, and increasing, Soviet strength. Not only were there more Soviet troops at the front; they were more experienced and more effective. The Soviets had more weapons. The Germans had seen ample evidence of American and British equipment that had been sent to Russia. Even more numerous Soviet-made weapons were much in evidence, particularly T-34 tanks, new aircraft types, and more effective pilots.

The Germans had learned in the past year that the Soviets could now hold a position if they wanted to. A hasty, impromptu attack was no longer sufficient to cut through the Soviet lines. Hitler decided that if supermen were no longer sufficient to smash the Russians, then superweapons would help do the job. Thus a planned May offensive against the still-forming Soviet positions was postponed until new tanks and troop reinforcements could be rushed to the Kursk front.

The German generals in Russia knew what their troops could do, were well aware of the Soviets' increased capabilities, and didn't believe any new weapons were going to make a big difference if the Soviets were allowed to dig in. Soviet deception did not bother them, yet. But Hitler and his toadies back in Germany were calling the shots. The new weapons and reinforcements were supposed to be in place by June, but they weren't. So the attack was delayed until July, and by that point, the Soviets had used their shovels, and a great deal of deception, to make the last German offensive in the east a catastrophic failure.

The key factor in this month-long series of battles in the Kursk "bulge" was deception, and it worked because the Russians were more realistic than their opponents. The Germans felt, with some justification, that their superior training would enable them to pierce the Soviet positions and allow another of the mobile battles which had been so costly to the Russians for the past two years. On the face of it, the Germans certainly had a lot going in their favor. Numbers of troops and weapons were about equal, and many times in the recent past, the Germans had been able to win while outnumbered. German troops were largely veterans, and the Wehrmacht now knew how to operate in the unique geography of Russia. In these two particulars, the Germans were correct. What they failed to deal with was an array of deceptions the Soviets employed to defeat the German advantages.

The Soviet deceptions took several principal forms:

1. The most important ploy was good old-fashioned camouflage and concealment. As the Soviet troops began digging in during the spring of 1943, they did more than just create trenches, bunkers, and tank traps (trenches tanks would get stuck in). Once the first line of fortifications was created (the one Germans could raid with infantry patrols), additional lines of fortifications were created behind the first one. While it was a common practice for most World War II armies to create multiple fortified lines, the Soviets took great pains to conceal these additional works. Camouflage, false positions, and concealment were all used to deny the Germans any accurate knowledge of these supplementary lines. This meant that once the German troops pierced the first line of trenches, they would encounter one unpleasant surprise after another. Minefields and antitank guns were the biggest problem for the German offensive, and these two items were the ones the Soviets paid the most attention to hiding. Equally important was the use of camouflage by the Russian reserve units, the ones that would move up and attack the Germans once the Nazis had wasted their vigor against Soviet defenses. The Germans knew the Russians had reserves, but Russian skill at deception prevented the Nazis from knowing the extent and location of the Soviet reserves (which were, in fact, immense).

2. The second major deception was the false airfields, supply dumps, and troop concentration areas the Soviets created throughout their rear area. The Germans were still able to control the air in early 1943 and they depended on this to enable their bombers to destroy Soviet rear-area targets. But many of the targets hit were clever fakes that cost the Soviets little. Particular care was taken to create fifteen false airfields filled with fake aircraft. Meanwhile, thousands of new Soviet fighters and bombers were kept at bases to the rear, out of range of German aircraft. While German intelligence experts were tallying all the enemy aircraft, airfields, and ground units they were "destroying," the actual Soviet strength was growing, and cleverly hidden. Entire armies (roughly equal in size to German corps of be-

tween forty and sixty thousand troops) were moved around the area without the Nazis being aware of it. Huge amounts of supplies (thousands of tons per location) were brought forward undetected. At the same time, false troop concentrations and supply dumps were created, to make the Germans think they knew what the Russians had, where it was, and what the Soviets were going to do with it. The Soviets knew that it wasn't enough to hide the troops; you had to hide all the support units, supplies, and equipment needed to support the combat formations.

3. The Soviets hid their plans effectively. While hiding troops, equipment, and facilities was important, it was vital for them to deceive the Germans about what Soviet plans and intentions were. Nazi reconnaissance aircraft and intelligence analysts were continually taking a long-distance look at what the Russians were up to. German doctrine stressed taking the initiative, which meant enemy plans were often irrelevant. But it was always useful to have an idea of what the other side might be up to. And so the Germans did have intelligence officers who reported regularly on what Soviet plans appeared to be. But a major problem the Germans had, and the Soviets exploited, was a bad case of arrogance. It wasn't until 1944 that the German senior commanders finally began to show the Soviets a little respect. But in 1943, too many of the senior Germans were inclined to believe the Soviets incapable of clever, and deceptive, operations.

4. Concealed movement was attempted when Soviet units, especially large ones, moved into the battle area; they were usually hidden from German view. This was not easy to do, and it took the Germans a while before they caught on to what was happening. Moving at night and remaining hidden during the day, the Russians were able to surprise the Germans with massive attacks by units the Germans thought were far away. Part of this hidden mobility was made possible by the construction of nearly two hundred hidden bridges (either underwater or pontoon spans that were moved to shore during the day). Being able to move large quantities of troops, equipment, and supplies undetected over long distances was something the Germans didn't consider possible. Until they were the victims of it.

The Soviets used a wide variety of stratagems to hide the size, location, and apparent intentions of their forces. This began with extreme secrecy in developing plans. Only a few Soviet officers even knew of plans being worked on, and all communication was face-to-face or via courier. As these plans developed, and more units and officers had to be involved, secrecy remained tight. For example, messages would be sent by radio to distant units getting ready for an offensive. But the replies would be sent back, more slowly, by telephone or courier. That way, the Germans could not use radio locaters and traffic analysis to pinpoint units receiving all these messages (and heavy message traffic was characteristic of a planned offensive). All of these messages would use cipher, and even if the messages were deciphered, they would mostly consist of code words (from an ever-changing code book). The Soviets intentionally changed the designations of units simply to confuse the Germans. The circumstances were further muddied through the use of disinformation. Rumors would be spread among Soviet troops (some of whom would always be captured during German raids) and behind German lines (via the partisans). These rumors were concocted to conform to events the Soviets were staging to make the rumors seem plausible. For example, a rumor would be started that a specific Soviet tank army was moving into an area. The Soviets would establish radio stations in that area, broadcasting messages indicating the presence of a large tank unit. While the messages were coded and such, many of them were about fuel and spare parts, which would indicate a tank unit. The partisans would spread rumors that they were being ordered to move to a certain area, which was usually indicative of preparations to support a Soviet offensive (led by a tank army). These preparations would usually include partisans sabotaging railroads, bridges, and the like. If these rumors were made plausible, the Germans would conclude that a Soviet offensive was going to occur in a specific area and would redeploy Nazi forces to counter it. This was precisely what the Soviets wanted—to get German reserves away from the area where the real attack was to take place.

Many Soviet units were kept far away from front lines, but these outfits had worked out plans to get to the battle area quickly. The aircraft superiority the Soviets were hiding from the Germans would make it safe for these Russian units to travel in daylight, if need be. However, the Soviets preferred to move troops at night, and then, carefully conceal them before daylight. The Soviets developed camouflage techniques that enabled them to hide entire armies (over thirty thousand troops and several thousand vehicles) in forests. Even in the open, weeks of camouflage work done at night would transform a flat, featureless area into a hiding place for thousands of troops, along with their artillery, tanks, and supplies. Since there were no sensors for ground reconnaissance at night in World War II, Soviet movements would go unseen if they were done carefully and left no signs for the German recon aircraft that were out and about at first light. When the sun came up, these Russian units were under orders to be invisible from the air. Soviet reconnaissance aircraft would fly over areas where Russian units were resting during the day, to make sure the ground troops and their equipment were well concealed.

A key Soviet advantage was their ability to extract a high degree of discipline from all the troops involved. Punishment for failure in "camouflage discipline" could be severe. Russian officers had the authority to execute troops on the spot for serious infractions. While this rarely was used outside combat situations, offenders could also be sent to "penal battalions." These suicide outfits were used to lead high-risk attacks. Need a minefield cleared quickly? Call in a penal battalion to run across it before your own troops do. The Russians also employed military police and secret police troops to ensure that deception measures were carried out. Such incentives and policing activities enabled the Soviets to perform impressive feats of camouflage and concealment.

Air bases were more difficult to hide, mainly because of the highly visible landing and takeoff areas. But even these could be hidden to some extent. The aircraft, crew quarters, fuel, and ammunition dumps could be concealed. By using fake airfields with more visible facilities and dummy aircraft, the Soviets made Germans think they were bombing the real Soviet airfields to pieces. What the Germans thought were smaller, secondary air bases, were in fact the ones that held most of the Soviet aircraft.

The Soviets never stopped looking for, and experimenting with, new deception techniques. During the 1942–1943 period, they developed and perfected the following deception routines:

- Air attacks in support of ground forces, particularly the ones launched before the tanks and infantry troops began moving, would be directed at areas that were not targets of the ground offensive. These deceptive air attacks would be a final indicator to the Germans that the offensive was going to hit where the Russians wanted the Nazis to think the action was. Thus assured that they had figured out where the Russian offensive was going to be, the Germans would begin deploying their reserves to the wrong place.

- The Russians would not allow any noticeable additional reconnaissance in the area to be attacked. Before a major attack, it was customary to increase the use of recon aircraft and ground patrols. The success of an offensive was directly related to how much the attackers knew about enemy units, and their positions. Normally, additional recon aircraft flights would be made to gather as much photography of enemy positions and activity as possible. Additional recon aircraft over the area would be an indicator, to the enemy, that something was up. Likewise, ground-patrol activity would increase to inspect enemy positions more closely and to capture prisoners. Again, this would alert the enemy that a major attack was on the way, and they could beef up their defenses accordingly. Another aspect of reconnaissance for an attack was the increasing presence of officers at the front, usually in small groups, all looking at their maps and generally sorting out who would do what in the coming attack. This, too, was a normal form of reconnaissance. The Soviets eliminated all external manifestations of these attack preparations in front of the area to be attacked. Instead, they made sure that the usual number of recon flights were made, but sent their best planes and pilots to ensure that the maximum quality photos were obtained. Patrols were kept at normal levels, and Russian troops doing the patrolling, of course, knew nothing of the coming attack in their sector. This was important, because members of these patrols were sometimes captured, and the Germans had ways of making the unfortunate Russian captives tell all they knew. But the reports of those patrols went back to the officers planning the attack (normally, only summaries of patrol activity would reach higher headquarters). If officers from the attacking units visited the front, they did it secretly at night. But often, even this was not done, as there was always the risk that the Russian troops would notice (and figure out what was going on) and some would later get captured and interrogated by the Germans. While the Soviets lost an advantage in not having the officer of the attacking units getting up front before the assault, it was considered worth it. The element of surprise was a tremendous benefit. The Germans could be very thorough and deadly in preparing for what they expected the Russians to do. But if the Soviets suddenly came out of nowhere in a surprise offensive, German effectiveness declined noticeably. Thus the Russians were able to even the odds quite a bit and, eventually, win more and more battles.

- To further conceal the coming offensive, the Soviets refrained from the usual registration fire for their artillery. For as long as artillery has existed, it has been customary to send a few shots at the enemy position to make sure your aim was correct. This is called "registration fire" and continues until each artillery unit has made corrections in its aim to ensure accuracy for the big barrage to follow. When defenders noticed this registration fire, they knew they were in for a lot of enemy artillery fire soon thereafter. That, in turn, was usually followed by the appearance of masses of enemy tanks and infantry. The Soviets, by skipping the registration fire, rendered their artillery somewhat less accurate. But as Soviet guns were used in large quantities (over a hundred per kilometer of front being attacked by their ground forces), the effects of initial inaccuracy were not critical. The surprise element was much more important. Since the Germans were not expecting a huge amount of artillery fire, they were not prepared for the rain of shells that caught many of them out of their fortifications. Thus German casualties were heavier. More important, the surprise element meant that German reinforcements had not yet started moving toward the threatened sector. This gave the Russians more time to chew through the German line without running into more Germans.

The Russians had carefully studied how German intelligence gathering and analysis was performed, and just as carefully fed the Germans pieces of a puzzle that existed only as a Soviet deception. The Germans caught on later in the war, but they never caught up.

The biggest shock the Germans got when they attacked the Kursk bulge was not so much all the concealed fortifications and defenders, but the hordes of attackers who later entered the fray. The Soviets had concealed the size and readiness of their tank and air forces. Soviet deceptions made the Germans think Russian forces were massing where they weren't, while the real Soviet assault units secretly moved into position for devastating surprise attacks on unsuspecting Germans.

The Germans knew that there would be some Soviet attacks, if only to slow the advance of the Nazi assault troops. In the event that the German offensive failed (something the Germans were prudent enough to consider), they expected some kind of Russian counterattack to recover lost ground. What the Germans didn't expect was a major Soviet offensive. This Russian offensive was used, for the first time in the war, to beat the Nazis during the warm weather. In fact, for the rest of the war, it was the Russians who were attacking and the Germans defending (while retreating), no matter what the season. And in large part it was made possible by some clever, and relentlessly applied, deceptions.

Deceiving the Germans: Overlord's Bodyguard

"Operation Overlord," the amphibious landings in Normandy which would be forever known as D-Day, was the most complex and risky undertaking in the history of warfare. To ensure that the invasion had every chance of success, the Allies developed an elaborate series of deceptions, all, in the words of Winston Churchill, designed to provide the truth with "a bodyguard of lies." While several of these often elaborate deceptions will be treated separately elsewhere in this book, it seems convenient to summarize them here.

"Ferdinand": a plan designed to convince the Germans that the Allies intended to make their principal thrust in the Mediterranean in Italy, to the exclusion of a "secondary" invasion in southern France or the Balkans.

"Fortitude North": an amphibious invasion of central Norway. Since Hitler was particularly sensitive about the security of Norway, which was an important source of iron ore and a valuable base in the war against the Russian convoys, this was a fairly elaborate deception, and was quite successful.

"Fortitude South": an amphibious invasion across the Strait of Dover against the Pas de Calais, the most direct and logistically easy invasion route, albeit the riskiest, since the Germans could read maps as well. As a result, this was the most elaborate of all the cover plans, and the most successful. The deception was maintained for nearly two months after D-Day as a way of convincing the Germans that the Normandy landings were actually a diversion. In effect, the Allies managed to convince the Germans that the enormous Normandy operation was one huge feint.

"Glimmer": a simulated assault landing against Boulogne, near the Pas de Calais, on D-Day itself, to pin down German troops during the initial hours of the invasion.

"Ironside": a threatened landing against Bordeaux and southwestern France, intended to keep German troops away from Normandy for as long as possible.

"Royal Flush": a series of diplomatic deceptions designed to suggest that various neutrals (Sweden, Spain, and Turkey) might be contemplating joining the Allies, or at least allowing them air-base rights or similar benefits.

"Taxable": a simulated assault landing against Fécamp, just north of Normandy on D-Day itself, to pin down German troops during the crucial first hours of the invasion.

"Vendetta": an invasion of southern France. Since the Allies actually intended to land in southern France (Operation Anvil/Dragoon), this was a truly tricky trick. First the Allies had to mount a convincing threat, and then, once the D-Day landings had actually taken place, convey the impression that this threat was for a second landing, to occur within a week or so of the Normandy operation. But then the Germans had to be convinced that it had all been a ruse after all, so that they would begin to strip away troops to support the fighting farther north, in time for the real landings, which occurred with considerable success on August 15, 1944, about a month after the Vendetta cover plan was blown by the transfer of troops involved from North Africa to Italy.

"Zeppelin": a series of deceptions designed to suggest that the principal Allied thrust would be in the Balkans, either against Pola, at the head of the Adriatic; on the Dalmatian coast; in Albania, Greece, or Crete; against Romania; or some combination of several of the above simultaneously.

Note that these are not all of the cover plans designed to protect the secret of D-Day. There were others, not all of which were as elaborate, such as "Copperhead," in which an actor who greatly resembled British Field Marshal Bernard Law Montgomery was flown to Gibraltar and thence to Algiers, where he held staff conferences with various prominent officers from the Mediterranean Theater on the eve of D-Day, in the hope that the Germans would lower their guard. And then there was "Titantic," four airborne landings simulated by thousands of half-size dummies which were scattered over a wide area of northwestern France, designed to confuse the Germans in the opening hours of the actual invasion.

The Balkans Deception: "Operation Zeppelin"

The object of "Operation Zeppelin" was to pin German troops in the Balkans so that they would be weaker in France in anticipation of the Normandy landings. This was a very elaborate scheme, involving not merely the United States and Britain, but also Russia. The deception actually hinted at several Allied offensives, to be conducted either singly or in concert. In summary, the various operations involved an Allied offensive against

A. Crete
B. Peloponessus, the southernmost peninsula of mainland Greece
C. Albania
D. The Dalmatian coast of Yugoslavia
E. Mainland Greece
F. The coasts of Romania and Bulgaria (in cooperation with the Russians)
G. Pola and the Istrian Peninsula at the head of the Adriatic, assuming the success of one or more of the earlier operations (items A through E)

The object of the plan was to keep major German formations in the Balkans for as long as possible. The idea was to convince the Germans that a major Allied move in the Balkans could be expected in mid-March of 1944, possibly in conjunction with a new Soviet offensive. As the Normandy operation drew closer, the details of "Operation Zeppelin" were modified.

As developed in February of 1944, "Zeppelin" suggested an Allied offensive against Crete, the Peloponessus, Albania, and Dalmatia, or some combination of these, on about March 23, 1944, around the new moon, to be supported by an Allied-Soviet assault on Romania and Bulgaria, plus an Allied landing on the Greek mainland in late April. As March 23 neared, "technical difficulties" forced the postponement of operations against Crete, the Peloponessus, Dalmatia, and Albania until late April, with the assaults on mainland Greece and in the Black Sea therefore "necessarily" postponed until late May. But those pesky "technical difficulties," mostly as very real shortage of landing craft (the available landing craft being hoarded in Britain for the main event, D-Day) forced a further "postponement." As a result, going into late April, all assaults were rescheduled for late May. And yet again there were problems, so that finally, the operations against Crete, the Peloponessus, and in the Black Sea were "canceled" so that those against mainland Greece, Albania, and Dalmatia were to take place in mid-June. By then, of course, the Normandy landings had taken place.

"Operation Zeppelin" seems to have been a success, keeping the Germans just sufficiently anxious about their situation in the Balkans to prevent the transfer of one or two mobile divisions to France. So the deception was helpful militarily. In fact, politically, it was remarkably successful, in an unanticipated way.

One of the formations "committed" to "Zeppelin" was the Polish III Corps, consisting of two Polish divisions: the 2nd Armored and the 7th Infantry. Concentrated in southern Italy, this impressive force was supposed to conduct an amphibious landing against Durazzo, the principal port of Albania, with the intent of driving on the capital, Tiranë, while other Allied forces (i.e., the U.S. Seventh Army, whose real objective was southern France as part of "Operation Anvil/Dragoon") landed in Dalmatia and Istria. In reality, the Polish III Corps did not exist, save for one tank brigade, which was hardly capable of a major amphibious undertaking. The rest of the corps consisted of notional formations created by British specialists in deception. So successful was this bogus threat to Albania, that it had serious repercussions in high diplomatic circles. By early 1944, the Allies had committed themselves to supporting Marshal Tito's Partisans in Yugoslavia, rather than the Royalist "Chetniks," who seemed as willing to fight Tito as they did the Germans. And Tito wanted nothing to do with Polish troops in the Balkans. A little elementary geography will explain why. Although not part of the Balkans, an independent Poland would be sufficiently strong to assert its will in that region, assuming no interference from even stronger outside powers.

Tito's objections to Polish troops "intruding" into the Balkans created a major problem for "Zeppelin" planners. They could not inform Tito of the deception, since his headquarters was riddled with German sympathizers and spies. Nor could they "replace" Polish troops in the operation with others, more politically acceptable to the Yugoslavs, since building up the existence of notional units was difficult and time-consuming. In the end, the directors of "Zeppelin" decided to continue carefully to deceive the Germans as to the presence of Polish III Corps, while deceiving Tito as to its absence. Fortunately for inter-Allied unity, they did not have to do so for long, as "Zeppelin" was abandoned after the landings in southern France.

The Birth of Electronic Warfare As We Know It

It was over Europe from 1942 to 1945 that electronic warfare as we know it was created. The use of electronics to perpetuate these deceptions arose from the fact that it's rather difficult for aircraft in flight to hide. Flying at night had some advantage, but this limited what an aircraft could do. The frequent night bombing missions flown during World War II were later found to be of often marginal usefulness. Not only that, but the widespread introduction of radar early in the war stripped away much of the invisibility the bombers gained by flying at night. But the electronics engineers responded with an ever-increasing array of electronic gadgets to fool the radar, and other electronic devices. And the Germans in turn developed still more electronic items to even the balance or gain an advantage.

The British use of radar in the Battle of Britain (summer 1940) was a key element in winning that campaign. By the time the British began bombing the Germans in earnest during 1941, they themselves had to deal with radar (the "Freya" system). Thus from 1940 on, "electronic warfare" as we now know it was prosecuted in earnest. While the 1940s' devices were crude by 1990s standards, they did the job and were directly responsible for the huge growth in combat electronics after 1945.

The British were the first to install an elaborate radar based on an early warning system. The Germans weren't far behind them. But in other ways, the Germans were ahead of the British in the electronics department. In early 1940, the Germans introduced Knickebein, an airborne navigation system using signals from ground transmitters. This enabled the Germans to carry out accurate night bombing. By late 1940, the British had developed Asperin, a jammer that blocked the use of Knickebein and deceived German navigators as to their actual position.

During the summer of 1940, the Germans introduced Wurzburg,

an improved ground radar (compared to their earlier efforts) with a 40-km range. This system could plot altitude and was thus able to control the fire of antiaircraft guns. Later in the year, the Germans began using Freya, a radar with a 120-km range, which could not measure the altitude of aircraft it spotted. However, the long range gave early warning of bomber approach and provided time for night fighters to be sent aloft, and air raid warnings to be sounded. At about the same time, the Germans developed Wurzburg II. This was actually a pair of radars, one to track bombers and another to track German interceptors. This was the beginning of "positive ground control," wherein ground controllers sent the night fighter directly to the bomber target, providing supervision nearly all the time. This proved to be a deadly combination.

The Germans continued to improve their ground radars as the British bombing of Europe became more common in late 1941. At that time, the Germans introduced Wurzburg Reise, an improved Wurzburg radar with 65-km range. These constant improvements in German radar during 1941 put the British under a lot of pressure to develop ways to deceive the electronic tools their opponents were using to such good effect.

The Germans also sought ways to make their night fighters more effective. To this end, in early 1942, they introduced Lichtenstein. This was an airborne radar for night fighters. Its reliability was a sometime thing and its range varied from 200 to 3,000 meters. But it was the beginning of lethal night fighters that didn't have to be directed (by ground controllers) to a near collision with enemy bombers. At the same time, the Germans fielded Mammut. This was a more powerful early-warning radar with a range of about 330 km. It could still not plot altitude, but it did give even more warning for interceptors and antiaircraft gunners. A month later, the Wassermann radar went into service. This early-warning radar had a range of 240 km and could plot altitude.

As you can see, the rate at which new technology was introduced was moving along at an unheard-of pace. A piece of equipment might go through three or more major upgrades in a single year. As with all cutting-edge technology, the new stuff was usually not very reliable. Maintenance technicians often spent as much time on the new gear as did the operators.

The Allies were not idle during this period. In early 1942, the British started using Gee. This was an airborne navigation system using signals from ground transmitters. An aircraft some 600 km from transmitters would know its location to within 10 km. This made it much easier to send out night bombers and be assured that they would find their target. A few months later, "Shaker" tactics were introduced. This involved Gee-equipped pathfinder aircraft flying ahead of bomber formations to find the targets and drop some bombs to provide aiming points for the masses of night bombers following right behind them. In the summer of 1942, the Germans introduced Heinrich transmitters, which jammed Gee signals and made Gee unseeable by the end of 1942.

In late 1942, the electronic war of deception and counterdeception went into high gear. First, the British brought out Moonshine. This was a device aircraft could carry that detected Freya (German radar) signals and increased the strength of those radar signals bounced back. This made it look like a larger bomber formation was present, which, in turn, caused the Germans to send interceptors after the wrong groups of bombers. About the same time, the British also started using Mandrel, an electronic jammer carried in the lead aircraft of a bomber formation to jam Freya radar.

By the end of 1942, the British were also using Tinsel, another electronic jammer that disrupted ground-to-air communications. This made German night fighters less effective. At about the same time, another device was added to aircraft that amplified bombers' engine noise so as to confuse ground observers who tracked bomber formations by their engine noise. The frequent unreliability of early radar equipment made it useful to continue employing this older method of using ground observers to detect night bombers. That is, have a network of trained human observers standing by with telephone and compass in hand. When aircraft were heard, the observers would call in the bearing of the sound and an estimate of its loudness. At a central location, these reports would be collected, and a good idea of what was in the air could be put together.

At the end of 1942, the British replaced their now compromised Gee system with Oboe. This was a 430-km range ground radar device that calculated bombers' precise location and sent a signal when bombs should be dropped. It could be used day and night. In early 1943, the Allies began introducing H2S. A ground mapping airborne radar, it could distinguish between water, cities, and rural areas, thus allowing bomber navigators to pick out prominent landmarks as waypoints on the way to their target, even if cloud cover or fog made it impossible to see the ground. This system was not fully debugged until late 1943.

The British were getting hammered by German radar-equipped night fighters, and retaliated in early 1943 with two new gadgets. Monica was a tail warning radar for night bombers which enabled the bombers to detect and then to lay some heavy machine-gun fire on German interceptors that thought they were creeping up on an unsuspecting bomber. Monica would alert crew when another aircraft was within 1,000 meters of their bomber and, in effect, allowed for a deception that often led to the enemy fighter being ambushed. The second device was Boozer, a radar receiver that alerted aircrew when they were being detected by Wurzburg or Lichtenstein radars. This enabled the bombers to brace themselves for a German attack. Without Boozer, they never knew exactly when the Germans had spotted them.

In response to increased German use of night fighter operations, the summer of 1943 saw the Allies introducing the Al Mk 9, a much-improved radar for their own night fighters. At the same time, the Allies began using Serrate, a radar receiver for night fighters, which detected German Lichtenstein airborne radar. This allowed Allied night fighter pilots to determine where a German night fighter was and engage it. Since all of this usually took place above the clouds, there was enough starlight and moonlight to allow engagements once radar had brought the aircraft close enough to each other.

The summer of 1943 saw a major advance in electronic deception when the Allies introduced Window. This was nothing more than tinfoil strips, cut to the right length to cause German radar to see a "wall" of what appeared to be huge objects but were, in fact, nothing more than tinfoil strips. Bundles of it were tossed out of Allied aircraft and this, in effect, created an electronic smoke screen behind which anything could be happening. Also called "chaff," this deception could have been used earlier, but the British knew the Germans would quickly figure out what it was and how it worked. The Nazis could then resume bombing British cities hidden behind their own clouds of chaff. By mid-1943, the Allied bombing campaign was going into high gear, losses were horrendous, and the Germans seemed less capable of resuming their bombing, so the chaff was let out of the bag. This was to be a typical problem with electronic deception—the danger that the enemy would quickly figure out what you were doing and use it against you.

About the same time, the communications war went another round as the Allies deployed Special Tinsel. These were updated jammers to deal with new German aircraft radios (designed to operate in spite of the original Tinsel jammers).

In the fall of 1943, the Germans began using Naxburg. These were radio receivers that could detect the Allied H2S ground mapping radars over 300 km away. This clever bit of deception enabled the Germans to track Allied night bombers accurately without using radar. About the same time, the Allies introduced ABC. This was an airborne transmitter that would jam new "jamproof" radios in German fighters and make it difficult for the fighters to get information from the German ground radar and control system. Thus Naxburg was somewhat compromised by ABC, although the Allies didn't know it at the time.

Late 1943 saw a virtual flood of new electronic warfare devices, making the aerial combat over Europe a rather chaotic undertaking. Among the many devices first used during the last few months of 1943 were the following:

- Corona, an Allied device, was a modified Special Tinsel jammer that, instead of jamming radio signals, sent out false instructions to German fighters. This was a pretty classic deception delivered via the latest in high tech.
- SN-2, a German night fighter radar that was immune to Window and had a range of 400–6,000 meters, was one of the more common examples of electronic deceptions; that is, a change in the equipment to defeat an enemy deception causes the foe to change its deception, which causes the cycle to continue.
- Wurzlaus was a modified German Wurzburg radar that could sometimes differentiate between stationary tinfoil clouds and nearby aircraft that were, of course, moving. The British were right about chaff—it did work. Not as well as they had thought, but enough to handicap the German radar system. But then, as so often happens, a desperate victim soon came up with countermeasures.
- Nurnburg was another modified German Wurzburg radar that gave an electronic sound to the operator, as well as the blip on the radar screen. After some training, an operator could use his ears to tell the difference between the radar signal coming back from a chaff cloud and one coming back from moving aircraft. The Allies continued to develop improved chaff, while the Germans remained right behind them in countering the latest wrinkle in those foil strips.

- Flensburg, a German airborne receiver that told pilots when they were detected by the Allied Monica tail radar, alerted the night fighter pilot to be prepared for some resistance from the bomber he was stalking. The Monica device had been a clever and effective deception. Coming up behind British aircraft at night, the Germans had only to fear an exceptionally sharp-eyed British tail gunner. With Monica, the British almost always knew the Germans were coming and inflicted heavy losses before the Germans figured out the problem.

- Dartboard was an Allied tactic of jamming German radio stations that were used to send coded messages to fighter pilots (whose normal radios were now frequently being jammed). Not strictly speaking a deception, except for the first times it was used, when the Allies still thought they were jamming the German fighters' radios.

Going into 1944, the air war unleashed another horde of electronic devices, including:

- Oboe 2 was the old Oboe, but it had a new type of radar signal so that Allied bombers could drop their bombs more accurately via the navigation signals from ground-based transmitters. This type of device had to be used carefully, lest the enemy figure out how to send false signals to Allied bombers. This wasn't easy to do, but it was always a possibility and made for a great deception.

- Naxos, an improved German airborne receiver, detected the Allied ground-mapping radar transmissions. This allowed night fighters to home in on Allied bombers without having to rely on increasingly jammed communications with ground controllers. This sort of deception only worked for a while, because the victims eventually figured out why the Germans were so easily finding Allied bombers. At that point, the Allies had a choice of either continuing to use the mapping radar and take losses, or turn the radar off and use less effective navigation methods.

- Jagdschloss, a very modern ground-based radar that could switch among four different frequencies and thus be more resistant to jamming, had a range of 150 km. It was, and still is, the most effective answer for an opponent who jams your radar signals with electronic noise. In a word: Switch to another frequency. Fooling the other guy as to exactly which frequency you are operating on is one of the fundamental deceptions of modern electronic warfare.

- Egon, a German fighter radio that was more resistant to jamming and enabled ground controllers to continually guide fighters, had a range of 200 km. Jamming was a widely used deception in the air, as it was a form of camouflage. You threw a mass of electronic garbage into the air and then hid behind it. Enemy radar and radio signals had a hard time getting through this static, but Egon was one of the several German devices that found ways to make their radios function in spite of the Allied jamming.

- Jostle was an Allied response to German attempts to use many different frequencies to avoid jamming. This was an airborne "barrage" jammer that jammed a large range of frequencies simultaneously. The main problem with barrage jammers was that they had to spread their power over a wide range of frequencies. A jammer that concentrated on a single frequency put all its noise into that one frequency for a much better effect, To date, however, no better deception for "frequency hopping" has been developed than barrage jamming.

- Window 2 was the designation for new tinfoil strips ("chaff") cut to a length that would jam airborne SN-2 radar. Window 2 had to be cut to the right length to jam a specific radar frequency. The Allies continued to experiment with different types of chaff to deal with different types and frequencies on German radars.

- Serrate 4 was a new Serrate radar detector that could locate the new German SN-2 airborne radar. Much to the Germans' dismay, the Allies continued to develop detectors for new German radars. This reduced the effectiveness of these radars, because the radar signals could be detected before the radar itself detected anything, giving the aircraft using the detector a marked advantage in the deception department.

- Perfectos was an Allied device that took advantage of German IFF ("Identify Friend or Foe") devices. The Germans were now using electronic ID (IFF), and Perfectos could trigger the IFF and use the subsequent ID signal to locate German fighters. The Allies used IFF starting in 1940, primarily to avoid having their returning bombers mistaken for enemy aircraft by Allied radar. It took the Germans several years to catch on to the advantage of IFF and develop their own. This was a very effective deception, one that still makes pilots nervous today (even though the IFF devices are supposed to be more resistant to this sort of thing).

- Micro-H, an Allied alternative to Gee, was used once the Germans discovered a way to jam the original Gee and other navigation devices. There was a constant battle in this area, with the Allies introducing new navigation devices and a few months later, the Germans figuring out how to jam it. Sending false signals to the enemies is an ancient deception that has flourished in the age of electronic warfare.

When the war in Europe ended in early 1945, both the Allies and Germans had even more electronic devices in the works. Most of these gadgets were for deceiving the enemy, and they did just that until the other side came up with a countermeasure. All of this had a profound effect on Allied air forces, and from this World War II war of wits sprang a flood of new devices for the subsequent Cold War. Vietnam and the Arab-Israeli wars saw, in effect, combat testing of all of this expensive gear and proved that the concepts were still viable. The 1991 Gulf War was a massive application of the techniques pioneered over Europe in the early 1940s. And the end is not yet in sight.

Surf Surprise

Amphibious landings are always tricky undertakings. Most of the complications come from just coping with the sea and the weather. But the troops on the receiving end of these assaults developed some lethal deceptions to confound the attacking troops still more.

Both the Germans and the Japanese quickly figured out that they could further complicate, and perhaps compromise, amphibious landings by adding hazards patterned after those nature had already provided. The most effective was the underwater obstacle. These were often nothing more than poles driven into the underwater surf. The height of these poles was calculated so that they were just below the water during high tide (when landing craft would most likely approach the beach). Since these poles were not visible at high tide, the landing craft would run into them and sink, taking the troops or vehicles they were carrying with them to the bottom of the sea. Unfortunately for the defenders, these underwater obstacles were visible at low tide. But then, the defenders didn't think this mattered, because the obstacles were covered by gunfire. The solution to the underwater obstacle problem was solved by yet another deception. The Allies trained underwater swimmers equipped with scuba gear and explosives to go in before a beach assault and destroy enough of the obstacles to allow for a landing. This usually worked, although the Japanese managed to ambush some of these frogmen anyway. The Japanese also planned to take this underwater obstacle method a step further by training divers to swim underwater toward the landing craft and destroy them with explosives. This was rather akin to underwater kamikazes, and never really worked.

Another favorite tactic was to plant mines near the high-tide line. Troops coming off the landing craft would walk right into an unseen minefield. Shelling the mined beaches before the landing was one solution to this problem, but you didn't want to shell the beach too much so that the ground was so torn up that it became difficult for troops to get across. And even this shelling was not guaranteed to get all the mines, leaving the attacking troops to deal with one more unknown as they strove to work their way inland.

This lethal beach game between defender and attacker went on throughout World War II, and continues to this day. For the attacker who is not careful and diligent in his preinvasion reconnaissance, fatal deceptions can still stop the assault at the water line.

Airborne Deception

It might appear absurd that an aircraft could hide while in the air, but hide they did, and often to very good effect. Aside from the obvious hiding place (clouds), experienced pilots learned to use the sun to their advantage. Coming "out of the sun" was a tactic discovered during World War I and is still valid to this day. By the time World War II rolled around, many pilots carried a pair of sunglasses with them, to wear if it was a cloudless day. The Ray • Ban–equipped pilot often owed his life to this bit of protection from sun glare.

Even though in the air, pilots often stayed close to the ground in order to avoid detection. Once camouflage (or at least dark-colored paint) was adopted for the top part of aircraft, it was possible to fly low and frequently escape detection from above. This was especially important for slower ground-attack aircraft that had to worry about getting attacked by high-flying enemy fighters. From two or three miles up, a fighter pilot had a hard time making out what was on the ground, or flying a hundred feet above it.

By the time World War II began, it was realized that other forms of deception were possible. While an aircraft's paint job might not seem like a useful form of concealment, pilots eventually noted that a different paint scheme on their aircraft could make a difference. This began during World War I, even as brightly colored planes were still popular. If aircraft are below you, and the aircraft is painted with the same type of camouflage scheme as, say, tanks, that aircraft will be difficult to make out against the ground below. If an aircraft is above you, and its underside is painted a pale blue, that plane will be more difficult to spot against the pale-blue sky. By the end of World War II, it was a common practice in most air forces to paint combat aircraft in this manner: ground camouflage on top, light blue underneath. The light-blue scheme was particularly useful for ground-attack aircraft, as they were vulnerable to ground fire.

Many other painting techniques were used, including no paint at all. The B-29 bomber was found to perform better without the weight and drag created by a paint job, and some were used in a bare-metal condition. Some air forces, especially the Soviets, often didn't bother with anything more than a standard green paint job. Air forces that had air superiority most of the time, or operated largely over the ocean, didn't bother with fancy paint jobs. Bombers that operated largely at night would adopt a dark paint job that would make them less visible to enemy searchlights or night fighters.

"Dazzle" and zigzag paint patterns were used because research found that this would spoil the aim of enemy gunners. It was later discovered that a regular camouflage pattern would serve nearly as well. Pilots also found, through experience, that a dark color, without camouflage detail, would work about as well to hide a low-flying aircraft from being spotted from above. After all, at a certain altitude, all you saw of the ground below was a lot of dark colors.

Ground-attack aircraft pilots learned to use the ground in other ways. A favorite deception was to fly down the middle of valleys and into open areas surrounded by trees (as along a forest road). The slower ground-attack aircraft had a maneuverability advantage, which allowed them to fly into tight situations that would find faster fighter pilots flying into the side of a mountain. While this spectacular event happens mostly in the movies, the thought of it made fighter pilots cautious when chasing after a low-flying ground-attack aircraft.

Fighter pilots also were quick to learn other deceptions. Although plane-to-plane combat was fast and seemingly chaotic, the better pilots used a full array of deceptive maneuvers. Experienced pilots knew that the other guy would think his opponent was going to go in one direction if the other aircraft tilted this way or that. But wait; the other pilot could be going into the early stages of a certain maneuver with the intention of deceiving his opponent. When one pilot thinks another is going to make a move, he must immediately begin to react. If the other pilot suddenly changes his maneuver, it's often too late for the less-experienced pilot to do anything about the deception he just fell for. As quick as aerial combat was, there was always time and opportunity for deception. The better pilots were the deceivers, and their prey rarely learned quickly enough to avoid their fate. All of this was old news by the end of World War I in 1918.

Helicopters took up where the World War II–era ground-attack aircraft left off. Even more maneuverable, helicopters typically hide behind trees and hills, "popping up" to get off a shot and then hiding again. In this way, helicopters could use cover and concealment to protect themselves from ground units (as well as a good camouflage paint job). Helicopters can be even more lethal against fighters, as some helicopters now have air-to-air missiles. As always, it's a matter of who spots whom first. For this reason, the two-man crews of helicopter gunships have to look up, down, and all around when in enemy territory.

Friendly Fire: The Tragic Deception

During the 1991 Gulf War, a usually hidden aspect of warfare attained unaccustomed prominence. It was noted that some 20 percent of American casualties were from friendly fire. That is, United States troops were killed accidentally by fire from other American troops. What was amazing about the reaction to this is that few people reporting this event were aware that friendly fire is as old as warfare itself. During our century, as long-range rifle and cannon fire have become more common, it has gotten more difficult to determine exactly at whom you were firing. Moreover, the projectiles would keep on going and hit whatever was at the end of the trajectory. The bullets and shell fragments did not distinguish between friend and foe.

Before gunpowder-based firepower became common, it was not unusual for troops, in the heat of battle, to hack away at their own comrades. This was particularly true for those many armies that did not dress up their troops in uniforms. While distinctive national (or tribal) dress might serve to identify troops, once the warriors got to brawling with each other, it was easy to lose track of who was who and take a fatal swipe at one of your own.

Friendly fire has never become a major issue with the public, aside from the media rediscovering it from time to time and flogging their new revelation for all they can. The press generally has little sense of history, especially in military matters. Not only do the media regularly fail to remember that friendly fire has been around for a long time, they also miss the fact that it's also a form of self-deception and, at times, a means of deceiving the enemy.

The troops have long known about friendly fire, lived with it, and trained to deal with it. Although one thinks of the battlefield as a chaotic place, that is not what soldiers try to make it. Most of a soldier's effort on the battlefield goes toward creating as much order as possible. The troops devote a lot of energy to sorting out where everyone (friendly and enemy) is, where the weapons are pointed, and what everyone is supposed to do when the shooting starts. The chaos comes in when the two sides mix it up. Coordination of plans between the two sides is rare (very rare), and the clash of carefully prepared plans is, well, chaotic. Moreover, there is a lot going on. The troops at the front not only have their personal weapons (rifles and machine guns), but also attached mortar units. Then there are the separate artillery units, some miles to the rear. Add the armed helicopters and bombers, and you can begin to appreciate how easy it is to hit someone you didn't intend to. Despite all the precautions taken, battles have a habit of occurring at night. Even in broad daylight, the considerable amount of firepower on the battlefield creates smoke, dust, and fear that ruins one's ability to put your firepower exactly where you want it.

This last item, fear, is not to be underestimated. Fear of death or mutilation makes even the most levelheaded soldier a bit distracted.

During battle, dangers are magnified. What might have been an innocent noise or movement in the bushes before combat becomes a threat once the shooting starts. It's not for nothing that soldiers are "disciplined" during their training. Ideally, this "mindless discipline" will cause the soldier to do what he was ordered to do despite the mind-numbing fear. This fear is so overwhelming that soldiers often can't remember exactly who did what to whom during the heat of battle. The training imparts a number of automatic reflexes which will countermand the urge to turn your machine gun in another direction during battle (and likely as not, shoot some of your own troops).

When one or both sides is on the move, the chaos is even worse, and the likelihood of friendly fire losses even greater. This was the situation during the 1991 Gulf War, with thousands of friendly and enemy armored vehicles rolling across the desert at the same time. This was not the first battle of this type, the first ones having occurred during World War II in North Africa and Russia. The Arab-Israeli wars also saw hundreds of armored vehicles maneuvering, and shooting, within a few-score square miles. In all of these battles, there were cases of friendly-fire losses.

All of this mobile firepower did provide opportunities to pull new kinds of deceptions that exploited the higher risk of friendly fire during high-speed operations. Since mobile units needed fast communications to deal with a rapidly changing situation, the enemy could jam its foes' radios or send false signals. This would result in the affected units either going where they weren't needed, or ending up in a fire fight with friendly units. Although World War II tank guns had a relatively short range (most shots were under five-hundred meters, but the shells could be effective out to a thousand meters), the smoke, dust, or night would provide ample opportunity to shoot up your own people. This was especially true when radio communications were jammed. Friendly infantry were always at risk, as infantry were much feared by tank crews and unidentified infantry was often subject to a "shoot first and identify later" treatment. Modern armies make elaborate plans to minimize friendly-fire losses during mobile battles, and endeavor to work up deceptions to get the enemy to shoot at their own people.

On the battlefield, there were often very clever techniques for exploiting friendly fire to gain an advantage. This happened quite a lot in Vietnam, where the undergunned North Vietnamese would strive to "grab the Yankees by the belt" when things got too hot. This technique involved moving as close to U.S. troops as possible, so that the Americans would be reluctant to use all their firepower for fear of hitting friendly troops. The Americans learned that those "suicidal charges" the Communists made were not suicidal. The Vietnamese knew that if they kept their distance, they would be gradually chopped to pieces by American artillery and bombs. But if they moved up close to the Americans, even if they took high losses doing so, they would be on the receiving end of fewer bombs and shells. The Americans were reluctant to bring down a lot of firepower when their troops were mixed in with the enemy (although it was done at times). Moreover, the Communists found that there was always a chance that the Americans would be defeated quickly by the seemingly reckless infantry attack.

A more common variation on the "get close and let the enemy get hit by his own artillery" routine is the "false signal" gambit. During battles, it is common to rely on flares or colored smoke to mark where shells or bombs are to go. It is always assumed that the radios may get knocked off the air and flares and colored smoke, as primitive as they are, can always be relied on. But if the enemy has found out what your signals are, they can play havoc with your plans.

If, for example, your troops on a certain part of the line were to use various combinations of green and red flares to indicate where their supporting artillery fire was to fall, the enemy could simply send up its own flare signals. These would either cause the artillery fire to fall where it wasn't needed, or thoroughly confuse the artillery commanders to the point where they refrained from firing at all. Either way, by use of this deception, the enemy would have eliminated a lot of punishing artillery fire.

If the enemy was really sharp (and lucky), he would send the signal to the opponents' artillery to "fire on our (the friendly troops') position." This action was taken when the position is about to be overrun and the friendly troops have retreated to their bunkers. If this fire comes down when the friendly troops are *not* in their bunkers, a lot of friendly casualties result. And the position is more easily taken by the enemy. This has happened all because of a clever deception and the use of friendly fire.

During World War II, friendly fire (after a fashion) was used by the Germans to deceive the enemy. In Russia, retreating German units would have their own aircraft fly low over them on the road. This was to make the pursuing Russians think that the columns of vehicles they saw in the distance were Russian, not German, and were being attacked by the low-flying German fighters. But it didn't always work.

Another example of deceptive friendly fire was the handful of times the Germans got a crashed Allied fighter operational again, and then used it against Allied bombers. This did more than just cause some damage to a few bombers; it also made guarding bombers a lot more stressful for Allied fighter escorts. The gunners in the bombers became trigger-happy if they saw a friendly fighter flying toward them. The fighter escorts generally stayed out of range of the bombers' .50-caliber guns (to avoid friendly fire), and sought to intercept enemy fighters as far away from the bombers as possible. But there were times when enemy fighters were seen coming at the bombers with the escorts still in hot pursuit. At other times, the escorts would cut across a bomber formation on their way to intercept oncoming enemy fighters. Once the word got around that the Germans had a few Allied fighters on their side, it became more dangerous for American pilots to fly too close to their own bombers. In fact, the Germans had very few of these U.S. fighters in flyable condition, and very few attacks were made. But the psychological effect was much greater. Bad news travels farther, faster, and with more impact than its reality.

General Kenney's Navy

Lieutenant General George Kenney, commander of the Fifth Air Force under Douglas MacArthur during the New Guinea Campaign in World War II, was an unusually imaginative airman with a penchant for deception. Noticing that by late 1943, the Japanese were increasingly reluctant to commit their air forces in the face of growing American air power, he sought ways to entice them into a major air battle, one in which his fighter pilots could have a field day. The problem was that in order to do so, he had to offer the Japanese a prize worth the risk of their precious airplanes. After thinking on the matter, he came up with an interesting scheme, one with which he approached Army Chief of Staff George C. Marshall when he ran into him at a high-level strategy conference.

Would it be possible, Kenney asked, to "let me have" an "old or new boat fixed up with a painted wooden or other cheap deck and smokestack so that she looks like a carrier, put some dummy aircraft on deck?" Kenney went on to explain that he would run this "aircraft carrier" around in Japanese-patrolled areas, with several layers of fighter cover stacked high above her, and ambush any Japanese aircraft that attempted to attack her.

Marshall's reaction was that he had no ships to spare, and that Kenney had best try "to get the boat locally." Kenney next tried the Army Service Forces, which controlled a lot of shipping, but they had none to spare. He then approached the navy, but Seventh Fleet Commander Admiral Thomas Kinkaid demurred. It is easy to see why the navy was reluctant to take part in the deception. Not only would it have to provide the ship, at a time when shipping was very tight, but it would also have to provide the crews, who might be exposed to considerable danger. And for Kenney's trick to be convincing, the "carrier" would have to operate with a realistic escort, destroyers, and the like, which were also in short supply. With that, Kenney dropped the idea.

Kenney's proposed dummy aircraft carrier reveals an aspect of deception not always considered. A scheme may be eminently possible, but not practical, given logistical and technical problems. Of course, a lesser man might have been crushed by the rejection of his plan, but Kenney went right on scheming and came up with some brilliant deceptions during his career.

Footnote: *Britain's Wooden Battleships.* During World War I, the British laid on an interesting deception to suggest to the Germans that they had more battleships in certain areas than was actually the case. Several old merchantships were taken in hand, stripped of the upper works, and then provided with a new wooden superstructure, plus wooden turrets, from which protruded wooden guns. These were then sailed around in areas where the Royal Navy wished to create the impression that it had a lot of battlewagons. The trick might have worked, but for a slight problem. A German submarine torpedoed one of the dummy battleships. It must have been interesting to see the look on the U-boat's skipper's face as his expression changed from elation to something else entirely when he saw the turrets float away from his battleship.

The Phantom Army

Military deception attained unprecedented levels of sophistication during World War II. The British and Americans created what amounted to an entire "phantom army" in order to fool the Germans. So successful were they in this effort that the Germans were convinced that the U.S. Army had some 20 percent more divisions than was actually the case, and the British Army nearly 70 percent!

The United States, for example, conjured up one army group (the "First United States Army Group," or "FUSAG," about which see page 209), one army (the "Fourteenth Army"), three corps, one armored division, five airborne divisions, and fourteen infantry divisions during the war for the specific purpose of deceiving the enemy. In addition, a further airborne division and nine infantry divisions which had been officially activated but not actually raised were more or less incorporated into this notional army, so that there were a total of thirty nonexistent divisions officially part of the U.S. Army—a paper augmentation of about a third above actual divisional strength.

The creation of a "phantom" unit was surprisingly difficult. Initially, it was almost impossible, due to a peculiarity of U.S. Army regulations, which required the activation of new divisions on American soil. These restrictions had to be lifted before the creation of notional units for purposes of deception could be undertaken. Things did not actually get going until 1943, by which time the British had been at it for years. Deception divisions were sometimes assigned numbers that fitted quite logically into the existing order of battle. For example, the Germans were aware of the existence of armies numbered from "First" through "Tenth," as well as a "Fifteenth," so that a "Fourteenth" sounded reasonable. And the notional "46th Infantry Division" followed neatly upon the existing 45th Infantry Division, while the "22nd Infantry Division" did appear to be logical, given that there were already 23rd, 24th, and 25th Infantry divisions. However, some ghost units were given designations similar to those of existing formations, so that there was a notional "6th Airborne Division" as well as a real 6th Infantry Division, and a "17th Infantry Division" as well as a 17th Airborne Division. The idea was to be as confusing as possible. And some of the notional units had impressively high numbers, such as the "157th Infantry Division," to suggest that there were a lot more U.S. divisions out there than was really the case (eighty-nine, plus six divisions of marines).

Of course, in order to be confused, the Germans had to find out about these outfits. Many techniques were used to assist them in discovering the new formations. Double agents proved very useful for passing false order of battle information to the Germans, and even real enemy agents could be tricked into serving the Allied cause in this way. Other techniques included false radio traffic, "lost" documents, "accidental" slips of the lips in bars, and ploys like marriage notices in newspapers, wherein the groom's assignment was carefully noted.

A Sampler of "Phantom Army" Shoulder Patches

The three insignia shown here are typical of those designed for the notional units "raised" by the U.S. Army for purposes of deception during the Second World War. In every way they reflect the normal patterns of insignia design prevailing at the time. Thus the "Fourteenth Army" patch shows an "A" for "Army"—within an acorn, which represents strength, the colors white on red being those officially assigned for an army headquarters' flag. Similarly, the "6th Airborne Division" patch shows a parachute in white on a blue shield, the infantry colors, while the "131st Airborne Division" patch has a black spider on a gold background, symbolizing descent from above.

One particularly clever trick involved *National Geographic Magazine.* The magazine was given official assistance in preparing a lavish, full-color spread of American military insignia, including army shoulder patches. The army cleverly arranged to have inserted among the legitimate insignia properly designed patches for most of the notional formations. Then, after only a few copies of the issue had been printed, the army had the presses stopped. Within a few days, a revised version of the insignia issue (June 1943) was issued, with several interesting deletions. Some copies of the original version of the magazine were, of course, allowed to get into circulation. This was already a pretty clever ruse, but then, a bit of luck occurred to reinforce it.

Since the army's heraldic-design people were themselves unaware that the formations for which they had been requested to design insignia were fraudulent, they not only designed quite attractive and symbolic patches, they also issued proper manufacturing specifications for them. Needless to say, an enterprising capitalist down on Seventh Avenue (New York City's "Fashion Avenue") quickly produced samples of these insignia in the hopes of securing the contract! And the army promptly bought some thousands of them, which were actually issued to troops at times: Troops heading overseas were sometimes issued these false patches at their ports of embarkation. So one way or another, the Germans soon got wind of the "existence" of these formations. In passing, it might be added that copies of the magazine in question are worth a considerable sum today, as are the insignia, some of which were actually worn, when units bearing the appropriate designations were later raised in the postwar years.

In their creation of notional units for the purpose of deceiving the Germans, the British used several techniques not employed by the Americans. They occasionally "created" a new division by the simple expedient of redesigning an existing one. Thus their 77th Division became the 45th. While a small staff maintained the fiction that the original formation was still in existence, the new one began to appear in German intelligence reports. They also designated training and holding formations as divisions, which was, in fact, the case with both the 77th and its successor. In addition, they would sometimes designate a territorial command that had few or no combat troops as a division. For example, they created a "7th Infantry Division" on Cyprus with three notional brigades, two of which had the same numbers as outfits on distant colonial postings.

Finally, they would sometimes shuffle units between the British and Indian armies, so that the 36th Indian Division was transferred to the British Army as the 36th Infantry Division.

All this trickery turned out to be quite effective at fooling the Germans. The British were particularly successful. In fact, they were so successful that it can be difficult to trace formations even now, some fifty years later! On a worldwide basis, the Germans overestimated Western military strength by some sixty divisions. They overestimated American strength by 20 percent and British strength by a remarkable 70 percent, representing something like a million additional combat troops. In terms of D-Day, this meant that the Germans believed there were over 40 percent more Allied divisions in Britain than was actually the case: some eighty-five to ninety infantry and armored divisions, plus seven airborne divisions, when the actual figures were about thirty-five and three. It was an effort well worth the expense.

STRENGTH OF THE ALLIED PHANTOM ARMIES

	NATIONALITY				
TYPE	AMERICAN	BRITISH	GREEK	POLISH	TOTAL
Army Group	1	0	0	0	1
Armies	1	4	0	0	5
Corps	3	4	0	1	8
Airborne Divisions	6	1	0	0	7
Armored Divisions	1	3	0	1	5
Infantry Divisions	23	17	1	1	42
Armored Brigades	0	3	0	0	3
Infantry Brigades	0	8	0	0	8

Arguably, the total of notional Allied divisions could easily be increased. The U.S. 2nd Cavalry Division, a mostly black outfit which went to North Africa in early 1943, continued to be counted in German intelligence estimates until the end of the war, despite the fact that it had been disbanded some months after the end of the Tunisian Campaign. Likewise, the Germans appear never to have caught on to the fact that the United States disbanded a number of National Guard cavalry divisions shortly after mobilization. And then there was the British 18th Infantry Division, which continued to turn up in German estimates of Allied strength, despite the fact that in early 1942, it had been captured by their Japanese allies at Singapore.

Footnote: *The Russians.* The real masters at creating deceptive formations were the Soviets, who at times managed to convince the Germans that they had up to twice as many units as was actually the case. But that's another story.

The Marines Invade Martinique . . . or Maybe It's the Azores?

When France fell to the Germans in June of 1940, most of the French Empire fell into line behind the new pro-Nazi Vichy regime. This included the islands of Martinique and Guadeloupe in the West Indies, where there was a considerable French garrison, an important naval squadron, and over a hundred new American aircraft still in their shipping crates, which had been purchased by France shortly before the disaster.

Over the next few months, the United States government engaged in careful negotiations with the commander of French forces in the Caribbean, a Rear Admiral Robert, as to the disposition of these forces. While Vichy leader Marshal Henri Pétain and Admiral Robert both assured the United States that they would never allow French property to fall into the wrong hands, there were grounds for doubting their sincerity. Some American political and military leaders, not to mention numerous columnists, suggested that we send in the marines, a course that President Roosevelt was loath to adopt for fear of offending Latin American sensibilities. So the situation more or less simmered along for another few months. Admiral Robert appeared to be keeping his word, but doubts continued to fester.

Then, in early June of 1941, the 2nd Marine Division, based at San Diego, received orders to send its "best equipped and trained regiment" eastward, in order to cooperate with the 1st Marine Division on "amphibious maneuvers in the Caribbean." Soon the 6th Marine Regiment, reinforced to nearly four thousand men, was making its way to Charleston, South Carolina, partially by rail and partially by ship. The phrase "amphibious maneuvers in the Caribbean" was generally regarded as a euphemism to cover the occupation of Martinique, which was expected momentarily.

Meanwhile, the 1st Marine Division, based at Camp Lejune, in North Carolina, was alerted for duty. The scuttlebutt in the division was that the Martinique operation was actually a cover for a landing in the Azores, an undertaking for which it had received some training ("Operation Gray"). The Azores were controlled by neutral Portugal. Since the beginning of the war, the British had been trying to get Portuguese permission to use the islands to improve aerial surveillance of the central Atlantic, which would be of enormous help in defeating the German submarine offensive. Although friendly, the Portuguese had been very uncooperative about permitting the Allies to use the Azores, and it was an open secret that many prominent British military and political leaders had suggested taking the islands by force. So perhaps the Marines were actually bound for the Azores?

As soon as men of the 6th Marines reached Charleston, they began to smell a rat. Perhaps the first clue that they weren't going to any tropic clime was the distribution of woolen underwear as they boarded their transports. Then, within a few days, they found themselves at Argentia Bay, in Newfoundland, as part of a convoy bound for somewhere. And on July 7, the 6th Marines landed in Iceland, to take over garrison duties from British troops, who were desperately needed elsewhere.

The Martinique-Azores-Iceland ruse was a classic one, albeit little known. The ostensible mission, the Martinique operation, was a credible undertaking, as was the alternative plan, the occupation of the Azores. But as each could also be viewed as a mask for the other, both were actually a mask for the occupation of Iceland, even more vital to the war effort.

Footnote: For the curious, the French squadron down in Martinique was neutralized by some hard diplomacy (i.e., the United States agreed to supply food if the French agreed to remove certain vital portions of the ships' machinery and store them ashore). And the Portuguese eventually relented, allowing Allied aircraft to make patrols from the Azores.

The President Goes Fishing

On August 3, 1941, Franklin Roosevelt and two aides sailed from New London, Connecticut, aboard the presidential yacht *Potomac*. Over the next few days, *Potomac* cruised the waters around Buzzard's Bay and Martha's Vineyard, and, passing through the Cape Cod Canal, Cape Cod Bay as well. From time to time, people (some of them journalists) on passing boats could see the president and his associates fishing over the stern or lounging in deck chairs, sometimes pausing to wave a greeting to the curious passersby. Several newspapers reported the presidential holiday, a few of them with photographs, in which FDR could be faintly made out, wearing his white boating togs, with his cigarette holder jutting from his jaw at its customary jaunty angle.

In fact, the president was very far from his yacht. Unbeknownst to all but a handful of people (not including the head of the Secret Service!) was the fact that one night, off Martha's Vineyard, *Potomac* had rendezvoused with heavy cruiser U.S.S. *Augusta,* to which the president and his party were transferred in greatest secrecy. With a small escort, *Augusta* then steamed for Argentia Bay, in Newfoundland. And there, on August 10, the presidential task force anchored alongside battleship H.M.S. *Prince of Wales,* aboard which was Winston Churchill. For five days, the president and his principal military and political advisors (who had previously boarded *Augusta* and the escorting vessels amid the greatest secrecy) conferred on policy and strategy with the prime minister and his principal advisors. Out of this meeting came, aside from agreements on Lend Lease and convoy escort responsibility, the Atlantic Charter, a general statement of Allied war aims, which was announced to the world a few days later, after all parties had returned home.

There were several reasons for the deception that accompanied the presidential trip to Argentia Bay. The Atlantic was alive with German submarines, and Churchill (whose departure from Britain was accompanied by even more elaborate ruses) was particularly at risk: *Prince of Wales* was a very fast ship, capable of outrunning any submarine, but even the fastest ship could have been ambushed had the Germans foreknowledge of its movements. In addition, however, there were the president's domestic political problems. His modest efforts to prepare the nation for war and to aid the countries fighting Nazism had met fierce resistance from a broad spectrum of Americans, ranging from the far right to the far left. The political storm that would have resulted from advance notice that he was planning to meet with Churchill might easily have scuttled his efforts. By convincing Churchill to agree to the Atlantic Charter, the president came away from the meeting with something that most Americans could agree to, thereby muffling isolationist sentiment.

And the genial fellow fishing off the quarter-deck of the yacht *Potomac*? He was a fireman who bore some resemblance to the president, particularly with the cigarette holder clenched jauntily between his teeth. The two "aides" were U.S. Navy seamen who looked a bit like Stephen Early, one of the president's closest advisors, and Brigadier General Edwin "Pa" Watson, FDR's military secretary.

Although not strictly a "ruse of war," President Roosevelt's little trick was a timely and valuable deception, which certainly furthered the war effort.

The U.S. "Q-Ship" in World War II: A Ruse That Failed

During World War I, the British appeared to have secured some success with "Q-Ships." These were innocent-looking merchant vessels that toted an elaborate armament. Q-Ships deliberately exposed themselves to German submarine attacks, by sailing independently or straggling from convoys. As most sinkings by submarine during World War I were done with deck guns rather than more expensive torpedoes, the idea was that when a U-boat surfaced to sink the straggling ship, a "panic party" would take to the boats, while other crewmen would secretly man concealed guns, cutting loose as soon as the unsuspecting submarine drew into range.

Early in 1942, with German submarines finding the waters off the East Coast a happy hunting ground, President Roosevelt and Chief of Naval Operations Admiral Ernest J. King proposed that the United States send some Q-Ships to sea. Seven vessels were soon converted: a tanker, two small freighters, a pair of fishing trawlers, and two schooners. Typically, they were given a relatively heavy armament, well concealed, usually one or more 4-inch guns, several machine guns, depth charges, and sonar—and sent to sea with a fairly large crew. Their record was rather dismal.

Four days into her first patrol, Q-Ship U.S.S. *Atik* (3,200 tons, four 4-inch guns) was attacked by the German submarine *U-123*. She took a torpedo, and, as her "panic party" began taking to ship's boats, the rest of the crew manned the still-concealed guns. The submarine surfaced, in order to finish the ship off with gunfire. Suddenly, the stricken steamer dropped her false bulkheads and opened a heavy fire on the U-boat, killing one crewman. Making its best speed, the submarine fled on the surface, the damaged Q-Ship being unable to pursue. Later that night, the submarine returned, putting another torpedo into *Atik*, which sank with all hands.

U.S.S. *Foam*, a fishing trawler (one 4-inch gun), encountered a submarine on her first voyage, and was torpedoed off Nova Scotia by *U-432*, which apparently did not even know her victim was a Q-Ship.

None of the other U.S. Q-Ships had nearly such exciting adventures as did *Atik* or *Foam,* albeit all of them had longer careers and no casualties. Despite more than a dozen patrols in submarine-infested waters, the other American Q-Ships managed to engage only three submarines, all of which escaped with no apparent damage. At the end of 1943, the project was abandoned. It had cost the lives of 141 sailors, in *Atik* and *Foam,* about a quarter of those serving in Q-Ships, making them the highest-risk branch of American military forces in the war.

There were several reasons why the American experience with Q-Ships during World War II was so dismal a failure. To begin with, during World War I, Q-Ships had by no means been as successful as believed, albeit they were much more successful than their Second War descendants. Expectations were certainly too high. In addition, the nature of submarine warfare had changed. World War I submarines carried very few torpedoes, and preferred to reserve them for "worthy" targets. As a result, most sinkings by submarines during the critical period of that war had been made by submarines on the surface, using gunfire. This gave the Q-Ships a much better chance at getting in their licks before taking fatal damage. That is, if they encountered a submarine. In fact, even in World War I, Q-Ships had been a failure. The approximately 180 Q-Ships that the Royal Navy and the U.S. Navy sent to sea managed to kill only five submarines. The great success which it was believed had attended their employment was largely derived from some post-war memoirs and sensationalistic popular treatments. What is surprising is that neither the U.S. Navy nor the Royal Navy had made any serious study of their effectiveness. While both navies employed Q-Ships in World War II, the British, who had tried Q-Ships in 1940–1941, had abandoned them as a bad investment even before the U.S. Navy decided to get into the business. So, obviously, the U.S.N. was not paying attention. And certainly Q-Ships were a deception that failed.

Deception As a Matter of Course

Some deceptions are so commonplace that they are not even consciously recognized as such. For example, for many centuries now, the routing of convoys has typically been done in such a way as to deceive the enemy about their destination. As a case in point, take the massive series of convoys that were required for "Operation

Torch," the Allied invasion of North Africa in November of 1942, while U-boats still prowled the Atlantic with relative impunity. Moreover, since the objective of this operation was the occupation of French-held territories on the northwestern and northern coasts of the Mediterranean, there existed the distinct possibility of intervention by the French fleet, based at Toulon, in southern France, and Dakar, in western Africa, only a day's sailing from the invasion beaches. And the French might be aided by Italian and German naval forces in the Mediterranean. So "Torch" was a risky undertaking.

Convoys for this operation sailed from ports as widely separated as Virginia and Gibraltar—Maine, Bermuda, and Britain figuring in between. Three major convoys sailed from U.S. ports: one from Bermuda, two from Britain, and one from Gibraltar.

The "Northern" and "Southern Attack Groups," consisting of troopships and escorts, sailed together from Hampton Roads, Virginia, on October 23, 1942, heading southeast-by-south, as if they were bound for the Caribbean.

The "Covering Group" (battleships, cruisers, and destroyers assigned to protect the troop convoys from surface attack and provide gunfire support for the landings) sailed from Casco Bay in Maine on October, taking a southeasterly course, suggesting it was heading for the Mediterranean.

The "Center Attack Group" (more troopships and escorts) sailed from Hampton Roads on the same day, taking a northeasterly course, as if it were a routine convoy bound for Britain. Also that day, the "Northern/Southern" convoy altered course northward, its new course suggesting to observing submarines that it was headed for Newfoundland.

The "Air Group" (several small aircraft carriers with their escorts) sailed from Bermuda on October 25, taking an easterly course, as if headed for the Mediterranean. At the same time, the "Covering Group" altered course southward, so that observers in passing submarines might think it was heading for the Caribbean.

The movements of the first three convoys resulted, on October 26, in a rendezvous some five hundred miles north of Bermuda. The combined force now assumed an easterly course, as if it were bound for Portugal, a friendly neutral whose neutrality gave Hitler some worries. Meanwhile, the "Air Group," about seven hundred miles southeast of the combined invasion fleet, altered course sharply to the northeast, so that had it been observed, it would have seemed to be bound for Britain.

On October 28, the "Air Group" joined the main convoy, and the whole detachment altered course once more, this time to the southeast. This course, if followed, would have suggested a possible movement against Dakar, in French West Africa, which the British and Free French had once before tried to capture. Over the next nine days, the general trend of the convoy's course was altered significantly twelve times: from southeast, to south-southeast, to northeast, to south-southeast, to east, to north-northeast, to northeast, to east, to northeast, to east, to north, to east, so that on November 7, the three principal attack groups had arrived about a hundred miles off their individual objectives in Morocco.

While convoys from the United States were making their way across the ocean for the Atlantic coast of North Africa, other troop convoys were outbound from Britain for Africa's Mediterranean littoral. A "slow" (making an average advance of seven knots, allowing for zigzagging) convoy departed Britain on October 22, 1942, followed by a "fast" (nine knots) one on October 26. Both took a southerly course, which could imply many things: an attack on Spain or French North Africa; reinforcement of Malta; reinforcement of Egypt, where the British Eighth Army was locked in combat with Rommel's Italo-German forces on the Alamein Line; or even a destination in the Far East. On the night of November 5–6, the convoys passed through the Straits of Gibraltar, recombining into the two task forces which were separately to assault Algiers and Oran. At Gibraltar, the Royal Navy's "Force H" (two carriers, four battleships and battlecruisers, plus many other combatants) joined up to cover the two invasion forces. With "Force H" as an escort, the two invasion convoys proceeded on easterly course, "hugging" the Spanish coast, as if trying to avoid observation from French North Africa, a common ploy adopted by Malta-bound convoys. Not until the last few hours of its movement did the Algiers task force shape course for its objective, altering from east to southeast under cover of darkness on November 7. Meanwhile, the Oran task force actually sailed past its objective, to double back during the night.

And when the Allies deposited some hundred thousand troops on African shores on the morning of November 8, 1942, it came as something of a surprise. The word "deception" probably never crossed the minds of any of the staff officers charged with planning the invasion of North Africa. Their routing of the convoys was rooted in experience—for the British, dating back to the Dutch Wars of the seventeenth century. It was all a matter of course.

The "First United States Army Group"

One of the most important deceptions of the "Bodyguard of Lies" with which the Allies protected the Normandy landings was "Operation Fortitude South." This involved a notional formation under the command of George S. Patton himself, the "First United States Army Group," known as "FUSAG" for short. FUSAG was supposedly concentrated in eastern Britain, perfectly located for an invasion of the Pas-de-Calais, that part of France lying closest to England.

Altogether, about two dozen divisions were theoretically under Patton's command. Some of these were real outfits which just happened to be stationed in Kent. Most of his divisions, however, were fakes. On the eve of the Normandy landings, virtually all of his divisions were notional ones, for as the real ones moved out, new false ones were created. As a result, on D-Day, June 6, 1944, FUSAG had under command thirteen and a half divisions, of which only two and a half were real (including Britain's Guards Armored Division, retained on the theory that the Germans would expect so prestigious a unit to take part in the invasion), the balance consisting of three phantom airborne divisions (one British), two phantom armored divisions (one British), and six phantom infantry divisions (two British). These were organized into several notional corps, and FUSAG also had under command the headquarters of U.S. Ninth Army, a real formation awaiting transfer to the Continent.

Each of Patton's notional divisions had a few hundred men assigned. These included about two dozen signal corps personnel, assigned to simulate a division's worth of radio traffic. Aside from establishing the existence of the division, many of their messages were designed to suggest (sometimes blatantly) that the invasion was scheduled for the Pas-de-Calais ("Are we supposed to take all these VD cases to Calais with us!"). Most of the troops assigned to each notional division were engineers. They maintained and operated elaborate dummy installations (such as dummy barracks and workshops), weapons (rubber tanks, guns, trucks), and equipment (noise- and dust-making machinery, etc.) that helped strengthen the ruse. A lot of set designers and other folks in similar trades were recruited from the theatrical and motion picture industries to serve in these units.

FUSAG was an extremely successful ruse. Although a few senior German officers believed otherwise, virtually all of the senior officers, including Rommel and Rundstedt, overall commanders in the west, were convinced that the Pas-de-Calais was to be the principal Allied objective. The FUSAG ruse was maintained for weeks after the Normandy landings, in order keep the German Fifteenth Army pinned to the Pas-de-Calais. Since Patton was scheduled to command the Third Army as soon as sufficient maneuvering room had been secured in Normandy, an even more senior officer, General Lesley J. McNair, chief of Army Ground Forces, was brought from Washington to "assume" command of FUSAG. As luck would have it, McNair was accidentally killed by "friendly fire" (actually "friendly bombs"; all that could be found of him was his West Point class ring), while observing the preliminary bombardment for "Operation Cobra," Patton's offensive designed to break out of the beachhead to the south and spread out over France. Since by then the FUSAG ruse had pretty much run its course, it was quietly abandoned.

FUSAG was perhaps the most important single deception of the war. Although the Allies had deceptions hinting at a variety of alternative landing sites (see "Deceiving the Germans: Overlord's Bodyguard," page 178), there really were only two wholly practicable places, Normandy and the Pas-de-Calais. Had the attempt to deceive the Germans into believing the invasion was to come at the latter place failed, the Normandy landings would have been a much tougher proposition.

Snow Job

Among the toughest missions of the war were the convoys to northern Russia, essential to keep the Soviet Union supplied with critical war materials. It was a long and grueling run, even in summer, from Iceland or Scotland, up across the Norwegian Sea, into the Arctic Ocean, and then south to Murmansk or Archangel. Most of the route was in very close proximity to German naval and air bases in Norway and Finland: Murmansk itself was only thirty-five miles from the nearest German airbase. The worst of the Murmansk convoy battles was that of PQ-17.

Convoy PQ-17, composed of thirty-six merchant ships with an escort of six destroyers and supporting force of a handful of light cruisers, sailed from Britain in late June of 1942, bound for Murmansk. The convoy soon came under heavy attack by eighteen

German submarines and numerous aircraft. Despite serious losses, the convoy pressed on. Then, in early July, German surface forces—including the battleship *Tirpitz*, which outgunned everything the convoy and its supporting force had, plus two heavy cruisers—sortied from the Norwegian fjords. On July 4, the Admiralty ordered the convoy to disperse, each ship to make its way individually.

After the convoy dispersed, it was happy hunting for the Germans, as aircraft and submarines hunted down the fleeing vessels, twenty-three of the merchantmen being sunk. Only thirteen made it to safety, several by using a ruse thought up by George J. Salvesen, skipper of the merchantman *Troubador*. With two other merchant ships and an armed British trawler, *Troubador* had managed to elude the Germans, by steaming as closely as possible to the pack ice. But this was only a temporary solution, as the ships would certainly be spotted by the first German submarine or airplane that happened by. Then Captain Salvesen had a brainstorm. The ship's cargo included a surprisingly large allotment of white paint. Salvesen ordered the ship's starboard side (the right, the one most visible from the open sea) painted white, and convinced the other skippers to do the same. In addition, sheets and tablecloths and anything else white were spread over the ships' most exposed features, to enhance the Arctic camouflage. They then proceeded slowly through the ice to the relative security of Matochkin Shar (strait), which divides the two halves of Novaya Zemlya (we're talking very far north here, about sixteen degrees from the Pole), arriving there on July 6. Over the next few days, the little convoy was joined by other survivors of PQ-17, until there were seven altogether. The new arrivals quickly adopted Salvesen's camouflage scheme. Despite being overflown several times by German reconnaissance planes, the ships were never spotted. After about two weeks of lying to in the strait, the ships made their way into the White Sea, keeping close to the pack ice on their port sides (that is, the left), and making Archangel within a few days.

Ironically, the Germans thought up the same deception for the small weather ships they periodically sent to northern waters off Greenland. Without weather information from that area, the Germans would have had a very difficult time getting accurate weather reports for most of Europe (and the Allies would always have accurate weather reporting). The weather ships had to sneak past Allied aircraft and ships to get on station, and one of the stratagems used was to paint the ships white and stick close to the pack ice. Like the PQ-17 ships, the snow-white German weather ships proved very difficult to spot.

General Kenney Builds an Airfield

In furtherance of Allied operations in northeastern New Guinea during mid-1943, Major General George Kenney wanted to advance his forward fighter bases closer to Wewak, on the north central coast of the big island, some four hundred miles from Lae and the other Japanese bases on the Huon Peninsula—Wewak being the Allied objective as the principal Japanese air base in the area. The idea was that, if he could put his fighters closer to Wewak, he could escort heavy bomber missions against the major Japanese base, thereby heightening the isolation of those farther to the east. As there were a number of old airstrips (left by miners and missionaries) in the region, there was no shortage of potential sites at which a major base might be developed. But there were some problems nonetheless. First, the possible sites for air bases could not be reached by overland movement. Construction engineers and equipment would have to be flown in. However, since the Japanese still retained considerable air strength in the region, and since they could reasonably be expected to object to the Fifth Air Force setting up shop on their doorstep, it was also reasonable to assume that the airfields would be subjected to devastating enemy air attacks from the start. How to resolve this dilemma occupied the Americans for some time. However, Kenney soon came up with a creative solution.

Not very far from Lae, there was an old airstrip at a place called Bena Bena, once used by gold miners but now so overgrown, the Japanese had not thought it worth the effort to rehabilitate. In short order, and with some carelessness, airborne troops were dropped at Bena Bena. Rapidly securing the area (which was wholly undefended), the troops quickly cleared enough of the strip for construction engineers and equipment to be flown in. With that, work began in earnest. Of course, the Japanese soon got wind of this, as the work was ill-concealed from aerial reconnaissance. In short order, the emperor's eagles began paying regular calls on Bena Bena, bombing and strafing to their hearts' content. But despite the devastation, the tenacious construction troops kept at it. Although the regular news communiqués issued from Kenney's headquarters put a positive spin on the raids ("Seventeen enemy bombers attacked Bena Bena yesterday. No damage was reported."), the Japanese were pleased to note that their attacks seemed to have virtually halted progress on the air base. This was precisely what Kenney wanted them to think, for Bena Bena was never intended to be an air base at all. In fact, it was a ruse, a trick designed to draw the attention of the Japanese from the real air base, which was just then under construction forty miles from Lae. Only a handful of engineers had actually been landed at Bena Bena, and they were instructed to do the minimum amount of work necessary to attract the attention of the enemy, most obviously by clearing the airstrip of brush and by creating lots of dust.

While the Japanese were blasting Bena Bena, construction troops had been less spectacularly inserted on June 16, 1943, into Tsili Tsili, another abandoned airstrip, some forty miles from Lae. Working quickly, and with excellent camouflage discipline (they left the clearing of the brush-covered airstrip for last), the troops managed to have the strip ready to receive aircraft within ten days, a remarkable achievement. Kenney immediately laid on a series of heavy bomber raids against Wewak, with fighters from Tsili Tsili providing escorts. The destruction wrought at Wewak (a possibly optimistic estimate of over four hundred aircraft destroyed or damaged in the first two days alone) greatly interfered in the Japanese ability to support their bases on the Huon Peninsula, thereby furthering the Allied war effort. And, to add insult to injury, Bena Bena was eventually developed as an air base, after all.

Hiding Behind Distance

Distance can be used to conceal troops, equipment, and facilities. Moreover, if one force is known to have poor spotting ability, the other side can use this knowledge to stage surprise attacks. Surprise is the most decisive battlefield deception, as it is a deception that immediately pays off in battlefield advantage and, often, victory. "Sneaking up on the enemy" only works if the other guy doesn't see you coming. If you are spotted, your surprise attack turns into an ambush of your own troops. A risky business, and not the sort of thing for amateurs to fool around with.

For centuries, wise commanders sought to put lookouts on high ground (or simply up a tree or on a tall building) to spot the enemy before the enemy thought he would be spotted. When aircraft were first used for this kind of reconnaissance some eighty years ago, it was thought that a new age in military operations had dawned. Such was not the case. At higher altitudes, the troops and equipment being sought out were harder to identify. If the aircraft flew lower, they were subject to energetic fire from enemy troops wishing to protect their anonymity. Air reconnaissance did prove a valuable technique, but insightful commanders learned that these long distances could be used as a deception. Spotters in aircraft were under a lot of pressure to find something out there, and to identify it. Since these flyers didn't want to get killed either, there was a temptation to maintain a high altitude (out of range of rifles and machine guns). The airborne spotters would make mistakes, and clever ground commanders eventually learned to use these mistakes to their advantage.

To put all of this into perspective, consider how far military units and equipment could be seen and identified, depending on the altitude of the observer.

OPTIMAL SPOTTING DISTANCES IN CLEAR WEATHER

(all distances in meters)

WHAT'S ON THE GROUND	HEIGHT OF OBSERVER	SPOTTING RANGE	IDENTIFY RANGE
Vehicles in the open	100	5,000	3,000
Vehicles in the open	1,000	9,000	5,000
Vehicles in the open	10,000	15,000	NA
Vehicles under cover	100	4,000	2,000
Vehicles under cover	1,000	7,000	5,000
Vehicles under cover	4,000	8,000	5,000
Artillery in position	100	3,000	1,500
Artillery in position	1,000	4,000	3,000
Artillery in position	4,000	5,000	NA
Fortifications	1,000	3,000	2,000
Fortifications	4,000	7,000	6,000
Fortifications	8,000	12,000	10,000
Bridges, etc.	10,000	20,000	20,000

Referring to the above table, "Spotting Range" is the distance at which you can determine that something is out there; "Identify Range" is the distance at which you can determine if the object spotted belongs to the class indicated.

"Vehicles in the open" are trucks or armored vehicles moving on a road or cross-country. Moving vehicles are somewhat easier to spot, although a trained observer will catch stationary vehicles almost as easily.

"Vehicles under cover" are trucks or tanks in revetments or backed into a tree line. What usually gives these vehicles away are tracks in the dirt or other disturbances in the landscape (flattened grass, etc.). If these vehicles are completely covered or camouflaged, they are much harder to spot. In fact, camouflaged vehicles cannot normally be spotted unless photos are taken of the area and examined carefully by trained analysts.

"Artillery in position" is field or antiaircraft artillery (or missile batteries) in firing position. Of necessity, these weapons have to be out in the open so that they can fire.

"Fortifications" consist of trenches, bunkers, and other firing positions prepared by ground combat units.

"Bridges, etc." are large, man-made structures out in the open. Bridges are the easiest to spot, because they cross an easily seen river or other waterway.

The above values assume trained observers equipped with binoculars and perfect weather. Very hot weather or smog and the like will lower the spotting distances by up to 40 percent. Poorly trained observers can knock it down even more. You can compensate for weather conditions and poorly trained observers by having more people out there looking. This is a popular tactic, but it wears out the troops, and poorly trained spotters will still miss distant items no matter how many eyes you employ.

"Height of observer" makes a big difference, but not as much as you might think. At sea, where the sameness of the ocean makes any ship stand out, a lookout fifty or so meters up can see for about twenty thousand meters. But on land, there is much "clutter" (foliage, hills, buildings) to distract the observer. As a result of the clutter problem on land, maximum distance you can expect to spot something is about 30 percent less than at sea. An important part of a soldier's training is how to spot things in the distance. It isn't an easy skill to acquire.

The latter half of the twentieth century saw the introduction of more and more electronic spotting devices. Radar is the most common, first appearing in the 1930s. There are now radars all over the place, and many other kinds of "sensors" for detecting heat, metal, and electronic emissions. This has not made eyes obsolete, as there are never enough electronic gadgets to go around. The machines break down more often than eyeballs do, and soldiers will still have to rely on their eyes for some time to come.

Things you are looking for often make it easier for you to find them. The most obvious example is the smoke that a ship often emits from its stacks. If a distant ship is emitting a large amount of smoke, the lookout's job is much easier. That smoke drifts upward and tells everyone in the vicinity, "Here I am," even while the smoking ship is beneath the horizon (caused by the curvature of the earth). The same phenomenon has been noted in the air after jet engines were introduced widely during the late 1940s. Many jet engines left smoke trails some of the time (depending on how the pilot was handling the controls). Normally, a jet has to be within a few thousand meters before it stands out. With a smoky engine, you could easily be spotted much farther away. It was common, during the Vietnam War, for the smoky (and large) American F-4 Phantom to be stalked by the smaller, smokeless, and unseen Soviet jets that the North Vietnamese used.

As always in air warfare, whoever sees the foe first is often the winner, or at least has a substantial advantage.

Rubber Tanks

World War II (1939–1945) saw spirited reaction to the reconnaissance advantage introduced by the first use of aircraft during World War I (1914–1918). The people on the ground developed numerous ways to fool aerial observers. Aside from the usual camouflage efforts, now extended to conceal one from aerial observation, dummy weapons became much more popular and successful. Dummy weapons were nothing new; lighting additional campfires or painting logs black to look like cannons were old, and reliable, deceptions. What the World War II era brought was the use of modern materials to construct more elaborate fakes.

At first, dummy tanks and trucks were made of wood and cloth, but they were fragile and difficult to set up. Soon came the rubber, inflatable decoy. These were much lighter, easier to set up, and could easily be moved about once deflated and packed. A special trailer was built that created realistic marks on the ground, showing characteristic tank "track" marks. The trailer would be towed to where each rubber tank was to be set up. With that, the enemy photo analysts would be convinced that the rubber tank was real, for there, behind it, were the characteristic marks on the ground that only a heavy tank would leave behind.

The military advantages of this deception were substantial. If enemy recon aircraft spotted hundreds of tanks, enemy plans would have to be changed to deal with this threat. The Allies used the rubber tank ploy against the Germans many times and as a result, the Nazis always overestimated the number (already substantial) of tanks the Allies had. The Germans could have been a lot more aggressive against the Western Allies were it not for the rubber tanks. The presence of these "additional" tanks forced the Germans to hold back large numbers of their own armored vehicles as a reserve for when the Allies sent these "tanks" into combat.

Of course, the rubber tanks never saw combat (aside from being shot up by German aircraft a few times), and German armored units being held in reserve were often pounded by Allied aircraft before the panzers ever got to shoot at Allied ground troops.

Many an Allied soldier in 1944 and 1945 owed his life to rubber tanks, and most were never aware of it. For example, in preparation for the final Allied offensive in Italy in 1945, an entire dummy armored division was assembled in the rear of the 92nd Infantry Division (a mostly black outfit with a Japanese-American regiment attached), convincing the Germans that the blow was to come on the west coast, rather than in the center. Like any good deception, the use of rubber tanks was kept secret as much as possible.

Rubber tanks are still in use. In the late 1980s, the U.S. Army developed a decoy version of the M-1 Abrams tank. It costs only thirty-three hundred dollars and weighs about fifty pounds. When disassembled, the dummy tank is about the size of a duffel bag, while its portable generator is about the size of a twelve-inch television. When erected, which can be accomplished by two men in a few minutes, the decoy not only looks like a real M1 (at least from the front), but also simulates its heat signature, to fool infrared detectors. What's more, it can take several hits and remain standing, giving the illusion that you missed, or that there are more enemy tanks around than you thought.

While rubber aircraft were not a big item during World War II (tanks are easier to build and inflate to the right shape), they are today. And with the high speed of modern combat, attacking bombers often don't notice the peculiar effect their bombs and missiles are having on the aircraft they are hitting on an enemy airfield. During these attacks, the parked aircraft may be fake, but the antiaircraft fire isn't. While the aircraft lost on the ground are cheap imitations, the aircraft shot down are not. Rubber aircraft, then, can be quite lethal. Sort of like cheap bait.

A Shot From—Somewhere . . .

During all the wars of this century, troops often have lived in constant fear of attack from snipers, or groups of troops, hidden in the background. This is a development of this century, but as with most new techniques, it has an ancient pedigree. It's all because soldiers like to avoid danger.

Many soldiers contend that their most valuable piece of equipment is their shovel. Rarely does an infantryman go into battle without one, and it has long been known that the shovels of many troops, properly applied, can deliver some striking deceptions. Since ancient times, teams of diggers were used to secretly tunnel under the walls of enemy fortresses. When these tunnels were completed, supports under the enemy walls would be knocked out, collapsing those walls. While the defender was still recovering from that shock, an assault would be launched into the breach and the fortress often taken without further ado. When gunpowder arrived on the scene, it wasn't long before these tunnels were filled with powder and ignited at the proper moment. The collapsing tunnel might not always bring down the wall above, but the exploding powder nearly always worked.

When artillery became common five hundred years ago, the shovel was still an important tool for sieges. But deception was no longer the goal. Troops dug trenches closer and closer to an enemy fortress. Their digging protected them from the defenders' cannon fire and allowed attackers to move their own guns closer to the enemy's walls. If the attacker had enough artillery, and men with shovels, the fort would eventually fall.

In the past century, the shovel again became a tool for deception. When long-range (500–1,000 meters) rifles became common at the time of the American Civil War (1861–1865), clever soldiers sought ways to avoid bullets from afar. Although it took half a century for the idea to catch on, soldiers and their shovels eventually developed ways to dig for protection, and deception. By World War I (1914–1918), rifles and machine guns had made the battlefield a very lethal place. Digging afforded some protection, but for the attack-minded soldier, something else was needed. What appeared was camouflaged firing positions for snipers.

There had always been soldiers who were particularly accurate with whatever type of firearm was available. The term "sniper" came from the name acquired by British hunters of the elusive snipe in India during the nineteenth century. While snipers were common almost as soon as there were firearms available, the introduction of smokeless powder late in the 1800s marked the beginning of the sniper's reign as the stealthy scourge of the battlefield. If a sniper was well enough concealed, his shots would not leave a telltale puff of smoke. On a noisy battlefield, or in dense or crowded areas like forests or towns, the sound of a shot would not always reveal the location of the shooter. By 1900, the standard infantry rifle was very accurate indeed, especially the German Mauser '98, the American Springfield, and the British Lee-Enfield. In the hands of a marksman, no one within range was safe. But if the marksman fired, he would quickly be fired upon. So these stealthy marksmen, by now called snipers, learned to use their shovel and their wits to protect themselves.

During World War I, the millions of troops manning trench lines that stretched (literally for thousands of miles) across Europe considered snipers just one of many frontline dangers. Nothing special, just another reason to keep your head down. It was during this period that snipers began using telescopic sights on a large scale. But during World War II, snipers became a major battlefield presence no matter where the fighting was, in or out of trench lines.

The Germans and Japanese soon discovered that poorly trained American units could be stopped cold for hours by one well-concealed and resolute sniper. Better-trained troops immediately fired their weapons in the general direction of the enemy and fanned out when they encountered a sniper. Soon, the lone shooter was either killed or forced to withdraw. But American troops usually went into combat without realistic training. It was common for entire U.S. infantry companies (100–150 men) to hit the ground because of one sniper.

It would often take hours to get all the troops up and moving again, and by then, the sniper could have withdrawn to a new position.

A good sniper was a master of deception. Like a submarine, a sniper could dish it out, but was unable to absorb much punishment. So the efficient sniper paid as much attention to concealing himself as he did to his marksmanship. While movies like to put Japanese snipers in trees (albeit camouflaged), that was not a common practice. While it makes for a more spectacular scene when the GIs shoot the SOB down, the Japanese were more practical and less theatrical. A Japanese sniper would prepare a ground-level position, often a pit connected to a trench that he would use to switch to another position without being seen. The Germans and Russians also favored the carefully prepared position technique. The snipers usually had a good tactical sense and would quickly figure out the likely routes the approaching enemy would take. The sniper would then pick out the best firing positions and figure out a way he could switch from one to another without being seen or shot.

A successful sniper would pick a firing position that the enemy would not expect to receive fire from. In this respect, an expert sniper was an accomplished psychologist. Most troops, and people in general, had the same habits that a sniper could exploit. For example, if a shot rang out and a soldier was hit, the other troops would tend to look up and forward. A clever sniper would set up a firing position at ground level and try to get his shot in from the side or rear. Once the troops ran around a bit looking for the shooter, and the sniper sensed that no one had a good idea where he was, he would get off another shot.

In urban areas, snipers had a plenitude of hiding places. Unlike the movie cliché, snipers avoided roofs and sought basements instead. Again, part of this was psychology. Troops under sniper fire would first look to the roof. Moreover, it was easier for a sniper to escape from a basement than from a rooftop. In a built-up area, a good sniper had a wealth of inconspicuous firing positions to choose from. Many positions were fiendishly obscure. For example, a favorite trick is to set up a firing position at the end of a hall leading to an open (or blown-off) door to the outside. The field of view should lead to a large open area (like a town square). Such a position allows for a shot the enemy troops will have a hard time tracing, and ample opportunity to get away to another firing position. Basements, or even partially destroyed buildings also provide ample firing positions. A few experienced snipers can select dozens of firing positions and switch back and forth for hours while harrying hundreds of enemy troops and shooting scores of them.

A sniper's most lethal opponents were experienced troops who had dealt with snipers before. A good sniper would quickly realize if he was up against such troops, and act accordingly. In other words, under these circumstances the sniper would operate much more cautiously and end up inflicting far fewer casualties. Most snipers were not suicidal, and their commanders could not afford to have snipers with a death wish. Snipers were too valuable. Many expert snipers survived the war, thanks largely to their skill at deception rather than their outstanding marksmanship.

But it was in Vietnam that Americans met their most resourceful snipers. Because most of the fighting was in guerrilla mode, the Communists (Vietcong irregulars and North Vietnamese regulars) made the most of sniper tactics. Often, the majority of fire that American troops received was sniper fire. The Communists were pragmatic about how they used their snipers, as they quickly discovered that the Americans had far more firepower and were quick to use it. For example, each American trooper had an automatic rifle, and about every tenth man had a machine gun. Units in the field were supported by artillery, mortars, bombers, and helicopter gunships. It was not unusual for a sniper to get hit with many of these weapons, often with fatal effect on the hidden, solitary rifleman.

The Communists quickly adapted their sniper tactics to deal with American firepower. It was noted that the maximum U.S. firepower was available during planned operations. Under these conditions, snipers went for the long-distance shot, followed by a rapid retreat. Often this worked, although the Communists were always impressed by how quickly the Americans could bring down artillery fire and bombing attacks. While all this firepower may not have been accurate, there was so much of it that even the most artfully concealed sniper was likely to get hit. During these large-scale operations, snipers were often used to cover a retreat and, as such, were knowingly in a dangerous situation. But all the Communist troops knew that by sacrificing a few of their number as snipers, the American ground unit could usually be held up long enough for the bulk of the retreating force to escape.

A more lethal use of Communist snipers was at those times when Americans were not engaged in a major operation. U.S. base camps and other camps out in the bush were always subject to random sniper attacks. Often, the Americans would send out troops to try to snag the enemy shooter and, as often as not, they wouldn't find them. The Communists considered their snipers a potent political weapon. Having a sniper shoot at Americans from a village would often result in U.S. troops storming into the village to find the now-departed sniper. The villagers would be terrified, and often killed or injured. Net result: Some of the villagers would now be anti-American enough to join the Communists. This technique was not unique to the Vietnam War; it had been used for over a century.

Firing from among unarmed civilians is a deception technique as old as warfare. This method of avoiding detection, and to a lesser extent, retaliation, has become more effective with the introduction of "rules of warfare" and mass media in the past century. The rules of warfare (the "Geneva Conventions") forbid attacks on noncombatant civilians. The rise of mass media made it more difficult to conceal killings of civilians. As a result of these two changes, soldiers have become increasingly reluctant to open fire on civilians, even when it was clear that the unarmed civilians were being used as concealment for armed foes. American troops frequently encountered this technique in Vietnam when facing snipers. With the end of the Cold War and increasing use of American soldiers in conflict-ridden nations, being confronted with seemingly harmless, but actually lethal, civilians is becoming more common. It is well to remember that armed opponents who are outgunned will seek any advantage they can. The opponent may be uncivilized, but rarely will he be unintelligent.

Blowing Smoke

It has long been noted that fog, mist, and clouds can produce militarily useful concealment. Some parts of the world commonly have morning fog or mist and local troops have to learn to deal with it. Until artificial smoke devices were invented in this century, fog and the like would usually be a danger to defending troops, as it would allow attackers to get close without being exposed to missile fire (spears, arrows, muskets, cannon, etc.). Defenders derived some benefit from these natural obscurants, because the attacker had to find their way partially blinded. You could always hope that the attacking troops would get lost, and this sometimes happened. But well-trained attackers knew how to find their way in the fog and usually didn't cover much more than a few hundred meters under these conditions anyway. Sometimes fires would be set to provide a smokescreen, but this was rather obvious; the enemy would thus be alerted, and was never commonly used.

With the coming of gunpowder weapons in the past five hundred years, battlefields often became obscured by the smoke of thousands of muskets and cannons being fired. Since these primitive weapons weren't all that accurate, and were fired in the general direction of the enemy, this black powder smoke screen was rarely a major factor for either side.

In this century, right about the time smokeless gunpowder appeared and cleared the battlefield somewhat, it was discovered that smoke grenades, smoke shells, and smoke pots could provide copious amounts of smoke on demand. This led to a new form of deception: blowing smoke.

Rather than wait for nature to provide a convenient fog, artificial smoke could be used as needed. While a strong wind would rapidly disperse artificial smoke, average weather conditions were, quite naturally, the norm and no problem.

Artificial smoke is delivered by three means: smoke grenades, smoke shells, and smoke pots. Smoke grenades are thrown like regular grenades; smoke shells are like high-explosive shells, but contain smoke-producing material instead of high explosive; smoke pots are containers of various size that contain smoke-producing material. The characteristics of these various items are as follows:

ARTIFICIAL SMOKE WEAPONS

	GRENADE	SHELL	POT
Weight of Item (pounds)	1–2	15–100	5–100
Time to Form (seconds)	10–20	1–3	10–30
Max Length of Cloud (meters)	20–40	20–100	50–200
Duration of Smoke (seconds)	60–120	40–160	300–900

"Weight of Item" is in pounds; the variation results from different national designs as well as a variety of different sizes available for most items.

"Time to Form" is in seconds and is how long it takes before you have a useful amount of smoke.

"Max Length of Cloud" is in meters and is, on average, how long a smoke cloud you will get for each item. The smoke is designed to stay close to the ground, so most smoke clouds are less than ten meters high.

"Duration of Smoke" is in seconds and is the average time that the smoke cloud will last before dissipating into uselessness.

In most cases, more than one of these items is used at once. Several grenades are tossed out to make sure there is sufficient smoke created. Smoke shells are fired in volleys, the number of shells depending on whether it's a battery volley (four to six guns) or a battalion volley (twelve to twenty-four guns). The shells land in a line, quickly putting up a wall of smoke. Volleys may then be fired again and again to maintain the concealment provided by the smoke. Smoke pots have to be placed and ignited by hand, although in some rare cases, the pots can be ignited by remote control.

The normal use of artificial smoke is to hide troops from observation. If the enemy couldn't see your troops, he couldn't accurately fire at them. While smoke did not stop bullets and shells, it made this fire much less accurate.

The infantry is very fond of smoke grenades, as soldiers often find themselves in situations where they need some concealment, fast. Pop a little smoke and slip away relatively unharmed. But the infantry is also fond of using smoke as a pure deception. A knowledgeable enemy will expect something to be happening behind the cloud created by a few smoke grenades and will often fire into the cloud. For the smoke-using infantry, that enemy fire means the other guys are preoccupied with smoke, thus allowing for other maneuvers by the troops that tossed the grenades.

In Vietnam, smoke grenades of various colors were used to mark targets for air strikes. Desperate troops could hurl colored smoke at the enemy even if there wasn't any friendly aircraft around. The enemy, not knowing about the absence of aircraft, but seeing the colored smoke that usually preceded the appearance of bombs and rockets, would take cover. The American troops could then make their escape.

Armored vehicles also use smoke a lot. Most tanks now have smoke-grenade launchers mounted on the turret. One or more of these grenades can be launched by the crew, putting an obscuring cloud of smoke to their front, thus ruining the aim of enemy gunners. Tank battles are often rather like chess games, with individual tanks moving to and fro trying to get a favorable firing position against any enemy tanks that might become visible. Smoke thus becomes a key tool in hiding your own tanks and their location or direction of movement. Smoke is so important that many older tanks had the ability to automatically pour oil over the hot-engine manifold and create an impromptu smoke cloud that a tank could back up and retreat through. Along the same lines, some nations issue smoke grenades that put out black smoke, which is used to make a vehicle look like it's been hit by enemy fire. This usually fools pilots, and often ground troops as well. The most common use of black-smoke grenades is to make a fake tank parking lot (full of inflated rubber tanks) look like the real thing after the bombs hit. But this trick is even used on the battlefield. The enemy will usually stop shooting at a target once it appears to be hit. A clever tank crew can stop its tank, pop a black-smoke grenade out the hatch, and then wait for the enemy artillery or antitank fire to shift.

Although submarines have little need for smoke grenades, they used a similar deception during World War II when they would release some oil (and perhaps debris from the torpedo tubes) when under attack. The destroyers topside would conclude that the sub had sunk, and stop their attack. Unless, of course, the destroyer commander wasn't fooled. As always, a deception is only as good as its believability.

Smoke pots are most often used to hide rear-area targets from observation or attack. Protecting a vital target like a bridge from air attack often involved the use of smoke pots. When it is known that enemy aircraft are in the area, the pots are ignited and the bridge is covered with smoke. This will not stop the attack, but it will spoil the pilot's aim.

You would also use the pots to support major attacks. Crossing a river under enemy fire, for example, would often require the use of hundreds of smoke pots to hide the troops going across the water in assault boats. Most smoke pots can float, and can be released to flow with the current while emitting their protective cloud. Any attack across open ground benefits from smoke, and smoke pots are the best way to do it if you have the time and opportunity to get the pots up front and ignited.

In combat zones, smoke shells are the more common form of delivering large quantities of smoke—everything from mortars to heavy-artillery-fire smoke shells. A particularly favorite smoke-shell type for Americans is White Phosphorus (WP, or "Willie Peter"). In addition to emitting a large quantity of smoke, the phosphorus burns intensely, is difficult to extinguish, and provides a very unpleasant way to die. A little Willie Peter not only blinds the enemy, but does a number on his morale as well.

Smoke is not as protective as it used to be. Some top-of-the-line tanks (the latest models of the U.S. M-1, for example) have a heat sensor that can see through most types of smoke. These sensors can be used by infantry and aircraft, if you can afford the expense. A more expensive smoke formulation can add more protection, and more of this better (and pricier) smoke is being used.

Walking on Water

Water and armies don't mix. Getting across rivers is a major headache during wartime, especially in this age of motor vehicles and all sorts of things that can't swim. It's not just the relative paucity of bridges, but the tendency of these river crossings to be destroyed. Each side is always trying to destroy the bridges the other side depends on. As a result, there is always a market for new ideas on how to deceive the other side of the current state of your bridges.

One of the more innovative bridge deceptions of World War II was when the Russians, in preparation for an offensive, built a bridge at night. Actually, the construction took many nights. But during the day, German recon aircraft did not see a bridge. The bridge wasn't invisible, but was built so that its roadway was a few inches beneath the water. When the time came for the attack, the Russian troops appeared to walk on water, as their tanks rolled right over the river as well. The underwater bridge was a favorite Russian deception and they used it frequently.

The underwater bridge trick was not always possible, as it was highly dependent on the depth of the river, light, and color conditions (lest the underwater structure be detected from the air), but there are many other deceptions possible with bridges. A common stunt is to make an apparently wrecked bridge usable. Most bridges consist of several sections and a successful bombing usually "drops" only one section. Depending on how resourceful your engineers are, and how badly banged-up the wrecked section is, it can be repaired. But the repairs can be constructed in such a way that the section still appears, from the air, to be destroyed. Or more obvious ramps can be built, and only used at night. In either case, by day the bridge is left in an apparently wrecked condition, while at night, traffic continues to cross. The enemy, thinking the bridge is down, will not continue to bomb it.

A common alternative to the secret repairs is the parallel pontoon bridge. Since approach roads are an important part of a bridge, these routes can still be used even if the bridge is beyond fixing. This is done by building a pontoon (floating bridge section) crossing that runs right next to the bridge. The nice thing about pontoons is that, since they float, the sections can be unbolted and moved to camouflaged positions on the shore during the day. At night, the pontoon bridge is put back together and traffic continues to move. Again, if the enemy can't see it, he can't bomb it.

Stealing a March on the "New Romans"

In the wake of Germany's spectacular defeat of France in 1940, Italy decided to take the offensive from its North African territory (Libya) into British-controlled Egypt. Styling themselves as "New Romans," the Italians pushed their six infantry divisions (all of them nonmotorized and two of them consisting of ill-trained Libyan levies) over sixty miles into Egypt, confident that the two British divisions available in Egypt would not be able to resist this advance. While the British troops were better trained and equipped, they were greatly outnumbered. A surprise attack was needed to redress the situation, and the British proceeded to pull off the first of many deceptions they inflicted on the Italians.

The Italians cooperated by halting after advancing five days and a hundred kilometers inside Egypt and digging in. Rather than a continuous line of fortifications, the Italians built a series of independent fortified camps. The Italians had problems with logistics, and a lack of motorized troops. They also lacked boldness, which might have carried them to Alexandria if they hadn't halted on September 18.

The Italians were going to wait for more supplies to be stockpiled, and bring up more trucks and armored vehicles from Italy. Since the Germans' air force was filling the skies over England (the Battle of Britain), British forces in Egypt were unlikely to receive any reinforcements from home anytime soon.

This situation prevailed through October and November. But the British were not content to sit and wait for the Italians to get their act together. Throughout this period, British patrols ranged all over the "Italian controlled" desert to the west. The Italians actually controlled little beyond their fortified camps, and were content to sit and wait. But the British collected information, while the Italians relied on questionable information from spies and agents in the Egyptian territories to the east. Many of the Italian spies were on the British payroll, and the Italians were being set up for a spectacular deception.

The British had determined that the Italian system of fortified camps had left gaps that could be exploited in a surprise attack. The British had also determined that many of the Italian units were not really ready for combat and could be expected to break and run if hit by an unexpected attack. Thus the British decided to use their meager forces (one armored and one motorized infantry division) for a surprise attack. For surprise to be achieved, the plan had to be kept secret. While the Italian spies in Egypt were not very good, they were numerous, and much information found its way back to Italian intelligence officers.

The British began their deception by announcing a training exercise out in the desert, to take place in early December. British divisions were deployed as if on a training exercise. This made sense, as a little tune-up was to be expected from the customarily well-trained British. The Italians thought nothing of this, as their spies reported no unusual activity beyond preparations for this training exercise. Few British officers knew that the training exercise was to be turned into a daring night march, followed by an attack on the Italian positions. A special British deception team set up false radio traffic to lull the Italian eavesdroppers into thinking that the training exercise was real. But at night, the British units got new orders. They advanced straight through the Italian positions, attacking the fortified camps from all sides. The attack was surprising enough; that the British had found their way through the Italians' defensive system was shocking. The Italians reacted as the British had hoped they would (by panicking). Italian resistance soon broke, and the shattered units fled west. The British captured nearly forty thousand prisoners, as well as most of the Italian artillery. The British then pursued the Italians hundreds of miles through the desert into Libya. This offensive was only brought to a halt by the need to send some of the British troops to other fronts, and by the appearance of German divisions the following spring.

The End of the Line

Deceiving the enemy as to where the end of the railroad line is can be an important advantage.

The war in North Africa (1940–1943) was fought largely in a desert where everything, including water, had to be brought across sandy wasteland. Trucks took quite a beating, and only one side, the British, had use of a railroad. It wasn't much of a railroad, extending into the Egyptian desert along the Mediterranean coast toward the Libyan border. But when the Germans had pushed the British back to the Libyan border in late 1942, this single rail line became most important. The Germans had exhausted their supplies, as they had to bring everything forward by truck over hundreds of miles of desert roads. With their railroad, the British could quickly and easily move up large quantities of men, weapons, equipment, and supplies. With all of this, the British could fight a decisive battle and defeat the Afrika Korps once and for all. But the Germans knew where the railroad ended and realized that regular bombing raids on the rail terminus could cripple this transportation effort. The large quantities of material being unloaded at the railhead made an excellent target. Moreover, if enough of the locomotives and their cars could be hit, the railroad effort would be crippled.

Painfully aware of their vulnerable position at the railhead, the British developed a deception plan to avoid the German bombers. A fake ten-mile extension of the railroad was built, and a fake railhead was constructed at the new terminus of the railroad. British work crews laid down rails (but didn't do all the other work required to prepare a roadbed) at the same rate they would if they were actually building a rail line. From the air it did look like the British were extending their railroad. And at the end of this new line, the British constructed a fake locomotive and a string of rail cars. The Germans fell for it and bombed the fake railhead many times, but never the old one (which was reorganized to look less like a railhead). The buildup for the Battle of El Alamein went ahead, and the Germans were defeated.

Why They Called Him the "Desert Fox"

One of the few German generals to attain fame in the Allied nations, Erwin Rommel first became a media favorite while fighting the British in North Africa. From 1941 to 1943, Rommel managed to give the British fits with his Afrika Korps and Italian allies. Rommel became known as the "Desert Fox" for his uncanny ability to run circles (sometimes literally) around the British. A large part of Rommel's success was due to his use of deception, something he did from long habit and began as soon as he and the first German combat units reached North Africa. Landing at Tripoli in the spring of 1941, Rommel promptly arranged a parade to show off German military might to the local Italians and Arabs, as well as to the Allied spies he knew would be watching. As German tanks rolled by the reviewing stand, they turned a corner and entered an area that was sealed off by German troops. Moving behind the reviewing-stand area, the tanks joined the vehicles waiting to drive past the crowds and did it again, and again, and again. The Allied spies could count, and the number of German tanks they reported to their British masters caused consternation. Thus Rommel began his North African campaign with a psychological edge.

Rommel, like all German generals of the period, was highly professional and had extensive experience in World War I. From this came a deep appreciation in the use of deception, and much practical experience in that area. Rommel habitually used deception whenever possible and to good effect. One item the Germans were particularly keen at was security. Although they didn't know that their secret codes were being read by the Allies, the Germans habitually piled on even more security measures. Once he reached North Africa, Rommel quickly realized that his Italian allies were not as security-conscious and that any plans he shared with them would quickly get back to the British. Typically, Rommel turned this Italian sloppiness to his advantage. He would tell the Italians only at the last minute when there was going to be a joint operation. Otherwise, Rommel fed the

Italians deceptive information that he wanted to get back to the British. For example, before the Germans were going to launch an offensive, Rommel would inform the Italians that he was planning a retreat, or some other nonthreatening (to the British) maneuver. Thus lulled, the British would be less prepared for the German advance.

Tactically, the Germans had it all over the British in the deception department for most of the North African war. The Germans would frequently draw the British tanks into desert ambushes. Often the British would stumble into a well-hidden German minefield, then get shot up by concealed antitank guns. While all this was going on, German tanks would be maneuvering far out of sight and then suddenly fall on the unsuspecting British rear area.

The British were not dunces in the deception department, but during many of the crucial battles in North Africa, the Germans were better. Better was often good enough to create humiliating British defeats.

Foxing the Desert Fox

The battles in North Africa (1940–1943) between the British and their Axis opponents saw extensive, and often decisive, use of deception. The British were better at it than the Italians, but the Germans were more skillful than the British, at least until near the end of the North African fighting. The British learned as they went, and their German opponents were enlightening teachers.

Although the British had a special section at their North African headquarters to handle deception, it wasn't until late 1942 that these professional deceivers had a chance to really strut their stuff. Before that, the Germans simply moved too fast for British deceptions to have much effect. But once the Germans had pushed all the way to the Egyptian border (at El Alamein) during the summer of 1942, the British had an opportunity to reveal some superior deception skills.

When Rommel and his Afrika Korps appeared on the North African scene in early 1941, the British had already spent months beating on the hapless Italians. Moreover, at the same time Rommel showed up, the British North African forces were being weakened by units withdrawn for service in other theaters. For the next year and a half, the Germans pummeled the British severely. Once the two armies settled down on the relatively narrow El Alamein front (with the Mediterranean to the north and the virtually impassable Qattara Depression to the south), the British had the breathing space and resources to run a number of successful deceptions against the Germans for a change.

The major British deception was to fool the Germans about the time and location of the coming offensive. The Germans were in a difficult situation. Too far from their own bases to supply an offensive, they did have sufficient resources to stop the coming British attack. Although the British had won a battle (Alam Halfa) in stopping the Germans from getting past El Alamein, the Axis troops still held the psychological edge. The British needed a decisive victory in their coming El Alamein attack. Deception was to be the key in obtaining that victory.

The 60-kilometer-wide front line was quite narrow for a battle in which many mechanized divisions would take part. Moreover, it was pretty obvious that the best place for the British to attack was along the Mediterranean coast. Here was the rail line, the only hard-surface road and the opportunity for British warships and amphibious forces to add their might to the coming battle. The Germans had five armored and mechanized infantry divisions in reserve (three of them German). Holding the line were six divisions, all of them Italian except for a single German one in the north. There was also a German paratrooper brigade down south to "stiffen" the Italians.

Once it became clear which part of the 60-kilometer front the main British offensive was hitting, the Germans would send in their mobile reserve divisions and smash the British attack. This was what the British wanted to avoid, and their deception plan sought to make the Germans think the attack was coming in the south. If this deception could be pulled off, a feint could be made in the south; the Germans would send their armored divisions down there and then the real attack would open in the north. With the coast road open and the German mobile units down in the desert, the Axis forces would be destroyed and it would be a great victory for the British. But only if the Germans could be made to believe that the attack would be down south in the desert, away from the coast.

That the attack would come in the south was not out of the question as far as the Germans were concerned. Still feeling rather superior to the British, the Germans could understand the British trying to hit the southern part of the line if only because that's where troop density was the weakest. The Germans could not afford to thin out their frontline troops much in the north, for if the British were able to move freely along the coast road, most of the Axis units inland would be lost while trying to get away (more slowly) cross-country.

The desert made it difficult to hide anything, but easier to showcase false objects you wanted the enemy to see, and believe. Although the British had air superiority, the Germans were still able to sneak their recon aircraft in. So whatever the British had on the ground (real or otherwise) would be seen by the enemy.

In September 1942, the British began Operation Bertram, as they called the crucial deception plan that was to support the coming El Alamein offensive and assure that it would be a victory. The Germans knew a major attack was coming. They didn't know exactly where and they didn't know the day it would start. The British deception was to cause the Germans to guess wrong on both of these important points. The plan consisted of four separate deceptions.

- The attack was scheduled for October 23, when a full moon was present, providing ample illumination in the desert for night movements. The British wanted to make the Germans think that the attack was going to come on November 6, when the night sky was at its darkest. From a military point of view, both a full moon and no moon had advantages and disadvantages. The Germans knew that either date was likely. The British deception plan threw out several tantalizing bits of information for German spies and sympathizers to pick up. British intelligence had operatives all over the Middle East, and these agents knew who was working for the Axis, and who was simply someone useful to drop a stray bit of information on. Many bits of information were let out in order to provide a puzzle that, once assembled, said what the British wanted said. Some of these informational gems were true, which was necessary to make the false ones more believable. One of the false bits was that the new American Sherman (M-4) tank was having teething troubles and this was holding up the offensive, making a November date more likely. This was a choice morsel to throw at the Germans, as Nazi propaganda had made much of America's inability to make decent weapons, so this would play into existing German prejudices. There were several publicly, and semipublicly, announced items for the Germans to chew on. Parties and receptions were arranged in Cairo for the night before the real offensive, and it was made known that many of the senior British commanders would be in attendance. A conference for senior Middle Eastern commanders was planned (a deception, of course), to take place in Tehran on the day the offensive was to kick off. The many officers who were to "attend" were indiscreet in making their travel arrangements, giving the Germans one more piece for their seeming real mosaic of British intentions. None of these deceptions would work, however, unless the other subterfuges in the field were convincing.

- The supplies needed to support the attack had to be hidden. It wasn't so much a matter of hiding the mountains of material as it was making the supply dumps in the south look larger than the ones in the north. All units on the 60-kilometer front line had supplies piled up behind them. The additional fuel, ammunition, and trucks for the real offensive in the north were made to look less conspicuous, while those in the south were made more noticeable. This was less of a problem up north, where the Germans knew of the railroad and paved road constantly disgorging mountains of supply. It was easy enough to hide the additional three thousand tons of supply for the attack up north. But down south, over a thousand men and many tons of wood, camouflage netting, and engineer equipment were used to create the illusion (from the air) of eight thousand tons of supplies being piled up for an attack. The British were able to take advantage of the fact that the Germans expected attempts to camouflage these mountains of material. Thus the mere presence of a lot of camouflage netting on the desert indicated, to the Germans, that there must be something underneath. There wasn't anything underneath, but one could not discern that from the air. A fake water pipeline was extended to the south. A major offensive needed a lot of water. The machines, as well as the troops, got thirsty. And the field hospitals treating the flood of wounded from a major offensive required ample water supplies.

- The artillery posed a particular problem. The Axis positions were dug in, fortified, and surrounded by mines and barbed wire. For the infantry to have a chance of getting through all this, they needed a lot of artillery support. The British had collected four hundred additional guns and thousands of tons of ammunition for the offensive. The ammunition was easier to hide than the guns and trucks of the artillery units. Indeed, the units carried a lot of ammunition with them and were resupplied by trucks running back to the supply dumps for more as several days of battle went on. The guns were hidden by moving them at night and camouflaging them, and the tractors that pulled them, to look like trucks during the day. The British had to be careful that all these false trucks (and the real ones that normally accompanied them) were parked in a place where the Germans would expect to see a lot of trucks. Each side had thousands of trucks, and it was normal for them to be all over the place moving supplies and equipment. But too many trucks in one place would raise an alarm with the German intelligence officers. As always with deception, it wasn't enough to make something look different; it had to look plausibly different and in context. For example, it's obvious that in the desert, you don't use winter (white) or temperate zone (green) camouflage nets. Less obvious, except to someone trained for intelligence work, is the importance of everything being in its place. The absence of a large number of guns in the south would be OK until late October, when the Germans would expect them to be brought up for the offensive they were expecting on November 6. Even so, in the days before the actual October 23 attack, many fake guns were set up in the south. The Germans noted that, and knew that more guns were probably going to be brought up at night until all artillery was in place for the November 6 offensive.

- The tank divisions that would exploit the attack were difficult to hide. The two armored divisions and a tank brigade assigned to the offensive contained some seven hundred tanks and over four thousand other vehicles. To make their task easier, the British set up a training area to the rear that was halfway between the coast and the area in the south they were trying to make the Germans think was the location of the "November 6" offensive. Because of truck and armored vehicle movement in and out of this training area, the Germans could see tracks in the desert leading to all sectors of the front. The problem was, how to get these tank units to the front line up north without giving away the game. The solution was to locate parking lots for trucks just behind the front up north. This was plausible, because of all the supplies coming out from Cairo by truck and railroad. But a tank looks quite different from a truck, viewed from the air. The solution was simple. While the tanks moved northwest to the truck parking areas in the two nights before the offensive, the trucks moved down to where the tanks had been. The tanks were then

> camouflaged to look like heavy trucks, while the real trucks were camouflaged to look like tanks. When the Germans looked at their aerial photos the next day, they saw what they had seen the day before; the same number of trucks parked near the front up north, and the tanks parked out in the desert.

The British deception worked. The first inkling the Germans got that they had been snookered occurred on the evening of October 23 as the British began attacking in the south. The Axis commanders thought they had been deceived as to the date of the attack, but not the place. So they began moving their armored reserve to the south.

But at 9:40 P.M. on October 23 (under a full moon), the British infantry advanced in the north on a 10-kilometer front while nearly a thousand guns blasted the German positions. Four hours later, the British tank units moved forward. But because of the artillery barrage and infantry advance in the south, the Germans saw the action in the north as a feeble diversion. While the German panzers began moving south, the British tanks were crashing through the German lines in the north. The November 6 attack never took place, for by November 4, the Germans and Italians were defeated and on their way west.

Operation Bertrand was, up to that time, the largest deception operation the British had run. Several thousand special troops were employed to make it work, as well as thousands of tons of material. It worked, and thus encouraged, the British (and their Allies) went on to deceive the Axis several times again before the war ended.

The XX (Double Cross) System

One of the more remarkable deceptions of World War II was the British ability to compromise the entire German spy system within Great Britain. Throughout the war, with very few exceptions, every spy the Germans had operating within Britain was caught by the British and, in most cases, turned into double agents working for the British. Naturally, this situation presented the Allies with numerous opportunities to deceive the Germans, and this is exactly what happened.

Pulling off this feat was made possible by many factors. First of all, British counterintelligence was good to begin with, and throughout the 1930s, the British knew who most of the German spies were.

Rather than just arrest every German spy who was uncovered, the British allowed many to continue operating, but kept an eye on what they did and who they did it with. This enabled the British to stay on top of German espionage activities, as the Germans had no incentive to increase the security of their spy network. The Germans cooperated by not being as professional as they could have been.

One benefit (to the Allies) of the Nazis taking over in Germany was the ascent of many Nazi politicians to prominent positions in German security agencies. These Nazis generally had contempt for the professionals they now supervised, and that contempt was returned in kind. The strained relations between the Nazis and professionals in the German security apparatus lowered the efficiency of those organizations. This provided ample opportunities for the more professional British counterintelligence to unravel German espionage activities in Britain.

There was also the question of exactly who Wilhelm Canaris, the head of German Army intelligence (the "Abwehr") was working for. There is much evidence that Canaris was not totally on the side of his Nazi masters. In fact, MI5 (British intelligence) and the CIA were quick to see that Canaris' widow was well taken care of after the war. To this day, no one in either intelligence establishment will say whether or not Canaris was working for the Allies.

After the war began, the British decided to attempt to "turn" the German spies, rather than just arrest and execute them (the traditional way to deal with spies in wartime). But this did not happen all at once. When known German spies were rounded up in late 1939 (after war was declared), the British were content to simply have smashed any incipient German espionage activity in their country. Before hanging the spies, British intelligence attempted to extract as much useful information as possible from them. The British were in a strong bargaining position, as they could offer a prison term in lieu of the gallows to cooperative spies. During 1940, as the questioning of the spies continued, it occurred to some of the British interrogators, and their superiors, that it might be possible to "turn" many of the captured spies (have them pretend to remain German spies but actually work for the British). The British had not released a lot of information on the captured spies. At most, they announced that His Majesty's government had the German espionage problem under control. The Germans expected many of their prewar spies to get caught when war broke out, and most of the others to lie low until things calmed down.

In late 1940, the British finally decided to attempt "turning" as many of the captured German spies as possible and, in effect, to control all German espionage operations in Britain. Many of the captured spies were given an offer they could hardly refuse: become a double agent or be executed for wartime espionage. Most German agents accepted this offer and spent the rest of the war feeding their controllers in Germany false or misleading information. This became known as the "XX (Double Cross) System."

What made XX remarkable was that the Germans never caught on. This feat was accomplished largely because the British were also reading German codes (the ULTRA system) and were thus able to feed the Germans real secrets that would not be damaging to the Allied cause. For example, the XX agents could radio true information to Germany on where Allied warships were. But this was done only when the ULTRA crew knew where all German submarines were, and knew that none was in a position to take advantage of this. Other forms of harmless information passed on were things the Germans were going to find out soon anyway. Items like meetings of senior Allied leaders (which eventually hit the papers), the introduction of new weapons (which eventually hit German troops), and sundry (harmless) trivia that only someone inside Britain could find out (like what the Royal Family was up to). All of this was done to maintain the trust of the German spymasters for those situations when the Allies wanted to feed the Nazis false information in order to gain a military advantage. In other words, for when the Allies wanted to stage a deception.

It took several years to get the XX system working to everyone's satisfaction. This took a lot of effort on the part of the British. One mistake could have brought the entire system crashing down. The Germans didn't suspect something as outrageous as XX, so they proceeded on the assumption that they had a functioning spy system in Britain. This meant that new agents were sent in, and sometimes, they tried to bring spies back to Germany. This last item gave the British fits, as they could not trust any of their XX spies to get back to Germany and not expose XX. Thus an elaborate deception had to be mounted to convince the Germans that new spies could get in, but old ones (that the Germans had hired in Britain or sent in) could rarely get out.

Taking care of new German spies was much easier. They were made "the offer they couldn't refuse" and promptly became XX agents. Some did refuse, and were hanged. These executions were made known to captured German agents who had not yet agreed to join the XX system, as well as those who already had. It encouraged all concerned.

The turned German agents were a mixed bag. Some were highly skilled Nazi true believers (the kind who tended to prefer death before dishonor), but mostly, they were a rather eclectic collection of adventurers, misfits, idealists (who were willing to compromise), and the like. The Germans knew they had a mixed bag of spies and were happy to get anything at all useful out of them.

There were never more than about fifty XX agents to be looked after, and most of them were low-level operatives. That is, they were ordinary citizens, often working at menial jobs. They did not have access to the inner workings of the British government or the military and could, at best, simply report what they saw as they went about their daily routine. To make the XX system more believable, and the job of the British controllers easier, periodically, one of these agents would "disappear." This was natural in the espionage business. Agents got caught, or lost their nerve and stopped spying. The Germans also had their problems in meeting their payroll. Agents had to be paid, and to take care of this chore, couriers came in from Germany via the neutral nations of Portugal and Spain. The British did not make it easy for these couriers to make their rounds, and would arrest and jail them as it suited their purposes. For example, a courier coming through with the payroll could be nailed, thus causing one or more agents to go off line because they hadn't been paid. Sometimes, XX agents would send a last message saying, in effect, "They are on to me," and then disappear.

The information concocted for the XX agents to radio back had to be plausible. Because of the mundane lifestyles of most agents, they could only report on those things a waiter or office worker could have access to. Travel was restricted in wartime Britain, so agents couldn't wander far afield looking for military information. Agents were expected to read the newspapers and listen carefully to pub talk and whatever rumors were floating around the civilian population. In large port cities, one could observe the ship traffic and report it. A bicycle ride near some military installations might yield some interesting insights.

The British did manage to insert some of their own agents into the German network. These were often "new" spies that the original XX agents "recruited" in England. There were also people who escaped Nazi-occupied Europe or had language and other skills that made them suitable for espionage work. These people were "recruited" by the XX network of phony German agents and were allowed to travel back to Europe to meet with their German controllers. Naturally, few of these agents knew about the XX system. The details of XX were kept by a small circle of British intelligence operatives. But this was standard espionage practice. The less any individual knew, the less would be lost if that agent were caught and interrogated.

The Germans also turned British agents on the Continent, but because of the XX system and ULTRA, the British minimized the damage these turncoats caused. Sometimes, the British were able to turn these agents once more, making them British agents turned by the Germans but actually working for the British. It was a murky business, as espionage always was. It was one area of human endeavor where paranoia was a positive trait.

From the beginning of the XX system in late 1940 through 1943, the British were grooming their XX agents for bigger things. Until 1944, there was little opportunity to use the XX system for any really big deceptions, mainly because most of the fighting was going on in Russia. Oh, XX was used to assist the war against German U-boats in the Atlantic, and to support the relatively minor operations in North Africa. But when 1944 came along, so did the preparations for the Allied invasion of France. This undertaking was to involve massive deceptions, and XX played a major role.

Since the D-Day invasion was being launched from Britain, the Germans put their British spy network into high gear. The Germans also sent many more new agents to Britain, and this put a much heavier workload on the XX staff, as all these new spies had to be captured and "turned" into XX agents. All this had to be done carefully, lest that one fatal slipup compromise the XX system when it was needed most.

One of the risks the XX staff encountered was the normal espionage procedure that allows for a captured spy, who is forced to turn, to send a signal in one of his coded messages to indicate that he was now a captive. Armed with such knowledge, the Germans could have played a little game on the British. This was a risk the British had to accept, and it took time before they were sure the Germans were hooked. While the British were running the XX agents, they also had ample opportunity to uncover German operations on the Continent. Although the Germans were careful to tell their agents in Britain no more than they needed to know, there were ways the British could use the XX agents to reveal information about German espionage operations elsewhere. This was often as simple as passing some information about a certain fake Allied espionage operation on the Continent, and then noting who did what over there. The Allies also had the assistance of their ULTRA intercepts. Thus, while getting ready to use XX for the big invasion in 1944, the British were able to run a lot of smaller, but still valuable, deceptions on the German spy system.

One of the more valuable items picked up was the cipher used by German spies when they radioed their messages. Since the XX spies had to use the cipher, the Germans sent them the new ones every time the code was changed (periodically, to make it harder to crack). This saved the Allies a lot of effort in cracking this particular code and allowed the Allies to quickly read the messages of other spies (in neutral countries like Spain, Portugal, Switzerland, Turkey, and Sweden) and act on what they had read. By reading these messages, the Allies were able to glean massive amounts of information on what the Germans were up to and what the Nazis were thinking.

The information requests the Germans made of their agents were also quite revealing, as they indicated what they didn't know, and what they were interested in knowing more about. As these requests were in support of upcoming German operations, they gave the Allies a good idea of what was going on inside the minds of the German leaders.

Best of all, although some Germans grew suspicious at times, the Nazi high command was never convinced that their espionage network was compromised to the extent that it was. The Nazis were not simply being blind, for the Gestapo (German secret police) and Abwehr (German military intelligence) were having some success in the occupied countries. But this success would have been much greater had the Allies not possessed XX and ULTRA.

The capstone of the XX effort was the deceptions the Allies played on the Germans in support of the invasion of Normandy in June 1944. These deceptions attempted to do nothing less than make the Germans believe that the Normandy operation was not *the* main invasion, but rather a feint. Given the size of the Normandy operation, these deceptions would appear an impossible task. But the Allies managed to pull them off, in no small part because of the "trustworthy" network of agents the Germans had in Britain. For over a month after the Normandy invasion, the Germans held back many of their strongest armored divisions so that they could deal with the "second" invasion. The second invasion never came because it was a deception, one of the most successful in military history. Thousands of Allied lives were saved and the defeat of the German forces in France came sooner because of this deception.

The XX system also paid many other dividends. It assisted in the war against German submarines in the Atlantic and saved the lives of many Allied agents on the Continent. But as valuable as the XX system was, it was kept secret for over twenty years after World War II. This was partially because of the Cold War, and the need to use some of the XX tricks against the Soviet Union. But the Communists were no slouches in the espionage department, and knew all about XX long before it was revealed to the public in the 1970s.

Where Are Those Damn Machine Guns?

While artillery causes the most casualties among the infantry, the machine gun comes in a close second. The infantry can't do much about artillery, but they can fight back at machine guns. If they can find the little beasts.

The modern machine gun was first introduced about a century ago, and during World War I (1914–1918), a multitude of techniques were developed to make the machine gun more effective, and less easy to find. The Germans were in the forefront of cultivating new methods to conceal their machine guns. One of their most effective techniques was to use oblique fire. Whereas most people think of a machine gun as something you fire straight at the enemy, the Germans came up with the idea of having the machine gun fire at an angle toward the approaching enemy. This was a clever, and simple, technique. Instead of having the machine gun fire straight ahead, it was fired at an angle. Thus all the machine gunner had in front of him was a mound of dirt. This technique required several machine guns cooperating in order to work properly. Since each machine gunner was not firing at enemy troops advancing directly at him, he needed another friendly machine gun to cover his front. This required well-trained and disciplined troops, which the Germans had. The psychological effect of this oblique fire was substantial. Enemy troops were not as well prepared for fire from the flank. People, even soldiers, tend to look forward when advancing. To add to the confusion, when some of the advancing troops are hit, it is not automatically obvious which direction the fire came from. The German machine gunners were trained to fire a short burst (a dozen or so bullets) and then cease fire. The confusion generated among the advancing troops becomes yet another asset for the defender.

Troops with some experience fighting the Germans were onto the oblique-fire routine. But to make things a bit murkier, the Germans didn't always use it, or didn't always use it in the same way. But many Allied troops encountered German machine guns for the first time, and several times thereafter, without a clue. It was a deadly experience encountering a machine gun firing from some location you couldn't quickly identify. The American Army didn't take up this German tradition until the 1970s. Which just goes to show you that a deception can still work against you after you have first encountered it.

Running with (Almost) No Lights

When an army is faced with enemy air superiority, it is often not worth the effort (and losses) to move supplies by truck during the day. So the driving is done at night. But driving at night requires that drivers be able to see the road. If headlights are used, trucks can be seen from a great distance and hit by enemy artillery or night bombers. There is also the possibility of ambush if enemy troops or guerrillas are in the area. While the ambushers can hear the trucks coming, the brighter the headlights, the more accurate the enemy fire will be and the farther away the vehicles will be spotted.

The solution is to reduce the amount of light used by the trucks to the point where the drivers can find their way, but the enemy cannot easily spot this activity. By trial and error during (and after) World War II, several techniques were developed. The obvious method was to turn the headlights off. Under certain conditions, this worked. If it was a clear night with a full moon, and you weren't driving through a forest or jungle road, there was enough light to travel at a reasonable speed. In any event, military trucks travel at a slower speed than commercial vehicles, because the troops usually move in convoys. There is a certain amount of discipline in this kind of driving, and this translates into lower speeds (generally twenty to forty miles per hour).

If sufficient moonlight isn't available, the alternate movement is to have the trucks led by a man on foot. This is quite slow, even if the human pathfinder is moving at a brisk pace, the convoy won't be moving more than five miles an hour. As the trucks will still be driving in convoy, the vehicles will close up so that each can see the one in front of him even in the dark. A variation on this is to have a man sitting on the hood of the lead vehicle, signaling to the driver as the turns in the road appear. This is still pretty slow—ten to twenty miles an hour. The "man on the hood" also provides someone to look out for booby traps or mines if that is a danger. Neither the pathfinder nor the guy on the hood is really a practical solution, just an alternative when you have exhausted all the others.

A more practical solution was to put a device on the headlights to vary the amount of light displayed. This allows for two more modes of light: dimmed and blackout. The dimmed mode lets only a small amount of light shine on the road. This helps the driver, and provides much less light to be seen from the air or ground. Blackout mode allows an even smaller amount of light to shine through (usually by using an electrical device that reduces the current going to the light in the first place). The brake lights in the rear of the vehicles are dimmed in a similar fashion. These devices limit the speed that vehicles can safely travel, and increase driver fatigue (and the incidence of accidents). But with either device, near-normal driving speeds can be maintained. The main difference is that blackout lights cause more driver fatigue and accidents than dimmed lights.

A more recent gadget is the underbody light. This is a light under the truck, which can be seen by the truck drivers, but not much by anyone else on the ground or in the air. The underbody light obviously requires more work to install than the dimmed and blackout lights. The big advantage of the underbody light is that it is nearly invisible from the air. It is not a replacement for the dimmed or blacked-out headlights, but for the stoplights and turn indicators on the back of a vehicle. While these are visible for up to eight hundred meters, they don't show much more than a spot of light. The underbody light illuminates a portion of the road under the truck. The driver in the truck behind can get a good look at the illuminated bit of road for up to thirty meters. In this way, only the lead vehicle in a column need have its dimmed headlights on; the other vehicles in the column can follow the underbody lights. This is not an ideal way to drive at night, as you make only about ten miles an hour. The drivers have to stay very alert and will get tired quickly. But such a column is unlikely to be spotted from the air, and only enemy ground troops within half a mile are likely to notice the slow-moving column.

The observation range of the various lights shows the value of each form of deception.

Maximum Distance Lights Seen

TYPE OF LIGHT	FROM THE GROUND	FROM THE AIR	ROAD SPEED (in mph)
	(all distances in meters)		
Undimmed	8,000	10,000	40
Dimmed	2,600	3,500	15
Blackout	800	2,200	12
Underbody	1,200	0	15

The above chart indicates the maximum distance at which the various types of driving lights can be seen from the air and the ground, as well as the maximum speed the vehicles can travel at night with the different types of lights. Distances are in meters; road speed is in miles per hour. The actual distances and speeds will vary according to local conditions (amount of moonlight, weather conditions, what shape the road is in, skill of the drivers and air and ground observers, etc.).

The deception involved with vehicle lights was to minimize the chance of trucks being seen by the enemy while at the same time keeping the convoys moving at maximum speed. For the last sixty years, it has become commonplace for combat units to do most of their moving at night. When dawn comes, these units are not where the enemy thought they were the day before. Typically, these night moves would bring troops up to the front and position them for a surprise attack. Most of the successful offensives in the last sixty years have used this "move by night, attack by surprise" approach.

Even if you have air superiority, you cannot rule out enemy air recon flights at night. In fact, when one side is outgunned in the air, it will fly most of its recon missions at night to increase its chances of success. Not only will these recon flights more likely be at night, but many of the aircraft will be flying low—slowly and quietly to avoid detection. Thus the above chart takes on a more tangible meaning. The lead truck in a convoy, moving along with blackout lights, can still be seen twenty-two hundred meters away, viewed from the air. This is not a certainty, just a probability. But the lone truck with blackout lights (followed by, say, twenty other vehicles using underbody lights) is far safer than a column of twenty vehicles moving along with undimmed (normal) headlights. This spectacle can be seen ten thousand meters away, and there are twenty vehicles to spot rather than one.

By the end of World War II, all experienced armies knew the value of sneaking around at night with the headlights dimmed. It was a cheap and extremely useful deception.

In the last twenty years, sensors have become vastly more sensitive. Just dimming your headlights will not hide you from airborne scouts. But these new sensors are more expensive and not everyone can afford them. Indeed, the nations that can afford these devices are the ones that still can use blacked-out truck convoys to best effect. The less well-equipped nations are unlikely to have the new sensors, and thus likely to be deceived by enemy truck convoys running around at night with dimmed lights.

Appearances Were Deceiving

A major problem during strategic bombing campaigns was finding out how much damage, if any, was inflicted. The Germans first stumbled on this problem during the Battle of Britain (summer 1940). The British used several deception techniques to deny the Germans accurate data on what damage they did, or did not do, to British bases and factories. But it was during the far more massive bombing of German factories later in the war that this "target deception" reached new heights in trickiness.

One could not get a good idea of damage during the bombing itself because of all the smoke and, quite frequently, bad weather conditions. The Germans learned to use smoke generators in target areas to make the Allied view of their targets even more questionable. Thus it was common practice to send single reconnaissance aircraft over the target a day or two later, when the smoke had cleared, to take photos of the damage. For most of the war, it was assumed that if a factory building was hit with bombs, and especially if the roof appeared to have caved in, the factory was destroyed. After the war, it was found that this was often not the case. Many types of factory equipment (sturdy stuff like machine tools, presses, etc.) were not destroyed when the roof was blown down by bombs. In addition to keeping this development a secret from the Allies (lest bombers return to finish the job), the Germans would often repair the factory so that it looked like it was abandoned or still damaged. The debris was removed from the usually undamaged machinery and a false (apparently damaged) roof was erected. Thus it appeared, from the air, that the factory was still wrecked. But beneath this masquerade, production went on. Even when this deception was not carried out, the machinery was removed intact and installed in another (often underground) location.

The Germans also learned to camouflage their factories. You might think that a factory, especially one comprising a large complex of buildings, would be hard to conceal. Well, it would be hard to conceal completely, but not so difficult to change its appearance in order to make it look quite different from the air. Using paint, canvas, and even shrubbery, it was possible to make a factory appear as something else. This would often be enough to fool daylight bombers, and make them drop their bombs on something else. Some of these deceptions would merely change the shape of buildings, roads, or even rivers. At other times, false buildings would be constructed, to draw the bombers away from real targets.

The Germans also discovered, before the Allies, what the key targets were in a factory complex. In most cases, these crucial areas contained equipment that fed electricity, natural gas, or petroleum to the complex. Should these utilities be hit, the entire complex would be out of action due to lack of energy. Thus the Germans learned to create false pumping stations and electrical distribution sites. Eventually, the Allies caught on and these false utilities were sought out by the Allied bombers and hit. The real items were hidden and covered by much concrete and steel. It wasn't until after the war that the Allies realized how well they were being deceived in this area. The German plants were able to keep producing even when, from the air, they appeared to be a smoldering ruin.

Often, exceedingly simple deceptions were quite effective. Everyone was wont to use smoke generators to spoil the aim of high-level bombers. While the smoke could only be generated for a short time, with a proper early warning system, you could accurately predict when the bombers would be overhead and have the smoke generators going in time. The smoke could also be used to deceive the recon aircraft long after the bombers were gone, to make a functioning or lightly damaged factory look like it was still burning, or at least smoldering and unused.

These techniques were a major reason why, despite heavy bombing throughout 1944, German production peaked at the end of 1944. It was only after the war, when many German factories, and their records, had been examined up close that the Allies realized how successful the factory deceptions had been.

Deception via Hand Delivery

The last German offensive of the war (the "Battle of the Bulge") in December 1944, came as a surprise to the Allies. It shouldn't have, but it did, and the primary reason was a German deception that the Germans themselves were not aware of.

Throughout the war, the Allies had been able to read most of the coded German radio transmissions. German codes were thought to be unbreakable, at least by the Germans. The Nazis never suspected, or at least never seriously suspected, that the Allies were reading these top-secret coded messages. But the Allies did read the Germans' mail, so to speak, and were able to ascertain enemy plans in time to do something about it. The Allies used this information carefully, but it was still a key advantage. But in one spectacular case, this code breaking didn't work.

On December 16, 1944, the Germans launched their last major offensive of the war. Sending three armies through the lightly held Ardennes region (southern Belgium), the attack came as a complete surprise to the Allies. Although the Germans were stopped after two weeks of hard fighting, the Allies were understandably shocked. It was not just a failure of Allied code breaking (which most Allied generals didn't even know about), but an impressive bit of across-the-board deception for the Germans.

Naturally, the Germans wanted to achieve surprise, and they used many deception techniques to do so. Allied code breaking was foiled simply because the Germans insisted that all communications about the offensive be delivered by courier or passed over telephone lines. This was done largely because German practice was that if there was going to be an extremely crucial operation, any materials related to it were customarily transmitted by courier (often an officer) to add one more layer of security. Moreover, much of the hand-delivered and telephone communications were coded.

The Allies had become quite dependent on their code breaking by now, and never considered that the Germans would, for whatever reason, refrain from using radio communication for planning a major operation. The Germans had used the "hand delivered by officer" method in the past, but not frequently. Up until 1944, most German operations were against the Russians, so the Western Allies were not as acutely aware of all the German security techniques. As is often the case, the Allies discovered this one the hard way.

The Allies also suffered from a bit of self-deception. By late 1944, they thought the Germans were on their last legs and no longer capable of a major offensive. Moreover, a German offensive didn't make sense. If the die-hard Nazis wanted to go down fighting, they would be best served by conserving their dwindling forces for defensive actions. But just because the Allied generals didn't think an offensive was possible, or rational, didn't mean the Germans wouldn't try it. If the Nazis were fanatics, as the Allies believed, then fanatic gambits should have been considered.

The Germans used just about every deception measure in the book. Units gathered for the offensive were moved at night, and carefully brought up to the front line. Electronic deception was used, with groups of radio operators far away from the Ardennes sending out typical radio messages indicating that the units massing in the Ardennes were elsewhere. Until just before the assault, the second-rate German units holding the line in the Ardennes carried on as if nothing was happening, or was going to happen. So quiet was the hundred-mile-long Ardennes front that the Allies were sending battered units there to rest and recover. Because these Allied units could offer less resistance, the Germans were careful to do nothing to alarm them.

But enemy deceptions didn't stop there. The Germans also formed a special unit equipped with captured American vehicles, uniforms, and weapons. Each group of soldiers in this "deception unit" had among them someone who spoke passable English and would deal with any American troops they met on their way to Allied rear areas. The idea was to let these German troops in American uniforms loose once Allied lines were blown open. It was common, during World War II, for the front line to become quite fluid during an attack, and units from both sides would often wander around lost and seeking directions. The Germans dressed as Americans would head west, act as though they were lost, and attempt to spread rumors, commit sabotage and, in general, do as much damage as they could.

The Germans did achieve a high degree of surprise, and for over a week, they moved forward, creating the "bulge" that gave the battle its name. But the Allies were too strong and the Germans were not what they were in 1941. The offensive failed, even if most of the deceptions did not.

Oh—and the Germans dressed as Americans? They didn't do too badly. Most of them really did get lost, or were not believed and had to fight it out anyway. But the word got around that there was a bunch of Germans driving around in American uniforms. Sentries at roadblocks were ordered to question closely all vehicles they encountered. It was thought that a German who spoke good English would not be up-to-date on certain crucial aspects of current American culture. So the sentries often asked questions about baseball. And for the next month or so, many Americans who were not big baseball fans were mistaken for Germans who spoke good English.

Freezing Off Your Deception

Winter weather provides obstacles, and opportunities, to those who would execute deceptions. The most obvious deception involved having the troops wear white, and painting the vehicles white. As deceptions go, this one is pretty effective. But a lot more deception is required if you are going to survive winter warfare. While the white camouflage works nicely against a background of snow, anything that isn't white shows up starkly. Thus while it's easy to hide in winter, it's even easier to stand out more starkly than in any other season.

Winter deceptions involving camouflage require a different bag of tricks than in warmer weather. While the troops and vehicles can move around in their white attire, they will also leave rather obvious tracks in the snow. This is true even on roads, which in other seasons would not be so obvious. With a lot of effort, you can have the troops use branches from trees as brooms to cover over your tracks, but this isn't always practical. Covering your tracks is pragmatic if you are moving a unit into an area that provides some concealment (evergreen trees or a built-up area), but not if you are moving a long distance. Nature sometimes does the job for you by providing winds that blow snow over your tracks, or simply lays down a few more inches of new snow.

Fortifications and areas where troops are living show up clearly from the air in winter, unless precautions are taken. Troops should take to the forests as much as possible, digging in along the edge of evergreen forests, if these are available, or within built-up areas. Buildings are an obvious target, but at least the troops have an opportunity to conceal themselves and force the enemy to bomb all built-up areas if they want to hit the troops. This only works if the usual deceptions are used when moving into a built-up area. Vehicles have to be hidden, and fortifications made to blend in with the surroundings. Being able to do this consistently and effectively separates the amateurs from the masters of deception. Both Soviets and Germans became quite good at this during their World War II battles. Those who were not good at it didn't last long. Doing it right meant moving at night, having scouts going ahead to select hiding places, and finally, getting all the troops and vehicles under convincing cover before first light.

Silent Steel Giants

Tanks are often thought of as large, loud, and impossible to hide. Such is not the case, much to the consternation of those troops who have had a tank sneak up on them. While tanks are large, they are not so large that they cannot be hidden behind buildings, depressions in the ground, or among trees. While they have large engines, these power plants (especially the diesel ones most tanks use) are not that noisy, unless the driver is gunning the engine. The one item on the tank that makes the biggest racket (aside from firing its main gun) is the running gear. This is the system of wheels and tracks that the tank moves about on. This mélange of parts does make distinctive squeaking and clanking noises when the vehicle moves. But if the tank is moving slowly, there is much less noise from this quarter. On your typically noisy battlefield, it is quite common for tanks to creep around unnoticed. Experienced tank crews know of this capability, and strive to use this advantage on enemy troops who haven't yet learned to be alert for the silent steel giants.

Save the Banzai for Last

The image most people have of the Japanese Army during World War II is a mass of fanatical guys launching a hopeless charge led by officers waving *katanas* (samurai swords). This was the banzai charge, and these attacks did occur, and made a vivid impression on defending troops. But most of these attacks failed. Not only that, but the zealous banzai attack was not a pillar of Japanese infantry tactics, but rather an expression of frustration and despair. When the Japanese launched a banzai attack, they were announcing that they were in a hopeless situation and would rather die fighting than surrender. It took a while for Allied troops to realize this—not that it lessened the terror felt when on the receiving end of one of these assaults.

What the Japanese *did* use as their standard infantry tactics emphasized deception, not straight-ahead assaults. The Japanese were very fond of night operations and much camouflage. They believed that a battle won by stealth, subterfuge, and surprise was more worthy than a victory soaked in blood. The Japanese Army trained hard, and a lot of that training was at night. Of all the world's armies, none spent as much time training at night as the Japanese. Again, no other army was as strict as the Japanese, who enforced an extremely strict discipline in the ranks. Moreover, the troops tended to be enthusiastic about their soldiering. The Japanese made death and suffering a patriotic duty, and most Japanese got into the spirit of it all. So when it was time to pull off a deception, Japanese officers could depend on their troops to perform their tasks diligently, even unto certain death. The troops were hardworking, and did not think it unusual to be literally worked to death. The extent and thoroughness of Japanese fortifications was indicative of this. Not only were Japanese fieldworks stoutly built; they were generally very well camouflaged.

What made the Japanese somewhat less than supermen on the battlefield was a lack of logistical support, and overconfidence resulting from years of fighting the rather lackadaisical Chinese. Their early victories over the British also led them to believe that Western troops in general would be pushovers. Although the Japanese had been clobbered by the Soviets in two 1939 border battles, they excused this away as a "special case." Once the Japanese had acquired some firsthand experience with American soldiers and marines, they shaped up fast.

Their logistical problems were more intractable. The Japanese depended on superhuman effort from their troops, because Japan went into World War II a poor country—and things went downhill after Pearl Harbor. Japanese soldiers never had enough ammunition, food, or fuel. Thousands starved to death on Guadalcanal, for example, and many more suffered the same fate elsewhere. Yet, there were never any major discipline problems. The starving soldiers simply kept going until they dropped. But most deceptions do not require a lot of logistical support. In the Pacific jungles, there was ample foliage to use for camouflage. Operating at night was second nature to Japanese soldiers. If anything, Japanese skill at deception was one of their more potent weapons. During the war, Americans called the Japanese soldiers "sneaky." The Japanese would have considered that a compliment, as would any competent soldier.

But when all their deceptions and carefully prepared maneuvers failed, Japanese officers and soldiers were expected to go down fighting. Surrender was not something covered in Japanese training. Surrender was unthinkable. Enemy troops who surrendered were considered dishonored, and no longer worth the respect one gives a worthy foe. This is one of the reasons why the Japanese often treated prisoners so barbarically. For the Japanese, a hopeless situation required desperate measures, and surrender wasn't one of the options. The result was usually one final "give them all we've got" infantry assault. It was understood that this was going to be (literally) a do-or-die attack. Before the attack, officers and troops said their good-byes to one another, for none expected to survive. And few did.

The Deceptions of Pearl Harbor

For America, World War II began with a brilliant subterfuge. The Japanese attack on Pearl Harbor could not have succeeded without a series of successful deceptions. The Japanese were known to believe in starting their wars with surprise attacks. The Japanese were great believers in surprise, generally. This made any attempt at surprise more difficult, but the Japanese saw this as a worthy challenge. Moving a fleet across the Pacific undetected was no mean feat. The Japanese did not normally operate at that great a distance. American admirals knew this, and the Japanese took advantage of it. The planners of the attack realized that the distance problem could be turned to their advantage. The shortest route that could be taken covered some of the stormiest, and least traveled, waters in the world. This route would be a key element in the Pearl Harbor deceptions.

The first step the Japanese took to pull off this deception was to examine the shipping lanes between Japan and Pearl Harbor. Sure enough, the route that saw the least shipping traffic ran from the Kuril Islands (in northern Japan) to Pearl Harbor. If the fleet steamed north of Midway Island before turning south toward Hawaii, they would traverse an area that was rarely traveled by any ships. The main reason this route was bereft of traffic was not just because it was out of the way, but because the North Pacific, like the North Atlantic, suffered from wretched weather. In the winter, the weather was at its worst. That was when the Japanese planned to pass through this stormy region. As a test, the Japanese sent a merchant vessel along the course in October of 1941. Not a single ship was spotted. But the fleet's passage would not be entirely uncontested. Storms were fierce, and bad weather was the norm. While none of the two dozen Japanese ships was lost during the trip, some were damaged and many crewmen were washed overboard. Refueling was difficult. It could be truly said that the most difficult portion of the Pearl Harbor operation was the struggle to get through the stormy North Pacific in one piece.

The Americans knew what the North Pacific was like in winter, but they also kept track of the Japanese fleet by monitoring their radio broadcasts, which also yielded the general location of the transmitter. Americans were also breaking Japanese codes, but that took longer. Monitoring the coded (and sometimes uncoded) radio traffic of the fleet was called "traffic analysis," and most navies used this technique to good effect. But the Japanese knew that and used it against the American traffic analysts. In November, the ships involved in the Pearl Harbor operation went silent. Other radio stations began transmitting the messages that the now-silent ships would normally send. Individually or in small groups, these ships moved from the Inland Sea west of Japan and concentrated in a Kuril Island anchorage. On November 26, the six carriers of the Pearl Harbor task force, plus their escorts and tankers, set out into the murky, storm-tossed North Pacific.

As expected, the Japanese task force was not spotted crossing the North Pacific. They encountered one ship, but it was Japanese. Maintaining radio silence, even the sailors in the ships were not told what their mission was until December 2 (while still thirty-two hundred miles from Hawaii). The Japanese refueled the next day, sent the tankers home, and then picked up speed to cover the remaining twenty-five hundred miles to their launching position. Right on schedule, the Japanese arrived at a point five hundred miles from Pearl Harbor at 9:00 P.M. on December 6. The next day was a Sunday, and that was yet another part of the Japanese deception plan. The Japanese knew that Sunday was the one day that the U.S. military was least prepared—especially during peacetime.

American recon aircraft were scouting out four hundred miles from Pearl Harbor, covering an arc ranging from northwest to south, the direction a weather-wise defender would expect a Japanese fleet to come from. But the Japanese were coming from the north, after traversing the stormy North Pacific, an area Japanese and American sailors generally avoided at that time of year.

Knowing what was in Pearl Harbor was a crucial part of the attack. The Japanese had to know which ship was normally berthed where, as well as details on American combat aircraft in the area and the location of military facilities. Moreover, they had to know when it was likely that the maximum number of ships would be in port. Ships at sea on training exercises would be impossible to find, and they would be alerted by the attack on Pearl Harbor. Only the ships tied up in port could be hit by the surprise attack. The Japanese collected the information using local agents, plus a naval officer sent in as a civilian to make sure nothing was missed. Photos were taken, and the movement and berthing routines of the fleet noted.

One important item that made the Pearl Harbor attack possible was the 1940 decision to move the Pacific-fleet battleships from San Diego to Pearl Harbor. This was intended to "send the Japanese a message." That it did, but the message was "bomb me." Both Japanese and American admirals still considered the battleship to be the key naval weapon. In 1941, the aircraft carriers were still thought of as just another support system for the battlewagons. The Japanese were thus looking for battleships, and they expected to find nine or ten of them, two or three carriers, and dozens of cruisers, destroyers, and submarines sitting in Pearl Harbor on a Sunday morning. Except for the missing carriers (which were off on various missions), they were not disappointed. Their spies had done their job well. So well, the Japanese were able to build a scale model of the harbor back in Japan to aid in the attack planning. The attacking Japanese pilots knew more about the layout of Pearl Harbor than most of the people stationed there.

One of the technical issues that lulled the Americans into thinking such an attack wouldn't happen was the depth of the water in Pearl Harbor. For torpedo-carrying aircraft to launch their weapons, the water had to be deep enough for the torpedoes to sink quite a bit before coming to the surface and moving toward their target. The water depth in Pearl Harbor (only forty feet where the battleships were berthed) was too shallow for any known torpedo. What the Americans missed, and the Japanese didn't, was that in late 1940, the British had used modified torpedoes (and bombing techniques) to overcome the same problem when they made an air attack on Italian ships in Taranto harbor. The Japanese duplicated British techniques with their own aircraft and torpedoes. This was all kept quite secret, and even the sailors involved in the project were not told what that particular operation was for.

Another technical problem—how to get at battleships that were berthed between another ship and the pier—was solved by turning some sixteen-inch naval gun shells into bombs. These shells were designed to devastate battleships if they came down through the deck at a sharp angle ("plunging fire"), and it was straight into the ships' thinly armored deck that the bombers would drop their modified shells.

While the Japanese pulled off a magnificent series of deceptions at Pearl Harbor, they also stumbled into one. Thinking that sinking battleships would cripple the American fleet, they later found out that they had actually rescued these battleships from certain destruction early in the war. Only two of the eight battleships at Pearl Harbor were a total loss. Three others were sunk in the shallow harbor, but were raised and restored to duty. The damaged battleships could not be used in the first few months of the war when Japanese carriers reigned supreme. Any of these battleships caught by Japanese carriers on the open seas would have been sent to the bottom, beyond recovery. This is exactly what happened to two British battleships on December 10. The six repaired U.S. battleships later returned to service and survived the war.

A larger feat of Japanese self-deception was their decision not to bomb the fuel reserves at Pearl Harbor or the ship-repair facilities. Hitting these two targets would have caused fleet maintenance to be moved back to the West Coast of the United States for over a year until the damage could be repaired. The Japanese failed to appreciate the logistical problems of fighting a Pacific war, and the importance of the fuel and repair facilities at Pearl Harbor.

Even though America was reading much of Japanese secret radio traffic, United States leadership deluded itself into believing that Japan would not attempt something as bold as an attack on Hawaii. It was an embarrassing bit of self-deception. But even more fatal was the Japanese self-deception on what targets to hit at Pearl Harbor.

The attack was magnificent but, like the Charge of the Light Brigade, it did not aid the attackers' cause. America was in the war, it was angry, it was forced to depend on the weapon of the future—carrier aviation—and, worst of all, the Japanese thought they had won a victory.

The Other Way to Hit a Beach

American marines stormed ashore; Japanese "marines" sneaked ashore. And there were good reasons for both approaches.

One vivid image many people have of World War II in the Pacific is U.S. Marines coming from the sea against fierce Japanese resistance. This was indeed often the case, and the marines had to use such straight-ahead methods because the islands to be taken were often small and usually heavily fortified. The Japanese also had amphibious forces, as well as specially designed amphibious shipping. They did not have an elite assault force like the U.S. Marine Corps (USMC), but they did have sailors armed and trained as amphibious infantry. These Special Naval Landing Forces (SNLF) made the initial landing, followed by regular army troops. The Japanese made several successful amphibious landings early in the war (and many before Pearl Harbor). But Japanese landing methods were quite different from USMC tactics. Japanese amphibious operations depended on deception to succeed. Japan was a poor nation, especially compared to the United States, and habitually sought the cheapest, yet still effective, way to accomplish military goals. Deception was, of course, relatively cheap if you applied yourself to it. Moreover, most objectives for Japanese amphibious assaults were not on small islands, but the mainland of Asia or larger islands. The Japanese would scout the enemy shore thoroughly and find the area that was least likely to be defended or reinforced. If necessary, they would make a feint or demonstration somewhere else to draw enemy troops away from the actual landing site. What the Japanese were looking for was an unopposed landing, and they usually got one. Once the SNLF troops were ashore, they would fan out and ensure that no enemy troops were in the area. Then, more numerous army troops would come ashore and proceed to carry out the mission (usually taking a nearby port or city). The Japanese used warships to provide fire support, and carrier aircraft to obtain air superiority and air support for ground troops. Since Japanese army aircraft had exceptional range, the Japanese were often able to use land-based aircraft to support landings.

But the key to Japanese landings was deception. When they were not able to use deception, they tended to have problems. An excellent example of this was the first Japanese attempt to take Wake Island from the Americans. The Japanese bombarded tiny Wake Island on December 7, and on the eleventh, attempted a landing. The garrison was not large, but they had some artillery and some marines. The Japanese were driven off. On December 23, reinforced by more warships and two aircraft carriers (from the group that had attacked Pearl Harbor), the island was hit by a lot more firepower. Japanese troops got ashore, and soon the island was taken. This taught the Japanese that, where deception could not be used, overwhelming force was the only alternative. But as the war went on, it turned out that the Americans had the overwhelming force. The USMC used it, and were never driven off a beach they had landed on.

ULTRA, the Grand Deception

During World War II (1939–1945), the Allies were able to read German secret codes, and thus set the enemy up for a monumental bit of deception. The code-breaking project was called ULTRA.

For a quarter-century after World War II, one of the grandest military deceptions ever undertaken went unknown except to the small group of people who pulled it off. The secret was ULTRA, which was the code word for the successful British effort to break secret codes the Germans used for messages they transmitted. Such secret codes had become big business in the twentieth century as the volume of daily messages needed to keep modern armed forces going increased dramatically. Every major nation took care to make sure their codes were difficult to decipher, and that the codes were changed regularly. But secret codes are never secret forever. Any code that man devises, others can decipher. Codes are still viable, because it can take weeks or months, and sometimes even years of effort by highly skilled people to figure most of them out. Since the volume of coded messages is huge, and there are simply few people who have a knack for breaking codes, your codes are unlikely to be broken in time to hurt you.

Due to a set of fortuitous circumstances, the British were able to get a head start on breaking German codes, and then followed this up by developing advanced (for the early 1940s) technology to break the German codes rapidly. The "fortuitous circumstances" began in the late 1930s, when the Polish secret service managed to obtain one of the German code machines (it was basically a typewriter with a "code wheel" and other electronics built in) and, most important, had a mathematician on hand who was able to figure out how the machine worked. The Poles did not share this information with Britain or France since they did not want to risk a leak, which would have antagonized the Germans. However, during 1939, the Germans made it clear that they were likely to invade Poland. So the Poles approached Britain and France and rather shocked the code experts from these nations with the extent to which Poland had cracked the "unbreakable" German ciphers. The Poles passed on their knowledge to the British before the Germans conquered Poland.

The Germans called their code machine "Enigma," and Polish experts had already figured out a lot about Enigma before they passed the torch to the British. The Poles had discovered that Enigma codes could be broken, and developed techniques that allowed it to be done quickly enough to be of use. This speed was of vital importance, as the problem with breaking codes is that you must decode a great many messages before getting a general sense of what the enemy is up to. And you must break codes quickly; otherwise, you are merely studying what the enemy has done, rather than what he is about to do. While reading "old" messages is of some use, it isn't nearly as decisive as reading current stuff.

In practical terms, the Enigma codes could not be broken; the Germans calculated that conventional decrypting methods would require millions of man-years to decipher their messages. In theory, they were correct, but what can be done can be *un*done. In the case of the Polish cipher experts, Enigma was undone at high speed. The Poles took advantage of the tendency of the Germans to begin many types of messages with the same words, and this proved a sufficient hook for a clever Polish mathematician to find a means to take apart the Enigma ciphers. Having obtained one of the Enigma machines completed the "keys" the Poles developed to unlock Enigma.

Although the Polish techniques were adequate in 1939, the Germans continued to enhance the complexity of their Enigma system. Earlier additions to the Enigma system had driven the Poles to continue devising equally clever techniques to keep up with the German cipher efforts. But eventually, much more powerful methods were required. The British found a novel solution by building one of the first modern computers. Although more mechanical than electrical, the machine was built for the sole purpose of speeding up the breaking of German codes. It wasn't much of a computer by modern standards, but it was much faster at processing complex data than humans, and it could be run around the clock. While human decoders did the delicate stuff, the computer was able to apply the brute-force calculations needed to finish the decoding.

The Germans knew that they could lose some Enigma machines to capture (as some were), as well as the seizure of code wheels and code books. What the Germans never realized was how thoroughly the Allies understood the Enigma machine, and how much mental effort Allied specialists applied to developing ways to make Enigma codes easier to decipher.

Although the Germans would change code wheels periodically, and thus slowed down ULTRA, they never realized that a technique had been developed to break their new codes regularly. If they had suspected the extent of the Allied effort, they would have developed a new coding machine which the Allies might never (or at least not quickly enough) get their hands on and find a way to decipher. ULTRA's formidable resources could still have been brought to bear, but much time (a year or more) would be lost.

However, as a practical matter, it would have been very difficult for the Germans to introduce a new code machine during the war. To do so would have required the rapid deployment of thousands of new machines, the simultaneous training of operators, and subsequent disruptions to all their operations. The Germans did not want to do this, which added to their desire to believe that Enigma was secure.

The deception that ULTRA made possible had to be handled very carefully, lest it become known to the Germans. There were many Germans who *did* suspect that their codes were broken, but as long as senior German commanders still believed that Enigma codes were secure, the Allied deception could continue.

The Allies maintained very tight security around their code-breaking operation, and no one outside ULTRA personnel or the most senior political and military commanders even knew ULTRA existed. Information gleaned from ULTRA was carefully passed to Allied commanders, who were simply told that it was obtained via the usual intelligence-gathering operations. Allied generals and admirals came to depend on these visits from higher headquarters. But, except in a few cases, they did not become overly dependent on ULTRA information. The one exception was naval forces. The war against German submarines ("U-boats") was a desperate affair until the summer of 1943, and ULTRA information played a large role in locating German subs so that convoys could avoid them, or Allied ships and planes could go after the U-boats.

The German Navy (*Kriegsmarine*) was more security-conscious than the Army (*Wehrmacht*) or Air Force (*Luftwaffe*). Although all three services used the same Enigma machine, the navy used all its features, even though this made more work for the operators. The Kriegsmarine's greater security consciousness arose from the fact that they were more dependent on radio transmissions than the Wehrmacht or Luftwaffe. The army and air force could use telephones to contact most of their troops; the navy could not. German submarines had to use radios, and had to be careful in sending messages, lest nearby Allied aircraft and ships use radio direction-finding equipment to locate U-boats. The Allies let it be known that their radio direction-finding equipment was first-rate and constantly being improved. This was a deception to prevent Kriegsmarine commanders from realizing that Enigma was being deciphered. But using extra code wheels and other optional features of Enigma made Kriegsmarine codes harder to crack. Fortunately, the Allies were able to capture navy Enigma machines several times during the war, and thus were able to keep up with new twists the German Navy added to Enigma use. Subs, weather ships, and other warships were a source of captured Enigma machines and code books. Between the primitive computers crunching away, the Allied cryptanalysts burning the midnight oil, and a steady supply of captured Enigma equipment, enough Kriegsmarine messages were broken quickly to win the U-boat war before Britain was isolated.

Thus armed with German orders transmitted to their U-boats, German wolfpacks (groups of U-boats organized for mass attacks) regularly found Allied convoys suddenly taking wide detours. Worse, Allied aircraft increasingly came out of nowhere to sink U-boats that thought they were off the beaten track. German admirals blamed all this on Allied radio direction finding, airborne radar, masses of antisubmarine ships and aircraft, and the fortunes of war. Some German officers blamed all of it on traitors.

The Allies knew that the Germans would suspect that Allied spies and German traitors were causing all these leaks. Therefore, an additional, and crucial, deception was run by the Allies to keep the enemy from figuring out that Enigma had been compromised. The Allies made a big deal about all their "spies" within the German armed forces and government. This was a bonus during the ULTRA deception, as the Germans saw ample evidence of the Allies knowing Nazi plans. Spies within Germany were a likely source, especially since senior Nazis were a suspicious bunch to begin with. But in reality, there were not that many Allied spies within Germany. The Nazi secret police were rather efficient, and very ruthless. Many loyal Germans were caught in Gestapo (secret police) efforts to root out sources of all the information the Germans were getting.

The German high command never suspected ULTRA existed, nor did they see through the many deceptions that flowed from Allied code-breaking efforts. Indeed, ULTRA was not publicly revealed until the 1970s.

MAGIC, the Other Grand Deception

Completely independent of the ULTRA code-breaking project was an American effort to crack Japanese ciphers. This project was called MAGIC. Unlike the Germans, the Japanese did not use an Enigma machine, but rather a similar, but less capable, device. They used older encryption methods which could be (and constantly were) cracked by rooms full of clever and hardworking specialists. This was what American cryptanalysts did in the late 1930s and throughout the war. As with ULTRA, MAGIC obtained timely results by reconstructing the Japanese cipher machine and automating the cracking of coded messages. The Japanese, as a standard military practice, changed their codebooks from time to time, and this slowed down the deciphering of their messages by the Americans for several months. But, aside from that, secret Japanese radio messages were not very secret during the war. Some Japanese leaders were suspicious that their codes were compromised, but the American use of MAGIC information for deception was done carefully, and the Japanese never took measures to change their cipher system completely.

MAGIC made possible two of the most crucial deceptions of the Pacific war. First, in June 1942, there was the ambush American carriers pulled off at Midway Island. A large Japanese fleet was headed for Midway, with the dual purpose of taking the island and drawing the remaining U.S. fleet from Hawaii for a "decisive battle." Because of MAGIC intercepts (Japanese secret messages decoded), the Americans knew of Japanese plans and set up an ambush with their outnumbered carriers. As a result, the Japanese were defeated. Having lost four of their large carriers, the Japanese were no longer able to do whatever they wanted in the Pacific.

The second deception involved yet another ambush. MAGIC intercepts informed the Americans of a visit to the Solomons area by Japanese navy commander Admiral Isoroku Yamamoto, one of the most effective commanders the Japanese had. He had attended college in the United States and understood America better than most other Japanese leaders. Although opposed to war with the United States, Yamamoto nevertheless presided over the string of victories Japan enjoyed in the first seven months of the war. He was known to be a superb leader and naval commander. While there was a risk that MAGIC would be compromised if Yamamoto's aircraft was attacked, the removal of Japan's best naval commander was considered worth the risk. So on April 18, 1943, long-range American P-38 fighters found and attacked Yamamoto's transport and its fighter escorts over the Solomons. The admiral died, and with his death, Japan lost a naval commander who could have led Japanese forces in a much more effective defense. The Japanese did not suspect broken codes, but rather bad luck.

There were many other lesser deceptions to come out of MAGIC. Knowledge of Japanese plans in many battles and preparations for battles enabled Allied forces to fight more effectively. MAGIC was particularly useful in the prosecution of the American submarine offensive against Japanese shipping. MAGIC intercepts indicated where valuable Japanese shipping would be, as well as the location of antisubmarine forces and other warships. Since subs depended on stealth and surprise for their success, MAGIC allowed American submarines to undertake their deceptions with maximum effectiveness and much less risk.

Loot as a Deception

During the last few months of World War II in Europe, the front line was particularly porous. Agents for both sides could cross over for purposes of spying or sabotage. The Allies, however, developed one deceptive ploy based on the shortage of motor vehicles in the German Army.

After the last German offensive (the Battle of the Bulge, December 16, 1944 to January 2, 1945), many Allied trucks and jeeps had been captured by the Germans and immediately put to use in Wehrmacht units. A lot of winter clothing was also captured, which was another item in short supply in the German military. American counterintelligence operatives used this situation to their advantage several times. All the Americans had to do for some mission behind German lines was to put one or more of their German-speaking agents into a jeep. Mud was splattered over the jeep's markings (the Germans usually painted over the original U.S. Army markings), and the agents put on American winter clothing (with unit patches removed) over their regular uniforms. If caught, the agents were not, technically speaking, spies (and subject to being shot outright), but prisoners of war (therefore, eligible for gentler treatment). Thus equipped, the agents would drive through a lightly guarded part of the front line and do whatever they had to do. This usually involved obtaining some information, or bringing back another agent or defector.

At this point in the war, it was common to see German troops driving around German-controlled areas in captured jeeps, and wearing U.S. Army winter clothing. If American agents were challenged by Germans, they depended on their ability to speak German fluently to get them past any identity checks. Because of the chaos in Germany during this period, American agents were usually successful. They were, in effect, able to move freely among the enemy using language and "captured" equipment as camouflage.

IN THE SHADOW OF ARMAGEDDON

Deception in the Era of the Cold War (1945–1980)

World War II brought deception to a new peak of technical perfection. The generation after, the one "tempered by a hard and bitter peace," as John F. Kennedy called it in his Inaugural Address, saw the use of deception become increasingly common not only in "real" war, but also in "cold" war, the political struggle between the Capitalist and Communist worlds which dominated international affairs until the late 1980s.

The deception techniques developed during World War II were recycled and improved throughout the Cold War. The Soviets, in particular, did much to advance the state of the art. In America, electronic deception came of age from its turbulent beginnings during the air war over Germany. Curiously, these substantial advances in deception technology, and practice went largely unnoticed by the public. This wasn't unusual, but it did leave American voters and taxpayers out of the loop when it came to understanding and appreciating the emergence of deception as a larger component of warfare.

Nor were the two "superpowers" the only practitioners of deception during the period. In fact, they were often the victims of clever dealing at the hands of relatively minor powers with whom they had become entangled. This is a useful reminder that anyone can come up with clever, effective, and winning deceptions, and that even the most sophisticated and powerful nations can become their victims.

A Half-Million Chinese Sneak into Korea

Deceptions come in many forms, and the most lethal is the one you create in your own mind. One such case was the intervention of the Chinese Army into the Korean War (1950–1953). Senior American commanders decided the Chinese would not intervene in that conflict, despite substantial evidence to the contrary.

It all began when Communist North Korea, at the urging (to put it politely) of the Soviet Union, invaded South Korea in the summer of 1950. American troops had been withdrawn the previous year. A year earlier, the Communists had won the Chinese Civil War and were, that same summer, massing troops for an invasion of the last Nationalist Chinese redoubt on the island of Taiwan. Europe was still in ruins, as was Japan. America seemed uninterested in foreign affairs and it seemed, to the Communists, that their time had come.

Well, not quite. But in late June, seven of North Korea's ten divisions crossed the border into South Korea. The South Koreans had no army to speak of, although eight divisions were in various states of training and formation. Few of the South Koreans had any combat experience, while over twenty thousand of the northerners were veterans of the Chinese Civil War. By August, the Communists had taken all of South Korea except for a few thousand square miles around the city of Pusan, in the southeastern tip of the peninsula. The world was aghast. The newly formed United Nations condemned the aggression, and American and allied forces thereafter fought under the UN flag.

The battle for Pusan put the North Koreans at the end of a long supply line, and the Americans continued to pour in reinforcements. But the Communists thought they had only to reduce the Pusan position and victory would be theirs. The Americans did little to disabuse them of this attitude, as U.S. troops struggled to hold the line and even attacked in attempts to extend what little territory they held.

Then, in one of the most masterful surprise attacks of the twentieth century, two United States divisions (one marine, one army) landed at Inchon bay (a hundred and fifty miles northwest of Pusan) in September. The landing area possessed one of the most treacherous tidal systems in the world, and was thought impossible for amphibious operations. But the U.S. fleet, still the largest in the world and manned by World War II veterans, was up to the task. The landing demoralized the North Koreans, and those fighting around Pusan promptly fled north. Most North Korean units didn't make it, although many individual North Korean troops did.

With nearly all of North Korea's armed forces destroyed, the UN commander, General Douglas MacArthur, of "I shall return" (to the Philippines) fame, decided to unite Korea, and moved his units north. In early October, he reembarked the two divisions that had landed at Inchon (and retaken the South Korean capital of Seoul) and moved them around the Korean peninsula to the North Korean port of Wonsan, far in the northeast of the country. Disembarking there in late October, these two divisions proceeded to overrun large parts of North Korea. Meanwhile, American and South Korean divisions advanced northward from Pusan and past Seoul. By early November, most of North Korea was no longer under Communist control, although the two UN forces were physically split as a result of the landing at Wonsan.

Then came another massive deception, this one partially self-inflicted. The Chinese Communists looked with horror on the demise of North Korea's army, and the prospect of American troops on the Chinese border. China felt she possessed certain traditional authority over what went on in Korea, and had many ties with North Korea. For example, the combat-experienced troops of North Korea's army were Chinese Communist army veterans. There was (and is) a sizable Korean population in northern China, and many of these people joined the Communists during the civil war. Korea had, for centuries, been something of a protectorate of China, and in the minds of the Chinese Communist leaders, this had not changed. Thus China warned that it would intervene if UN forces entered North Korean territory.

Chinese trepidation sprang from several sources besides the traditional relationship between Korea and China and a fear of Western troops on their border. First of all, the civil war in China had ended only a year ago. There was still a fair amount of chaos in China. A U.S. fleet steamed off the Chinese coast, protecting Taiwan, the last bastion of the losers of the civil war (the Nationalists). America was seen as ready to undo the Chinese Communist victory by force of arms. Well, maybe not. But the prospect of American troops lining the border between China and North Korea had a disquieting effect on Chinese Communist leadership. Moreover, these were leaders of a revolution that had just won a great victory. What better way to establish the major-power credentials of the new Communist Chinese government than to defeat an American army. Hadn't the North Koreans smashed American troops during the past summer? For any number of reasons, when the Chinese Communists said, "Don't cross that line," they meant it.

America had a different take on the Chinese Communists. The Cold War had just begun, and the recent Communist takeover in China had not gone down well in the United States. Back in Washington, the "who lost China" finger-pointing game was still going on. It was overlooked that Chinese Communists had taken control because they were well organized and competent. But most American experts derided Chinese Communist military prowess, as the Communists had sat out most of World War II while the Nationalists did most of the fighting against the Japanese. The Soviets, the conventional wisdom went, were pulling all the strings in the Far East and were the power behind any local Communists. Moreover, the Chinese Communists, according to all reports, were basically a large mass of riflemen. They had few heavy weapons and could not possibly stand up to a modern army. While many Americans (military and civilian) had seen the Chinese Communist military effectiveness up close during the 1930s and World War II, their expert opinion counted for much less than that of Douglas MacArthur. General MacArthur considered it an absurd idea that the Chinese would make good on their threat. He felt that the lightly armed Chinese, who lacked any logistical capability, would not be so foolish as to take on the enormous firepower of a Western mechanized army. MacArthur was wrong, very wrong.

Realizing that their admonition, "Don't cross that border," was being ignored, the Chinese began assembling an army on their border with North Korea in late September, soon after UN forces entered North Korea.

MacArthur was right about Chinese forces relying on light weapons and not having much transport. But the Communists had learned how to get around these deficiencies. And they used these techniques to spring a monumental surprise on the UN forces advancing north.

While MacArthur was willing to fight the Chinese, the United States government was not. So American reconnaissance aircraft were not allowed to fly over Chinese territory. If they had taken a look, they would have seen a lot of activity on the Chinese side of the border. Although it was known that the Chinese Fourth Field Army (over four hundred thousand strong) was in the process of returning "home" to Manchuria now that the civil war was over, little attention was paid to what might be passing over the Yalu River from Manchuria to North Korea. The Fourth Field Army was one of the more experienced and successful units the Chinese Communists possessed, having just won several key battles in the civil war. While MacArthur and his staff made up more reasons why the Chinese would not intervene, over half a million Chinese troops were being moved to and across the Yalu River, which serves as the border between North Korea and China. Starting in September, these troops were sent across the Yalu at night and instructed to move only at night and to be in camouflaged positions by first light each day. This was nothing new to the Chinese, as they had learned to operate this way in the face of Japanese air superiority from the 1930s until the end of World War II.

Because the Chinese didn't have a lot of trucks, the UN aircraft didn't see a huge increase in truck traffic south of the Yalu, and what they did see could be explained away as North Korean trucking responding to the oncoming UN forces. The Chinese tended to move practically everything at night, and they didn't have large supply requirements in the first place. Because their infantry divisions were almost entirely foot soldiers, they didn't have to move on roads, although they often did so because they could march faster that way. In this fashion, nearly half a million Chinese troops literally sneaked into North Korea. Their scouts noted where the advancing UN units were, and Chinese generals planned multiple ambushes to kick off their counteroffensive.

With that many Chinese moving into the paths of advancing UN units, it was inevitable that some Chinese troops would be captured.

MacArthur and his intelligence staffs dismissed these prisoners as Chinese serving with North Korea's army, or simply denied that they were Chinese. MacArthur knew that nearly all of the combat veterans in North Korea had served earlier with the Chinese Communists, and believed that these "Chinese" prisoners were simply Chinese-speaking Koreans. MacArthur refused to entertain the idea that a Chinese army was going to contest his march to the Yalu River.

During October, the number of Chinese soldiers captured increased, and many Chinese units were identified. MacArthur's intelligence staff, ever eager to back up their commander's point of view, decided that there were perhaps a division's worth of Chinese in North Korea. Disingenuously, the intelligence experts proclaimed that the Chinese had obviously sent into Korea a single battalion from each of several divisions. This was self-deception of the highest order. American intelligence had received considerable information during World War II of how the Communists operated against the Japanese. It was known how the Communists coped with constant Japanese air superiority, but this knowledge was consciously suppressed to conform with what senior U.S. commanders wanted to believe.

Some of the UN divisional commanders knew better; they believed that there were large Chinese units lurking in the vicinity. This was particularly true for South Korean officers who had served in the Japanese Army during World War II. Many of these Koreans had commanded units in Manchuria during World War II and could speak Chinese. These South Korean division commanders would often interrogate captured Chinese troops personally, in order to confirm what was becoming increasingly obvious. The South Korean officers recognized the presence of large Chinese forces in North Korea in early October. But senior American commanders would have none of it. Their minds were made up. The Chinese would not intervene. It was thought that any significant force of Chinese troops would be quickly spotted by U.S. reconnaissance aircraft, and just as promptly destroyed by American bombers. Unfortunately, American field commanders had orders from on high to advance with all possible haste, so there wasn't much they could do to prepare for the disaster that was to come.

It was ironic that less than ten years after the surprise attack on Pearl Harbor, another such attack was to be launched on United States ground troops in the hills of Korea. The same combination of arrogance and ignorance had reappeared to set the stage for yet another needless American defeat.

During October, the Chinese managed to make it painfully obvious to MacArthur that Chinese troops were in North Korea. What MacArthur didn't realize was that the Chinese were going to pull yet another deception on the UN forces. The Chinese launched a limited offensive on November 1. They wrecked several South Korean regiments and roughed up an American regiment as well. The units hit were approaching the Yalu River, and the Chinese attack was meant to push UN units back. This was accomplished by what amounted to a large-scale Chinese ambush. Then, as quickly as the Chinese attacks had begun, they ceased. The Chinese continued to pour into North Korea and, for the most part, remained hidden.

What the Chinese now planned was a trap, a "false retreat" on a grand scale. MacArthur fell for it. He ordered UN units to prepare to resume their offensive as a "reconnaissance in force." After a few weeks of resting and resupplying, on November 25, the UN troops cautiously advanced to find the Chinese and drive them out of North Korea. Many of the frontline UN commanders knew the Chinese were still out there somewhere in great numbers, and despite MacArthur's orders to plunge forward to the north, ordered their troops to pull back the moment they encountered any Chinese resistance. This was good advice, and risky for the careers of the UN commanders who gave it—but it wasn't enough.

Two days after the UN advance began, the Chinese sprung their trap. With about two hundred thousand UN troops advancing through North Korea, the Chinese began attacking with over four hundred thousand of their own. Moving cross-country from their hiding places in the mountains, and still taking care to conceal themselves as much as possible, the UN troops had the impression that Chinese troops were popping out of the hills themselves. The attacks were devastating. UN units were roadbound, as they were heavily dependent on motor transport. The Chinese could easily encircle UN units, attack from all sides at night, and quickly move on. While moving on foot, the Chinese moved south faster than the UN forces had moved north on wheels. A lot of this had to do with the relative paucity of roads in North Korea. But the Chinese were accustomed to operating in a roadless environment, and just kept going. In the annals of twentieth-century warfare, the Chinese advance toward South Korea was actually one of the swiftest any army, mechanized or not, had accomplished.

By early January, the Chinese were in South Korea, and recaptured the capital, Seoul. However, at this point, the down side of Chinese tactics began to manifest itself. Advancing on foot in winter without much logistical support, the Chinese began to suffer heavily from lack of food, ammunition, and other supplies. The Americans controlled the air, and Chinese truck and rail movements were subjected to attack whenever they could be found. Although trucks moving at night were largely safe, bridges and tunnels could be attacked during the day to create obstacles to nighttime traffic. While the Chinese eventually overcame these air attacks on supply lines, the weather and stiff resistance by many UN troops led to enormous casualties (over half the attacking force).

By late January 1951, UN forces were counterattacking. By March 1951, Seoul was captured again by UN forces, and the Communists were pushed back into North Korea. There the war stalemated for two years, until a cease-fire was signed in 1953. Forty years later, troops are still there, with the war itself never officially ended.

The Chinese continued to use deception during the two years of stalemated fighting. The UN never lost air superiority, and always possessed much greater firepower. Although the Chinese (and the revived North Korean Army) suffered five casualties for every UN injury, they were prepared to trade lives for the ability to withstand a better-supplied army. For the Communist soldier, his shovel and his wits were his best weapons. So these troops dug entrenchments, and camouflaged them. UN forces were never able to overcome this tactic.

After the cease-fire, and when the Chinese left, the North Koreans kept digging and practicing deception. The demilitarized zone that exists in Korea to this day is riddled with North Korean tunnels, and the Communist troops to the north still avidly, and effectively, practice camouflage and deception.

The Korean War goes on into the 1990s, as do the deceptions that made it possible in the first place.

Mister Charlie's Bag of Tricks

Vietnam was notable for the vast number of deception techniques successfully used by the Vietcong (VC, "Victor Charlie," or "Mr. Charlie") and North Vietnamese units. The American troops entering combat in 1964 and 1965 were fighting Vietnamese guerrillas who had over twenty years of experience in battlefield deception. U.S. troops never really caught up, even though the Americans had vastly superior firepower and mobility. In pitched battles, the Americans usually won because of these advantages, and the determination of the American infantry. But in the long haul, it was Mr. Charlie's bag of tricks that won the war, or at least made it difficult for America to do so.

The United States was actually fighting two rather different Communist armies. Until 1968, the main opponent was the Vietcong, who were South Vietnamese guerrillas reinforced by North Vietnamese troops. After 1968, most of the fighting was done against North Vietnamese regulars, who were assisted by Vietcong troops that had survived the 1968 Tet offensive, and American firepower in general.

Both of these Communist armies used a wide array of deception techniques to survive all that American firepower. During the year of heaviest fighting (1969), the million Communist troops were faced by 1.6 million allied troops (a third of whom were American). In addition to outnumbering the Communists, the allies had thousands of helicopters and bombers, plus thousands of artillery and mortar weapons in action. Aircraft dropped more bombs on Vietnam (4.2 million tons) than on Germany and Japan during World War II (3.4 million tons). The impact of this firepower advantage could be seen in the losses both sides suffered. The allies had 260,000 dead (mostly South Vietnamese, about a fifth American) while the Communists lost about a million. The Communists would have lost considerably fewer if they had had adequate medical resources in South Vietnam, but even their best deception efforts did not allow them to do more than set up crude trails to evacuate some of their wounded, and similarly crude hospitals (often underground) to treat the injured who survived the trip from the battlefield. Over half a million civilians also died from all that firepower, and many of these noncombatants perished as part of Communist deception efforts to protect their troops.

The Communists used the entire gamut of deception techniques, but the overwhelming favorite was good old camouflage and concealment. Against American firepower, the Communists quickly learned that what the Americans couldn't see, they couldn't hit. As American involvement in the war increased, more and more of the Communist infrastructure went underground. Supply dumps, headquarters, barracks, hospitals, kitchens, and just about every other

Communist facility was located wholly or partially in underground bunkers by the late 1960s. Since Vietnam has a tropical climate, there were always abundant opportunities for concealment in the lush vegetation. But this would not protect the Communists or their facilities from bombs or artillery. United States forces soon learned that their most important chore was to find out where the Communists were. Once that was done, there was ample firepower available to destroy whatever was found. Going underground, and procedures like ensuring that cooking fires were not smoky, added another layer of concealment and a needed degree of protection against the abundant American bombs and shells.

The Communists in their man-made caves were still vulnerable to weapons like B-52 carpet bombing and Fuel Air Explosives. Groups of B-52 bombers would release hundreds of five-hundred-pound bombs at a time. As these hit the ground and exploded, three things happened. The most obvious effect was the explosion; anything above ground was blown up. The second effect: multiple concussions and overpressure of these bombs, which would often collapse underground bunkers and/or kill anyone in them. The third effect was the mess it left upstairs. The Communists climbing out of their caves after a B-52 attack found their cover, quite literally, blown. If the forest was dry enough, there would also be fires. They might find their water supply damaged or ruined. But at least they had a fresh supply of firewood.

The Fuel Air Explosive (FAE) was an even nastier weapon. Basically FAE is a container (often weighing several tons) of flammable liquid that bursts open to create a large mist that is then detonated. The resulting explosion is enormous, and that alone does considerable damage. But two other effects are even more damaging. First, there is the "overpressure," or shock wave. It is powerful enough to kill any troops underground, if an entrance to their tunnel is nearby. The B-52 bombings also cause this effect, but not to the extent of an FAE. Yet another effect, unique to the FAE, is the removal of most oxygen in the area. This can kill or render unconscious any troops nearby, above ground or underground. Fortunately for the Communists, FAE was a fickle weapon. If the timing between the formation of the gas cloud and the detonation were off just a little, the explosion would lose much of its power. Normally, FAEs were dropped out of the back of transport aircraft, or from a sling under a heavy helicopter. Because of their unreliability, FAEs were used primarily for clearing helicopter landing zones in the forest. However, when there was an opportunity to use them against Communist positions, the effect was usually devastating. Somehow, by the time the 1991 Gulf War took place, the press had so completely forgotten about the FAE that when Saddam Hussein hinted he had such a weapon, there were hysterical comments by some journalists as to how he was obviously ahead in military technology.

In Vietnam, American tactics soon came down to finding the Communist bases, particularly in the late 1960s when large (battalion and regiment) Communist units were operating in the forests and mountains. These units needed support (food, ammo, medical) and much of this support came from underground complexes hidden in the forest. High tech soon entered the fray for the Americans as they began to use infrared sensors to detect the heat escaping from underground kitchens. The Communists had to cook their rice, and a dozen or more fires (each cooking enough rice for forty or so troops) could be spotted from the air if the airborne infrared sensors flew within a few miles of the escaping heat. It was not uncommon for infrared sensors to spot the heat, and it then took a day or so to get a flight of B-52s over the area. Only the B-52s could really pound these underground bases effectively, and often, the bombing was off enough to cause little damage, or the Communist troops had moved on.

While camouflage and concealment were the first lines of defense against American firepower, the Communists also employed the entire range of deception measures not only to protect their troops and bases, but also to facilitate their military and political operations in South Vietnam. Following are some measures that were not used, as well as those that were.

- False and planted information were not widely used by the Communists, mainly because they had a very dispersed organization, and most of their units operated independently. There were no reliable communications with which to coordinate such deceptions. It was the Americans who planned most of the major operations, mainly because they had much better transportation capability. In addition to complete control of the air, the allies had much more access to the road system than the Communists.

- Ruses were used, especially against the South Vietnamese troops. Throughout the war, two thirds of the allied troops were South Vietnamese. While a lot of the fighting was left to Americans, the South Vietnamese units were more numerous and vulnerable.

The Communists knew that if they could hurt the South Vietnamese troops sufficiently, Americans would have to be brought in to take up the slack. And this would mean fewer Americans out in the bush looking for Communist base complexes. Sharing similarities in appearance and language (although North Vietnamese have an accent perceptible to southerners), it was still relatively easy to use ruses against the South Vietnamese. Ruses were most often used in support of commando operations against either American or South Vietnamese troops. Communist troops passing themselves off as South Vietnamese soldiers while staging a commando attack on U.S. bases did little to improve trust between American and South Vietnamese forces. There was little trust to begin with because of the rampant corruption in the South Vietnamese military, as well as the large number of Communist agents who had infiltrated all levels of the South Vietnamese government.

- Displays were not employed that much by the Communists, primarily because the Americans had sufficient firepower to shoot at anything that remotely resembled Communist troops or bases. But because Communist troops spent more time using their shovels than their weapons, there was no shortage of unused Communist bases for the Americans to bomb or shell. The use of multiple, fortified (usually underground) camps was a sound practice. It was not safe for Communist units moving around without an underground hiding place at the end of their march. It was also customary to move out of the way when American ground forces came near. Better to have an abandoned base found and destroyed than to see the troops that normally occupied it killed as well. And the Americans did find many abandoned camps, and bombed even more that were occupied when first spotted, but were empty by the time the bombers showed up. Realizing this, the Communists would sometimes even construct false base camps, if only to distract approaching American ground forces while Communist troops in the area were able to go about their business undetected.

- Demonstrations were rather more difficult for the Communists to pull off. Displays of unoccupied (or even fake) bases were one thing; exposing troops in a demonstration was rather more expensive. It was easier to build new bases than to bring new troops south, or recruit them locally. Despite heavy losses that the Communists took, they were, under the circumstances, quite frugal with their manpower. When necessary, they would send a small unit off through the bush to distract American troops, but the decoy unit would usually only attract a lot of firepower. Sometimes this was useful, if the weather was bad and United States aircraft were largely grounded. American artillery was less flexible than the bombers, as the guns had to move to within range of their targets. Artillery ammunition was also less abundant than bombs stacked up, often right off a ship, at some coastal air base. When demonstrations were used, they could be quite effective. The Communist infantry was expert at moving unseen through the forests. Normally, few men were spotted. If the small demonstration units were instructed to be a little sloppy, what the Americans saw would indicate a much larger unit than was actually present.

- Feints were popular when attacking American base camps. These installations were usually surrounded by barbed wire and minefields. But the garrisons could not be thick at every point of the perimeter line. A feint at one portion of the perimeter could draw much of the garrison, making it easier for the main attack force to break in somewhere else. Feints were sometimes used in the field, particularly when the Communists were trying to stage an ambush. Typically, one American unit would be engaged, and another Communist force would lie in wait across the most likely route that an American reinforcing unit would travel if it was coming overland. Although the United States had thousands of helicopters, there were never enough to support every ground operation all the time. Bad weather could also ground helicopters, or greatly limit their effectiveness. And often, the reinforcing unit could not be landed nearby the friendly unit it was rescuing. The choppers would disgorge their troops some distance away, and the final approach would be made overland, often through the area where the Communist ambush was set up.

- Lies were more part of the Communists' political operations. Communist insurgency in South Vietnam began in the late 1950s. Since many South Vietnamese had fled the north in the mid-1950s to escape the Communists, reunification under Communist rule was initially a tough sell. The Communists had an even harder time once American military and economic aid began to turn the tide against the South Vietnam unification forces (the Vietcong). The North Vietnamese began sending more of their own troops south (as guerrillas). Then, American troops entered the fray in 1964, which was the best thing that could have happened for the Communists. Now the Communists preached a "united front" of opposition to the "alien invaders," and political freedom after the foreigners were thrown out. That ("political freedom") was a flat-out lie, and it was only the first of many that were told to further the Communists' military agenda in the south. Initially, most of the fighting against the Americans was done by the South Vietnamese Vietcong. The North Vietnamese Communists convinced the Vietcong in late 1968 that the people were ripe for a mass uprising, and that an all-out offensive would win it all in one dramatic operation. The subsequent 1969 Tet offensive was dramatic, and a decisive defeat for the Vietcong. But the Communists kept lying, telling the world, their people, and the South Vietnamese that the Vietcong defeat was actually a victory. It was a victory for the North Vietnamese, as they could now take over operations in the south, so heavy were the Tet losses of the Vietcong. When South Vietnam fell in 1975, it was to North Vietnamese regulars, not Vietcong irregulars. And after the victory, the south found itself subordinate to the north, not reunited with it. And that was all done with lies.

The Communists had one tremendous deception advantage over their opponents, and that was the presence of spies and sympathizers within the South Vietnamese government and army. The Americans hired thousands of South Vietnamese to work at their bases and headquarters, and many of these turned out to be Communist agents or sympathizers. This was to be expected during what was, essentially, a civil war. It made deception difficult for the allied forces, and considerably easier for the Communists.

Foxing Mister Charlie

While the Vietcong and North Vietnamese made much use of deception, American troops also wrote a few new chapters (and drew heavily from old ones).

Most U.S. combat troops were conscripts, or short-term volunteers. This meant that Americans generally did not have the combat experience needed to deceive their veteran adversaries. But this was not always the case. A sizable minority of American troops (up to 20 percent) were veterans and long-term professionals. Such men were experienced and quite expert. These soldiers were often concentrated in special units that proceeded to beat Mr. Charlie at his own deception game. Experienced outfits were divided into two classes. The largest group consisted of special reconnaissance units, while the smaller number of elite troops were in Special Forces (mostly Army Green Berets, but including Navy SEALS—sea, air, land team).

The recon units were found in all American divisions. Some divisions had more recon troops than others, but all of these soldiers were highly trained, often quite experienced, and always highly motivated. The principal function of recon troops was to gather information about the enemy. To do this, they usually went in "Charlie's backyard." That is, they either went in on foot or via helicopter and looked around. Any enemy troops encountered were avoided and, if that wasn't possible, the Communist soldiers were engaged and, usually, wiped out. American recon troops took light casualties, especially considering the close proximity they shared with the enemy. To accomplish their mission, and avoid injury, U.S. recon units used a lot of deception. Among the deception techniques used were the following:

- Don't be where the enemy expects to find you. Only well-trained troops were able to travel off the trails and get where they were going. Special recon troops were able to do this, and the usual American infantryman was not. It took a lot of practice, and a good sense of direction, to move off the known trails. Troops had to be able to move through the bush without making a lot of noise and, most important, not get lost. They practiced silent (hand) signals so that when the enemy was encountered, no talking was necessary. This particular deception was vital. The Communists weren't expecting U.S. troops in their own territory, so this ceded the element of surprise to the Americans. As always, surprise is the most effective deception in combat.

- Be prepared. Recon troops traveled light, and often at night. When moving about in darkness, dark, camouflage face paint was used, as well as the usual camouflage techniques. Day or night, recon troops were difficult to see, even if you were looking at where they were in the bush. Whenever American troops stopped for the night, they, in effect, went to sleep in the form of a 360-

degree ambush. Their campsites were carefully selected to be out of the way and easily defended. Someone was always awake, and everyone knew what to do if they were wordlessly awakened in the middle of the night. An escape route was always available, and would be used rather than putting up a fight. If, in the course of the recon patrol, there was a fight, Communist troops from miles around would begin looking for the Americans. During the late 1960s, the Communists even formed special scouting units to chase after the increasingly effective U.S. recon units. Usually, these Communist scouts got more than they bargained for. American troops were trained to detect someone following them. They would often split their force and, while one group continued on, the other would double back and try to ambush their pursuers from the rear. Sometimes a plain old ambush would do, but it was more effective if the Communist trackers thought they were now dealing with two groups of Americans. The Communists would usually go after what appeared to be the larger group, until they discovered that the smaller group was stalking the trackers. The Communists eventually realized that special "antirecon" units would not be able to stop the Americans. By the end of the Vietnam War, the Communists accepted the fact that they were not safe in their own rear areas, and this had the additional effect of making life even more miserable for the North Vietnamese.

- Fake landing zones were set up. America had helicopters, and the Communists did not. Thus it was up to United States troops to devise deceptions that involved helicopters. The most common one, especially for recon troops going deep into enemy territory, was the old "LZ Fake." When helicopters flew into enemy territory to drop off a patrol, there was no way to hide the presence of the copters. They were noisy and, no matter how low they flew, they were up there for all to see. Normally, a patrol went in with four choppers. Two were cargo carriers (containing the troops), and two were gunships to provide firepower as needed. Normal procedure was for the landing zone (LZ) to be checked out by the gunships (they had some armor) and, if all seemed quiet and unpopulated, one of the cargo copters (usually a UH-1 "Huey") would land and unload its troops. Once on the ground, these soldiers spread out and made sure that the LZ was indeed devoid of enemy troops. If such was the case, a signal was given for the second cargo chopper to land its troops. The Communists knew this drill, and when one of their people on the ground saw a landing in progress, they would get the word out as quickly as possible. Communist base camps had radios, although they were not usually carried around by smaller ground units; thus it often took several hours before local Communist troops were aware of the Yankee interlopers. Then, Communist troops would be converging on where an American recon patrol had landed. This was worse than landing in a "hot" LZ (where the enemy was firing at you), because you could pull out the troops that landed and go somewhere else. Once American troops were on the ground and the choppers were gone, Communist ground troops became the major menace. The Americans would not know if they had been spotted coming in—at least not until later that day, when they found the area swarming with Communist troops looking for them. So someone came up with the LZ Fake.

The LZ Fake took two forms. One was to have the two cargo choppers drop to ground level as if they were discharging their troops. But the troops would stay on board, and the choppers would repeat the process again some miles away from the first fake landing. Perhaps several times the choppers would drop down. But on one of these "landings," the troops would jump out. Thus local Communist troops might get several reports of U.S. recon patrols landing. Each reported landing would have to be checked out, which would dilute the number of troops that could be sent to each one, giving American troops a better chance of getting away to do their business.

The second LZ Fake gambit was more aggressive. In this variant, gunships would let go with some firepower. The noise would attract Communist troops, who would assume that an American patrol had attempted to land and had encountered enemy fire. Not having as many radios as the Americans, many small (platoon, company-size) Communist units had no way of quickly reporting a firefight with a bunch of U.S. choppers, or even if recon troops had actually landed. This noisy procedure would send Communist troops off to investigate, leaving the recon troops to land somewhere else.

Often, both of these LZ Fake routines would be used in the same operation, complicating the situation for the Communists. Moreover, after a year or so of use, the word got around as to what the Americans were doing. This explained to the Communists why so many of their units rushing to a suspected LZ found nothing. But so diabolical was the deception, they couldn't afford not to investigate every suspected landing.

Sneaking Up on the Bar-Lev Line

The world, and the Israelis, were shocked in late 1973 when the Egyptian Army fought its way across the Suez Canal, through the fortified Bar-Lev Line, and into Israeli-occupied Sinai. Just six years earlier, Israel had handily defeated the armed forces of Egypt, Syria, and Jordan in less than a week (the Six-Day War). But in 1973, Egypt expertly used deception to turn the tables.

As with most successful deceptions, the Egyptians ran several programs simultaneously. Equally important, they identified and exploited self-deceptions that the Israelis had acquired since 1967. As a result of the Six-Day War, Israelis had developed contempt for Arab soldiers. The Arabs knew this, and the Egyptians exploited it by running their deceptions secure in the knowledge that most Israelis didn't believe what was about to happen was possible.

The Israelis were not oblivious of the danger from Egyptian armed forces. Between 1967 and 1970, there had been a low-level war going on across the Suez Canal. While Israeli forces had stopped at the canal in 1967, the continued fighting after 1967 caused them to build a series of fortifications along the waterway. This was the Bar-Lev Line. The canal was considered (quite rightly) to be like a river, and crossing rivers against enemy opposition has always been a formidable task. The Bar-Lev Line was actually a series of concrete observation posts, positioned every six miles along the canal. There were additional fortifications at the more likely crossing points. Each post held only fifteen men, and their primary task was to give warning of a crossing attempt and direct artillery fire on the enemy from batteries well in the rear. Behind the canal were small armored and artillery units, and farther back, bases holding the weapons and equipment of reserve brigades. These reserve units would be mobilized within twenty-four hours of any attempt to cross the canal. The Israelis expected their artillery and air force to keep any canal-crossing force busy until reserve brigades could get moving toward the canal.

The Bar-Lev Line was completed in 1969. In 1970, the Egyptians and Israelis agreed to a cease-fire in their "War of Attrition." This conflict, consisting of air battles, artillery duels, and raids by commandos and ground forces, had cost Israel 500 dead. To put this into perspective, the entire 1967 war had killed only 764 Israelis. While the Egyptians got their licks in, the Israelis considered themselves the winners of the War of Attrition. Not only had they defeated strenuous efforts by the Soviet Union to strengthen Egyptian air defenses; they had also crossed the canal and raided into Egypt at will. The Israeli attitude was that they could cross the canal in force whenever they wanted to and march on Cairo. It was thought that the Egyptians realized this and would act accordingly. That is, the Egyptians would not be so foolish as to cross the canal and attack into Israeli-held territory. After 1970, the Israelis got progressively more sloppy about the condition and manning of their Suez defenses. While Israeli intelligence forces remained alert in keeping an eye on the Egyptians, there was a profound change within the Egyptian armed forces.

In September of 1970, the Egyptian leader, Gamal Abdel Nasser, died at age fifty-two. Nasser had been Egypt's first leader after it became independent in the wake of World War II and deposed the Egyptian monarchy (in 1952). While Nasser was an able leader, he was not very adept in military affairs. In October, Anwar Sadat was elected head of the government. The same age as Nasser, Sadat had also come out of the army and had been active in the resistance against British colonial rule and the overthrow of the Egyptian king. Sadat took his military experience more seriously than did Nasser. Sadat was willing to try new techniques, and he learned from the failed War of Attrition. Sadat immediately began planning for the 1973 war.

From the beginning, the Egyptians knew that the key to success would be deception. Just as the Israelis had vanquished the Egyptians with surprise attacks in 1956 and 1967, the Egyptians would return the favor. The Egyptian plan was simple. First, the Egyptian armed forces would undergo unprecedented amounts of training. The Israelis would notice this, but the new training regime would set up the Israelis for the key deception. As part of the more extensive training for Egyptian troops, there were frequent large-unit exercises in the vicinity of the Suez Canal. This made military sense. The area was largely empty, but close to the bases where the troops lived. And, obviously, if the Egyptians went to war, this would be near the principal battlefield.

The Egyptians also sent patrols across the canal at night to inspect Israeli positions. It was noted that the Israelis had largely abandoned their efforts to upgrade, or even maintain, their Suez defenses. Meanwhile, the Egyptians kept the peace on the canal, further lulling the Bar-Lev Line troops.

It had been centuries since an Arab army had been able to stand up to a "Western" army, and the Israelis considered themselves Western in that sense. Noting the Egyptian training exercises, the Israelis didn't think it possible that the Egyptians could ever get to the level where they could clobber Israeli units. Yet, the Egyptians kept at it. Increasingly, their exercises came right up to the canal itself. Several times, the Israelis put some of their troops on alert. Sadat, meanwhile, used the higher readiness of his army to announce imminent "decisive action," and then backed down. To the Western mind, this seemed politically suicidal. But in Arab culture, such gestures were accepted for what they were—gestures, not commitments. In 1971 and 1972, there were major deployments of Egyptian troops to the canal. In early 1973, the Egyptians did it again. But the Israelis did not fully mobilize for any of these events. A complete mobilization of Israeli Defense Forces (IDF) meant that most of the civilian economy in Israel was shut down. The IDF was largely manned by reservists, which was the only way the Israelis could match the much larger manpower levels of their Arab opponents. Egypt alone, for example, had a population more than ten times Israel's. Thus the Israelis could afford to mobilize only when Egyptian troops were massed on the canal, when they were pretty certain that it was the real thing.

The Egyptians were aware that Israeli intelligence was very good, with an excellent network of agents in Egypt itself. More important, it was known that the Israelis could mobilize troops on the canal faster than the Egyptians could. So the deception had to delay Israeli mobilization as long as possible.

The Egyptians planned to attack on October 6. This was the Jewish holy day of Yom Kippur, the Day of Atonement. For Jews, it was the major religious commemoration of the year, and Israel would be shut down. As with the Japanese attacking Pearl Harbor on a Sunday, the Egyptian attack was to hit the Israelis when they were at a low level of readiness. On the other hand, it was also Ramadan, the holiest period in the Islamic calendar, so it seemed unlikely to the Israelis that the Egyptians would attack at that time. However, in the days leading up to October 6, disquieting reports flooded into Israeli headquarters. More and more, it appeared that this time the Egyptians were not just staging a training or propaganda exercise. But the Egyptian deception was working. Most Israeli leaders and intelligence officers thought the Egyptians were bluffing. A partial mobilization was ordered, but the Egyptians had already achieved their goal; their assault forces were in position and ready to launch their attack. At this point, the Israelis learned of a second Egyptian deception.

Anwar Sadat, having spent much of his adult life in uniform, had studied the last three Arab-Israeli wars (1948, 1956, and 1967). He concluded that the Israeli advantage was largely the quality of their officers and troops. Sadat saw that if he could markedly improve the quality of Egyptian troops, he could at least win a propaganda victory over the Israelis. The world expected little in the way of performance from Arab troops, and Sadat saw that he could use any marked improvement to Egypt's advantage. His battle plan was to get across the Suez Canal and let the Israelis take heavy losses attacking his new and improved Egyptian Army. If the Egyptians simply crossed the canal and were promptly thrown back by the Israelis, nothing would be accomplished. Thus the goal was to get across the canal in force, and then inflict significant losses on the Israelis while waiting for the superpowers and the UN to stop the war. If Egypt could achieve something resembling a draw before the inevitable superpower/UN–mandated cease-fire, then it would be an Egyptian, and Arab, victory. At that point, Egypt could deal with the Israelis from a more powerful diplomatic position.

To accomplish anything, Sadat needed better Egyptian forces. The Soviet Union, for its own diplomatic reasons, had offered the Arab nations vast amounts of military aid in the wake of the 1967 war. Egypt accepted, even though they found the Russians themselves arrogant and generally unpleasant. But the Soviets did know how to turn masses of (often illiterate) civilians into decent soldiers. Russia had been doing it for centuries, and the Soviets had done so with exemplary results in World War II. Soviet methods were simple, stressing simplicity and repetition. Thus Egypt's army was equipped with Soviet weapons (which were simple to use), and thousands of Soviet advisors flooded into the country. Although most Egyptian officers disliked the overbearing and arrogant attitude of the Russians, Sadat decreed that the harsh Soviet training program be followed. Egyptian troops buckled down and soon noticed that they were better soldiers. It was hard work, but most Egyptians were used to that.

Egyptian engineers practiced building bridges across bodies of water similar to the Suez Canal. They did it dozens of times. Day and night. With each repetition, they got better at it. Artillerymen, tank crews, and infantry practiced their jobs in the same way. The Egyptian troops not only got better at their military tasks, they also gained confidence. The 1967 war had been a severe blow to the morale and confidence of the Egyptian Army. Sadat knew that there were two things needed to restore that morale and confidence. First, hard training. Then, success on the battlefield. A successful crossing of the Suez Canal and capture of the Bar-Lev Line would be considered a success on the battlefield. At that point, the Israelis would be up against something they wouldn't expect: trained and competent Egyptian troops.

The Egyptian attack on October 6 was a success. The canal was crossed, the Bar-Lev Line breached, and the Israeli mobilization was delayed. The many Egyptian deceptions had caused the crucial delay in the Israeli mobilization, which gave the Egyptian troops time to prepare for their imminent battle on the east bank of the canal.

The first Israeli reserve units to reach the Suez Canal thought they would sweep away the Egyptian units and prepare for an Israeli crossing. That wasn't what happened. The Egyptian infantry was dug in and well equipped and trained with antitank weapons. Those first Israeli units were practically destroyed. At that point, two things happened that almost ruined the Egyptian plan. First, the Israelis quickly surmised from the available evidence (the massive canal-crossing operation and improved fighting ability of the Egyptians) that they would have to deal with a rather more efficient foe than the one they easily crushed in 1967.

The Israelis adapted very quickly to the new battlefield conditions the Egyptians had created. The Egyptians thus could not depend on the Israelis to continually underestimate Egyptian troops. The second thing that hurt the Egyptians was their overly enthusiastic reaction to their own success.

The original Egyptian plan was to cross the canal, dig in, and let the Israelis bloody themselves against the well-trained Egyptian defenders. The Israelis had been stopped in the past by a resolute Arab defense, but not for very long. For the Egyptians to make a noticeable dent in Israel's military reputation required well-trained and well-led troops. The Arabs didn't have much of that in the past, but they did now.

However, the Egyptians decided to deviate from their original plan and move beyond the Suez Canal—which was a big mistake. While Egyptian troops had trained hard for making an assault crossing of the canal and defending against Israeli tank attacks, they had not drilled nearly as much to deal with mobile warfare. The Israelis were quite expert at mobile warfare, and it was when the Egyptians tried to tangle with Israeli tank units on the east side of the canal that things began to go very wrong for the Arabs. The Egyptians got chewed up quite badly in these fast-paced tank battles, and the Israelis used the opportunities obtained to make their own canal crossing and cut off most of the Egyptian troops still on the east bank.

But, as the Egyptians expected, the superpowers and the UN applied much pressure on all concerned to stop fighting. Both sides did so on October 22. Although Israeli troops were less than thirty miles from Cairo, and ready to march on Egypt's capital, the Arabs still had five divisions on the east bank of the Suez Canal. These five divisions were surrounded by Israeli troops and rapidly running out of water and ammunition.

But no matter; compared to the situation when the 1967 Six-Day War ended, it was much better. The Arabs, particularly the Egyptians, had fought the Israelis for two weeks. Although the Israelis were winning when the cease-fire was declared, they had not yet crushed the Egyptian armed forces as they had in 1967. To the Arabs, this was a victory, and it was celebrated as such.

The Egyptians had other things to celebrate besides still being on their feet (and on the east bank of the canal) after two weeks of fighting the Israelis. They also managed to deceive the Israelis as to the capabilities of their antiaircraft defenses. Using the latest Soviet weapons and equipment, the Egyptians inflicted unexpected losses on Israeli aircraft. The Israelis knew what the Egyptians were up to, but underestimated the skill with which the Arabs would use their new weapons. Because of this attitude, the Israelis decided not to buy the expensive electronic countermeasures equipment offered by the United States.

The Americans expected to face these new Soviet weapons and had been hard at work in developing electronic devices that would provide some protection. The Egyptians turned out to be quite skillful against Israeli aircraft with their new Soviet short-range missiles and radar-controlled guns. They had managed to conceal from the Israelis just how effective these new weapons were when used by Arab crews. Of course, the success of this deception relied a lot on Israeli arrogance regarding the skill of their pilots. The Israeli air force did eventually overcome the new Egyptian antiaircraft defenses, but only after taking heavy losses and not being able to provide the degree of support their troops were accustomed to. The Egyptian air force was never able to compete with the Israelis on equal terms, but the antiaircraft missile and artillery units put up a fierce resistance, giving the Egyptians (by their reckoning) claim to yet another victory.

The extent of Egyptian deceptions can be gauged by the degree to which the Israelis studied them after the war. The Israelis admitted that they were caught napping in the early stages of the war, and made strenuous efforts to prevent a reoccurrence. But, for the Egyptians, the fruits of their victory came in 1977, when they signed a peace treaty with Israel. The Egyptians got the Sinai back (along with its small, but significant, oil fields) and were able to reopen the Suez Canal. As part of the deal, Egypt became the recipient of military and economic aid from America. To Egyptians, their deceptions in the 1973 war paid off handsomely on the battlefield and at the peace table.

The Masters of *Maskarovka*

Russians, before, during, and after the Communists, were/are great believers in *Maskarovka*. Like most foreign words, this one is difficult to translate exactly. The word can mean two things: either "shield" (or, more commonly, the canopy on an aircraft) or a wide range of deception techniques. If the word has a lower-case "m," it means shield; if it's an upper-case "M," it means deception on a large, often national, and usually grand, scale. And, yes; the word does have the same root as the English word "mask." But this is "masking" on a huge canvas.

All nations use large-scale deception to some degree, but none have used it as energetically, or successfully, as the Russians. The Russians' fondness for deception has been known for centuries. The term "Potemkin Village" comes from an eighteenth-century deception whereby false, and seemingly prosperous, villages were set up so that the shortcomings of the local administration would not be revealed. When the Communists took over in the early 1920s, they added more technology and police-state tactics to the traditional Russian arsenal of Maskarovka techniques.

The extent to which the Communists had improved their Maskarovka did not become evident until World War II. The Germans were the first to suffer from the extent of the Soviet Union's application of large-scale deception, which was felt in the following ways:

- The Germans encountered unknown weapons—several new ones they didn't know about (multiple rocket launchers, heavy tanks, etc.)—and the experience was most unpleasant. The Russians accomplished this surprise by simply not announcing the existence of these weapons and, in most cases, not even issuing the new weapons to their troops until after the German invasion. This last practice reduced the initial effectiveness of new weapons, because the troops using them had to gain some experience. Still, the military advantage for the Russians was considerable. For one thing, the Germans had to start from scratch figuring out how to deal with these weapons. The new Russian tanks required new antitank guns and tactics; it took the Germans nearly a year to develop both. New rocket launchers changed the rules of artillery use, and the Germans took even longer to devise methods to cope with the rain of shells these launchers could deliver. Stalin's rule was accompanied by a good deal of terror, which caused nearly all Russians to keep their mouths shut. Although the Germans had spies in Russia, it was difficult to discover anything in what was, arguably, the most effective police state on the planet. Even the gestapo and other German security organizations could have learned something from the Russians in this regard.
- The pervasive secrecy in Russia during the 1920s and 1930s hid more than new weapons; it also masked industrial strength. During this period, Russia was going through a tremendous amount of industrialization. Most of the new factories were built with national security in mind. That is, vital plants mainly were built east of Moscow. While the Russians did disassemble and move many factories east in the early months of the war, most of the important stuff was never in danger of being overrun by the Germans in the first place. The Germans were unaware of this new Russian industrial setup and falsely believed that by their seizing the traditional centers of Russian industry west of Moscow, the Russians would be unable to supply their troops with weapons and equipment. By early 1943, the Germans finally became aware of the fact that the torrent of new Russian weapons was not in

any way affected by the Germans having occupied most of European Russia. While Russian secrecy played a large role in this deception, part of it was the Germans' fault. It was easier to hide new weapons than vast new industrial complexes, and the importation of foreign technology to set them up. If the Germans had not been so contemptuous of the Russians, they could have ferreted out the truth. Had they done so, they might not have invaded in the first place. Even with an invasion, a more accurate knowledge of Russian industry would have led to more effective German strategy.

- Germany greatly underestimated the extent of the Russian armed forces' military strength. The Germans knew that in 1941, they had about as many divisions as the Soviet Union had (a little over two hundred). By the end of 1941, the Germans had killed or captured nearly four million troops, while losing less than 10 percent of that themselves. By preinvasion German calculations, there should have been practically no Russian forces left. But what was the German situation in late 1941? They were falling back in the face of a fierce Russian counteroffensive. This situation never changed throughout the war. The Germans eventually killed eleven million Russian soldiers in combat (plus even larger losses from disease and privation), while losing about two million of their own. Yet, as the war went on, the number of Russian divisions, and troops, increased. What the Germans had missed, and the Russians had successfully hidden, was an enormous system of officer- and troop-training schools and facilities. Moreover, the Russians had been mass producing all manner of weapons since the early 1930s. They had made extensive plans to raise new divisions from men who had served as conscripts and been released, in addition to drafts of untrained civilians. While this technique was common in many European nations, the Russians had carried it out on a larger scale, and kept many details secret. The Germans hadn't planned on the Russians being able to make such a rapid recovery. But then, the Germans had not expected to do so much damage to the Russians in the six months after the June 1941 invasion. What fate giveth, fate taketh away, and then some.

- The Germans had trouble with navigation. Getting around Russia was never easy. Not a lot of roads were ever built; rivers were the preferred method of transportation. When railroads came in the late 1800s, there was even less incentive to build roads. And the Russians used a different gauge (distance between the rails) from that of other European nations, making it more difficult for an invader to use captured rail lines. This situation persists to this day, with air movement again providing new transportation alternatives that resulted in even less incentive to build roads. The Germans were aware of the lack of roads, but felt that they could make do by moving cross-country and along whatever dirt tracks did exist. The Russians added deception to the obstacles created by their lack of roads. Early on, the Communists declared that all maps were state secrets, and even their own troops had to treat maps as classified documents. Those maps that were published were riddled with intentional errors and misinformation. The Communists did not encourage tourism, and those foreigners who were allowed to visit were carefully screened and could only go forth in groups under the supervision of Russian guides (who were often counterintelligence agents). In a way, the lack of roads made the lack of accurate maps moot, even though things like rivers, streams, villages, and heights were also distorted on Russian maps. The Germans simply relied on compass and dead reckoning, and made their own maps as they went. This still gave the Russians an advantage, because the Red Army had its own accurate (albeit "top secret") maps. As these Russian maps were captured, it made it easier for the Germans to create their own maps. Not until the 1960s, with the advent of satellite photography, was the extent of the Soviet falsification of maps manifest: Virtually all published maps were found to be inaccurate.

- The Russians were masters of tactical camouflage, both in the field and in the rear area. They used every trick in the book, and added many new techniques. One could justifiably say that the Russians wrote the book on twentieth-century camouflage techniques. When American troops encountered clever German camouflage in the West, they were often seeing techniques the Germans had copied from the Russians. The Russians were particularly skillful at hiding masses of troops and equipment on the flat and featureless steppes. German troops received many nasty surprises when they encountered this skillful Russian camouflage. Their shock was made worse because they had overrun millions of Soviet troops in 1941 without ever running up against much decent camouflage, or effective resistance, for that matter. The more steadfast and better-trained Red Army soldiers came along in 1943. But most Russian camouflage techniques had been worked out in the 1930s. What the Russians needed was a stable front line and time to train their troops. They obtained both of these items by surrendering most of European Russia to the Germans in 1941 and 1942.

- The Russians considered any facility worthy of strategic camouflage. They would camouflage roads, bridges, factories, cities, and even rivers. In most cases, they were keen on deceiving German reconnaissance aircraft. This meant that if the shape of a town could be changed before an expected German bombing raid, the bombers were likely to hit the wrong targets in the town. Sheets of painted canvas, camouflage netting, smoke pots, and earth-moving equipment could all be used to deceive aircraft pilots. Dummy aircraft, vehicles, and railroad equipment were also used. If the effort seemed worthwhile, the Russians would try it. A few thousand civilians with shovels could change the appearance of the ground enough to fool pilots, and save valuable targets from bombing.

After World War II, the Soviets put more effort into strategic camouflage. This was particularly true after the 1950s, when American high-altitude reconnaissance aircraft began overflying Russia. First the U-2, and then the SR-71, made the Soviets acutely aware of their vulnerability to spies in the sky. From the late 1960s, they also had to contend with an ever-larger number of increasingly capable American space satellites. Gradually, the Soviets built up a huge Maskarovka bureaucracy, complete with a high-tech command center. From this bunker, the identity and path of all American satellites was plotted and plans made to deceive these birds when they passed over sensitive portions of the Soviet Union.

Foxing improved sensors became a major effort from the 1970s onward. Infrared sensors could detect heat from space, and this enabled American sensors to see through many previously workable deceptions. Smoke pots were a good example. Smoke pots were simply containers of various size that held material that produced a lot of smoke. These could make a dummy factory look active, or simply hide a facility you wanted to keep secret. With infrared, the presence of smoke pots could easily be discovered. The Soviets did improve the composition of their smoke chemicals, so that the smoke was more likely to mask heat. But the American sensors kept improving, and when they did see through smoke and detect the presence of factories, housing, or other facilities, they provided a wealth of information. Infrared photos allowed analysts to determine what was under the smoke (or natural fog) by the patterns of heat. Factories and laboratories had a different "heat pattern," and it was difficult (but not impossible) to hide. The Soviets applied a lot of resources to defeating infrared, and other, sensors. In addition to hiding heat, they had to conceal large masses of metal and electronic emissions that could also be detected by satellite (and high-flying aircraft) sensors. Most of the results of this contest between American sensors and Russian deception is still secret. What has leaked out seems to indicate that the Americans were more successful. But in the deception business, that isn't enough. A little bit of successful deception keeps the other guy guessing, and goes a long way toward hiding many things.

While American satellites were successful in discovering a lot of large things—such as factories, warships, airfields, and parking lots full of tanks—they had much less success in revealing small details. This was particularly true with weapons systems. Although America had extraordinary electronic reconnaissance (recording and, if need be, decrypting, messages), plus a steady stream of refugees and defectors, the secrecy in the Soviet Union was so pervasive that new weapons were often not known until they were actually seen with field units. Sometimes the first inkling of a new weapon was during a deliberate display by the Soviets. Parades that revealed previously unknown gear were a favorite opportunity to tweak foreign intelligence agencies and remind them that the Masters of Maskarovka were still in business. Americans could conceal things too, as witness the various "black" aircraft programs (F-117A, B-2, etc.). But the Soviets did it on a much larger scale. As a result, new Soviet weapons often took years to figure out. Even the intelligence agencies were often stumped. When the Soviet Union fell apart in 1991, many more "secret" weapons were revealed, as well as their capabilities. For now the Russians were trying to sell just about every weapon they have to keep their economy going. This full disclosure of known, and unknown, weapons proved a humbling experience for many in the West who had made a career of seeking out Soviet secrets.

After World War II, the Soviets continued their 1930s' program of creating "paper" divisions backed up by lists of former soldiers and storage areas full of older (and often quite obsolete) weapons. The

Soviets also continued other pre–World War II deceptions. For example, not all new types of weapons were immediately issued to the troops. Or not all of the new weapons were issued; some were held back to deceive potential enemies on how many of these new systems they actually possessed. Numbers of all sorts were secret in the Soviet Union, and military numbers were the most tightly held secrets of all. In the post–World War II era, however, many of these pre–World War II deceptions began to backfire. The Soviets laid up tremendous amounts of tanks, artillery, and infantry weapons after World War II. Each succeeding generation of new weapons sent many of the older weapons into storage depots. By the 1980s, even the Soviets had to admit that a lot of these old weapons were less than useless. Decades of indifferent maintenance and haphazard storage had rendered most of these weapons (and large quantities of ammunition) useless, and even dangerous. If Soviet troops were given these weapons to use, they would suffer considerable injuries as the weapons failed in action. This was seen in Afghanistan, when large quantities of ancient artillery ammunition failed to explode on the enemy, or detonated as they were being fired.

When the Soviet Union collapsed in 1991, these thousands of weapons and munition-storage depots were frequently plundered by various warring factions, or criminals who simply bribed the guards to allow some of the stuff to "disappear." Thus a deception that worked once in 1941 came back to haunt the Russians when they tried to keep it going fifty years later.

In addition to concealing weapons, and their performance, the Soviets also used satellite deception to mislead the West on how their troops would operate in the field. Several times a year, the Soviets would hold large-scale maneuvers. Each of these exercises would involve many divisions, plus hundreds of aircraft and helicopters. Satellite photos of these maneuvers were thought to reveal tactics the Soviets were going to use in future wars. But the Soviets knew when American satellites were coming over, and sometimes arranged displays of tactics they had no intention of using. Naturally, this made it more difficult for Western intelligence analysts to figure out exactly what the Soviets were planning. This, of course, was the sort of confusion the Soviets wanted to create with these little deceptions.

The Soviets used more of these "false demonstrations" to confuse those in the West who were using satellite photos to figure out just what Soviet industry was producing in the way of weapons. Again, because the satellites could not take photos around the clock, but only when their orbit brought them over a specific location, the Soviets would arrange special events when they knew an American satellite was overhead. For a factory, they would ask all workers to leave at a certain time, stay away for an hour or so, and then return, or go out of sight and then return, so that it appeared that shifts were changing. Since production at a plant varies with the number of shifts there are, or how long the shifts last, these "schedule deceptions" would mess up any weapons-production calculations based on satellite photos. To carry the deception further, weapons shipped out of the factory would go on covered railcars. Some would be real weapons, others would be nothing but empty railcars with the usual covering. To deceive metal mass sensors, the railcars carrying dummy weapons would have an appropriate mass of metal under the cover. All of this doesn't work all the time, but it doesn't have to. If deceptions are partially successful, they work. All you have to do is deceive your opponent part of the time to keep him guessing. And guesses are not as good as facts.

Although the Soviet Union is now gone, Russia is still there, as is the Russian penchant for secrecy and deception. Old habits die hard, as Western businessmen learn at their cost when they deal with Russian counterparts who can't escape their centuries of habitual secrecy and deception.

Inspiring the Revolutionaries with Disinformation

Spreading lies and rumors is a long-standing practice in wartime. But it was developed to a fine art by the Soviets during the Cold War. This conflict, which raged (or simmered) from 1948 to 1989 (or 1991, take your pick), never saw the main antagonists (America and the USSR) fight each other directly—rather, it was a war of proxies. Other nations provided the battlefields; other peoples provided most of the fighters. The Soviets were calling most of the shots during this conflict, even if they were not firing them. While both sides used the media and propaganda, the Soviets were enthusiastic users of a particular form of media deception called disinformation. This is the old "repeat a lie often enough and it becomes truth" routine distributed via press release and planted media stories.

Disinformation is an ancient deception technique, but never has it been used so widely and for such a long time to keep numerous small wars going, and generate such levels of hostility toward one's opponent. These fighters were not risking their lives for the Soviet Union, but for a myriad of local causes. The Soviet disinformation program was intended to keep everyone in a combative mood, pursuing goals that meshed with the Soviet Union's foreign policy.

Some of the disinformation was pretty outrageous, such as the planted story that AIDS was invented by American military researchers. Most of the disinformation was more minor, and locally relevant, in nature. The Soviets had a large bureaucracy, and equally vast budget, to buy the services of local journalists worldwide. The stories supplied would generally cast aspersions on the actions or motives of the United States government and Americans in general. While the Soviets were usually inclined to shovel out lies and half truths pell-mell, they also had specific programs to bring down governments friendly to the West or, more important, to prop up the morale of rebels, revolutionaries, and terrorists fighting for a Soviet-approved objective.

What made the Soviet program unique was its worldwide nature. The Soviets were quick to realize that the media in most countries were not as independent as in the United States. In fact, American media were the exception. In most nations the media are, like the first newspapers two centuries ago, the creatures of one special-interest group or another. It was in the United States that "independent" media were invented, and even the American media are not completely free of biases and favoritism toward special interests. In most countries, the bias and special-interest control is much stronger. Yet, in all countries, the local media are, like it (or agree with it) or not, the primary source of information for the populace. Compared to America, the rest of the world's journalists are not well paid (even by local standards). Thus it is common for journalists to accept "gifts" (or outright bribes) in return for writing certain stories or slanting their reporting a certain way. The Soviets took advantage of this custom, and their local agents (who were often not Russians) were liberally supplied with cash in order to buy the media attention they needed. The American CIA engaged in the same practice, but the Soviets were much more aggressive, and generous, in this area.

While many journalists worldwide admire the American model for media independence, the Soviets realized that they didn't have to buy a lot of journalists in order to give their agenda sufficient exposure. Most of the Soviet disinformation was purposely developed as sensational stuff. The Soviets knew what kind of stories played best in the media, and this is what they provided. This was the importance of the large disinformation staff back in Moscow. Stories that played on local fears were favored. For example, over the years, the CIA was played up as the cause behind just about everything that people feared, up to and including the weather and earthquakes (more malevolent American scientists on the CIA payroll). In typical Russian fashion, the Soviets would plant dozens of stories in different countries, all hitting the same invented idea from a different angle. That way, the press in one country could cite a Soviet story planted in another country to back up their local "reporting." The Soviets also made the most of some outrageous story appearing in the Western press (whether it was a Soviet plant or not), by planting more outrageous versions and elaborations via the more pliable journalists of other nations. The Soviets realized that the media had become a global system and that there was a great deal of "follow the leader" (or "steal from another newspaper," depending on how you look at it) going on. The Soviets also knew that correcting an inaccurate story was nearly impossible. Once the lie gets loose, you can never correct the misinformation that then forms in so many people's minds.

The "Big Lie" was something that was created in this century as media grew in importance. The Nazis get a lot of credit for starting it, but it was actually the Bolsheviks (the earliest incarnation of the Communists) in Russia who first used it so effectively at the end of World War I. Indeed, the term "Bolshevik" is Russian for "majority," a title the Communist minority among the Russian Socialists gave themselves as they set out to seize control of Russia at the time of World War I. The Communists kept repeating the term Bolshevik (even when it was obvious they were a small minority of the Russian Socialist party), and eventually, more and more people just took it for granted that the Communists were the majority, the Bolsheviks. And soon they were in control of the nation. And at that point, they were still a minority, which is why they kept on killing off Russians who actually or potentially thought differently (into the early 1950s). Western countries only slowly became aware of what the Soviets were doing. The Voice of America and the BBC World Service radio broadcasts were intended to counter Soviet disinformation. But these efforts met with limited success. Imaginative lies travel faster and more widely than does the more mundane truth. Politicians in all nations know and take advantage of this fact. "Negative campaigning" is often little more than a disinformation campaign.

The Soviets were also quick to support special-interest groups that, directly or indirectly, supported Soviet interests. From its very beginning in the late nineteenth century, modern Communism (and Socialism) found a receptive audience in academic and literary circles. These intellectuals were manipulated by Soviet disinformation specialists and provided more legitimacy for the artful lies emanating from Moscow. The many pacifist and antinuclear groups in the West were closely observed by the Soviets, and helped along when possible by Soviet agents, influence, propaganda, or cash. In the 1980s, the Soviets were keen to use Western pressure groups to gain an edge in the arms race the Soviets themselves had begun in the 1960s. When the West finally realized that the Soviets were undertaking a massive rearmament program, America took the lead in producing high-tech weapons that could match the Soviet buildup, and that the Soviets would have a difficult time countering. During the 1970s, America was preparing to deploy cruise missiles and Pershing ballistic missiles (which the Soviets had no defense against) as well as neutron bombs (which would negate the Soviet superiority in armored vehicles). The Soviets made little headway against the cruise and Pershing missiles, mainly because these weapons were being brought into Europe to counter the numerous new missile systems that the Soviets had already moved into Eastern Europe.

The neutron bomb was another matter, as it was a nuclear weapon. But it was a special type of nuclear weapon. Unlike existing atomic bombs, the neutron version made a smaller "bang," but generated a lot more neutron radiation. This form of radiation was also special in that it was short range and short lived. Unlike the radiation of existing nuclear weapons, neutron radiation would quickly go away. Finally, and this was what the Soviets disliked the most, neutron rays could penetrate the armor of tanks, kill their crews, and not leave the vehicles "hot" with lingering radiation. Because of the smaller blast and "cleaner" radiation, neutron bombs could be used to great effect against advancing Soviet tank divisions without totally trashing the surrounding West German countryside (and population). Thus the neutron bomb dramatically changed the military balance. It could more readily kill Soviet tank crews and allow North Atlantic Treaty Organization (NATO) troops to move quickly through the devastated Soviet units with less fear of radiation poisoning. While no one felt comfortable about it, both the Soviets and NATO were ready to use nuclear weapons in Europe, and the neutron bomb was seen as a more "humane" way to do it.

To the Soviets, however, the neutron bomb presented some excellent disinformation opportunities. The Soviets had already proclaimed that they would never be the first to use nuclear weapons. This was another lie, as their war plans, discovered in the former East Germany after the end of the Cold War, specified the early use of nuclear weapons, not to mention poison gas. But at the time, the "no first use" stand gave the Soviets some stature among antinuclear groups in the West. The Soviets used this edge to support a campaign to get the neutron bomb stopped. And it worked. A combination of disinformation and fear of things nuclear caused the neutron bomb project to be canceled in 1978. Work on the cruise missile and Pershing continued, and these did enter service in the early 1980s. A few years later, the Soviets agreed to the first treaty to ever withdraw and destroy nuclear weapons, mainly as a means to get rid of the cruise and Pershing missiles.

The only positive side of disinformation is that, eventually, most people catch on and no longer believe the lies. But this takes time, often decades. And the turnaround has to take place separately in each media area. That is, while people may begin to see through the local disinformation campaign in one area, people in a neighboring nation could still be under the spell of the clever forgeries. The Soviet Union and its East European satellite nations saw their web of disinformation come apart during the 1980s. The fall of the Berlin Wall in 1989 will long be seen as the moment when the tower of deceit came undone. But in actuality, the process of disintegration took place over several years. And for many years to come, there will still be people in those formerly Communist countries who will continue to believe the lies, even if the majority does not.

The deception technique of disinformation had a palpable effect on dozens of battlefields during the Cold War. Thousands of pro-Communist fighters believed, to the death, in the tangle of disinformation the Soviets had created. Without such motivation, many of these wars, rebellions, and uprisings would not have happened. Information, even false information, is power. And this translated into firepower for decade after decade.

It will happen again.

7

THROUGH A SHADOW, DIMLY SEEN

Deception from the Eighties to the Near Future

As rapid as was the growth and impact of deception in the twentieth century, the future appears to hold promise of even greater opportunities for the dark arts to flourish. Ever higher levels of education, and more technology to play with, portend a coming golden age of deception. We're already beginning to see signs of this, which is why we use the 1980s for the starting point of this chapter.

One important thing to keep in mind about the future of deception is that newer and cleverer ways to fool the enemy are certain to be invented. Just as the introduction of airplanes, radar, infrared detectors, and microwave sensors have in the past led to the development of innovative ways to fool the sensors, so it will remain in the future. Precisely what new technologies will be available to spoof the enemy will have to wait until they are invented. Not all of them will be invented by the good guys. But they will be invented. And the side that hits upon these new techniques of deception first will have a decided advantage.

Nor should the old standbys be forgotten. Maybe no one will ever again use the ancient "light many campfires" trick to make the foe think there are more numerous troops than there are, but modern versions of this hoary ruse exist, in the form of dummy encampments, dummy radio transmissions, and so forth. And don't be too sure that someone might not manage, occasionally, to get away with using that old campfire trick.

The changing nature of the world community will also undoubtedly lead to the development of new and more subtle ways to trick the enemy—certainly on a political and diplomatic level. Consider, for example, the influence of the media on the UN/US intervention in Somalia in late 1992, and NATO involvement in Bosnia in 1994. A clever dictator or politician with a hidden agenda could easily manipulate the global media with tales and pictures of human suffering for his own ends. Consider, for example, the attempt in late 1993 by the Haitian military dictatorship to suggest that the world embargo on trade with the island nation was costing the lives of over a thousand children a day. This was a lie which, in this instance, failed to convince the world community to go easy on the warlords. We can certainly expect more of this sort of thing in the future, along with other interesting manipulations of the global media. It's been a tricky world for a long time, and it's going to stay that way.

The Other Gulf War and Its Deceptions

The Iraq-Iran war of 1980–1988 was generally ignored by most of the world. It was certainly not covered as a major war should be. And a major war it was, being a conflict that involved several million troops and killed half a million of them. Part of the reason why the war gathered so little attention was that it was largely fought as a replay of World War I. That is, two entrenched armies launched futile infantry offensives at each other.

Actually, that was not quite accurate, as Iran was doing most of the attacking, and suffered most of the losses. Again, unlike World War I, there was a fair amount of deception used. While the combatants were not always very well trained, they were not stupid. Both Iraqis and Iranians quickly realized that one way to come out of the experience alive was to deceive the enemy. And deceive they did.

The Iraqis started the war with an invasion of Iran in 1980. This failed, and within two years, the Iranian counteroffensive had thrown the Iraqis back to their own borders and in danger of losing the war. The Iraqis went over to the defensive, using their shovels as their most important weapon. The Iranians proceeded to launch attack after attack until 1987. Iraq held back the Iranians in no small part because of a number of deceptions. Three of the more notable deceptions follow:

1. They created an artificial lake (Fish Lake), to cover a key portion of their front, by flooding low-lying land. The Iraqis realized that this body of water would not, by itself, stop the Iranians. So they filled the bottom of the lake with obstacles (objects just below the water to sink boats, barbed wire to snag waders, etc.), sensors, and even mines. As the Iraqis expected, the Iranians soon took to using small boats to move troops across the lake. The water was so shallow in places, troops could wade across portions of it. But the Iranians were never able to make full use of water transport on the lake because of the ever-changing array of obstacles the Iraqis placed under the water. This became something of a game, with the Iranian scouts searching for obstacle-free areas on the lake before an attack, and then finding (during the attack) that the Iraqis had secretly gone out and placed more obstacles in supposedly "clear" portions of the lake. The Iraqis used the "what's really in the water" deception elsewhere as well, since most of the heavily fought over battlefield was marshland, or close to it.

2. The Iraqis eventually realized that a single defense line, no matter how well built, could be scouted by Iranian patrols and then demolished with a well-planned attack. So the Iraqis listened to their Soviet advisors and built multiple defense lines. The most important aspect of multiple lines was that, once enemy troops breached the first one, they wouldn't know what to expect as they went on to attack the next one. While a single line would be more formidable if all resources were put into one line, the multiple-line approach had proven more effective in the past.

This Soviet style of fortification was developed from their extensive

World War II combat experience. In that war, the Russians had learned that it wasn't enough just to dig a hole in the ground, put some troops in it, and equip them with a machine gun. If the enemy could easily make out the design of the fortifications, they could plaster them with accurate artillery fire before getting within range of the machine guns.

The key to effective fortifications was multiple lines, and a lot of deception. The obvious positions are the ones that are unmanned and exist mainly to attract enemy artillery fire. The real positions are those that are not so easily detected and contain the lethal weapons. Positions in the second and subsequent lines cannot easily be scouted by the enemy and lend themselves to clever layouts. The combination of hidden positions and unfamiliar minefields causes much higher casualties among the attackers, and is often sufficient to stop an attack.

These techniques would have been even more successful had there been better morale among the Iraqis. Iranian troops were much more enthusiastic about combat. Iraqi troops, even though fighting from fortified positions, would often run away as Iranian assault troops approached. This was particularly true with the first line of fortifications, the ones the Iranians had had time to study carefully. The Iranians would prepare their attacks carefully, and exploit any weaknesses in the way the Iraqi positions were laid out. Firstline fortifications were also vulnerable to having their minefields secretly mapped and partially disabled at night by the Iranians. The always-shaky Iraqi morale was often unable to withstand the twin shocks of the Iranian attack and seeing that their minefields weren't working. Second and third lines of defense did not receive the "preparation" the Iranians could give the first line, thus leaving their Iraqi defenders somewhat less rattled.

This was not a panacea for the Iraqi morale problem, however, as the Iraqis were still shaken by the fact that the first line invariably fell quickly. Moreover, the Iranians, once halted at the second or third line, would often spend a few days scouting the new obstacle to their advance, and eventually deliver a more effective attack on subsequent defense lines. But the Iraqis needed every advantage they could muster to overcome the questionable morale of their troops. Deception in the construction of second and third defense lines turned out to be a decisive element in the Iraqis being able to hold on in the face of repeated Iranian attacks.

3. This third, and one of the more important, deceptions was not

directed at the Iranians, but at the Iraqi people. Saddam Hussein, the Iraqi leader, knew that the war was unpopular in Iraq and that he could be defeated from within even if the Iranians didn't get him first. While Saddam had plenty of oil money to toss around as a morale-building exercise, he had to do something else when there was a major battle and many Iraqis were killed all at once (as was normal with major battles). His propagandists made much of how effective Iraqi troops were at killing a lot of Iranians. But the big battles killed a lot of Iraqis too. To minimize the effect on Iraqi morale, Saddam built many refrigerated mortuaries. In them he stored the bodies of Iraqi dead after a big battle, and sent the corpses home to the families over a longer period of time. In this manner, the losses from major battles were seen to be lower.

The Iraqi people accepted the fact that troops were being killed by artillery fire and patrolling, even when a big battle wasn't going on, and were deceived into believing that the major battles were hurting Iraq less than Iran. A large number of casualties announced at once was (and is) a decided blow to civilian morale.

To further soften the blow to the populace, Saddam made substantial payments to the families of the dead, and made much of the "heroic sacrifice" of the Iraqi "martyrs." The fact that bodies were recovered from battlefields and returned to their families was itself unusual for Iraqis, who were long accustomed to much less thoughtful treatment of their war dead.

All of this was excellent propaganda for a government that was more feared than respected. The Iranians were always making much of Saddam's faults (like invading Iran in the first place), and it was no accident that Saddam had to survive at least one serious assassination attempt a year. But the refrigerated mortuary deception was a key element in keeping the Iraqi people behind him during this disastrous war.

Iranian Deceptions

The Iranians also learned to use a deception of their own to good effect. Although much was made of the Iranian "suicidal human wave attacks," many of these operations were neither suicidal nor failures. Iran had three times the population of Iraq, and a population that was much more enthusiastic about the war.

While most of the Iraqi soldiers were conscripts, most of the Iranian troops were volunteers. Iran had less money to spend than Iraq, which resulted in a large, enthusiastic infantry army of Iranians, versus a better-armed and decidedly unenthusiastic army of Iraqis.

The Iranians made the most of what they had. Before an attack (they were constant throughout the war), the Iranians spent a lot of time patrolling and gathering information on Iraqi defenses. Weak spots were noted. For example, there might be a piece of terrain in front that wasn't well covered by the Iraqi machine guns. This would be noted. Minefields were examined and paths were cleared at night.

When it came time to launch the "human wave" attack, the Iranians would move through the paths they had cleared in the minefields and take advantage of the ground they already knew was relatively safe from Iraqi fire.

From the Iraqi point of view, it looked liked their minefields were worthless and that the Iranians were coming out of nowhere onto Iraqi positions. It's no wonder many Iraqis promptly fled. And those who did run away lived to tell other Iraqis amazing tales of how Iranians walked unharmed over minefields and appeared immune to machine-gun fire. This made the normally shaky Iraqi morale worse.

Of course, it wasn't always a walkover for the Iranians. They would purposely send some of their troops into Iraqi machine-gun fire in order to allow other troops to sneak up on the enemy positions. Iranian losses were heavy, often two or three times what the Iraqis lost. But if the Iraqis had been as resolute as the Iranians, those fortified positions would rarely have fallen. The deceptions of clearing minefields and working out sheltered approach routes made all the difference. For if the Iranians had simply tried to walk across uncleared minefields and charged straight ahead, few Iranians would have made it to Iraqi lines.

Medieval Deceptions in Twentieth-Century Afghanistan

The war in Afghanistan that went on from 1979 to 1988 was begun with deception, fought with deception, and ended because of deception.

The Russians have, since the nineteenth century, considered Afghanistan within their sphere of influence. When the Communists turned Russia into the Soviet Union during the early 1920s, this attitude did not change.

After World War II, the Soviets steadily increased their influence in Afghanistan, particularly among the educated, city-dwelling portion of the population. A local Communist party grew apace during this period.

But like everything else in Afghanistan, the Afghan Communists were riven by factionalism and feuds. However, Communists eventually gained control of the government. It wasn't your typical Communist dictatorship; the government still had traditional tenuous control over powerful tribes and clans out in the countryside. These rural folk were quite conservative, and generally serious about their Islamic religion.

When the various factions of the Afghan Communist party took to settling their differences with guns, the Soviets decided that they had to intervene before the Afghan Communists completely discredited themselves and lost control of the country.

In a masterful bit of deception, the Soviets had their numerous military advisors and diplomats in Afghanistan distract the Afghan police, politicians, and troops while a rapid airborne assault was launched on the capital of Kabul. Key Afghan military commanders and political leaders were lured away from their posts by Soviet officials. Other Soviet military personnel assigned to train the Afghan armed forces disabled important Afghan weapons systems. Then, without warning, Soviet paratroopers and commandos landed at key airports and began taking over the country. Within days, the Soviets controlled Kabul and were quickly on their way to doing the same in other Afghan cities. The most reliable surviving Afghan Communists were installed as the new government. Things went downhill from there.

In the countryside, the Afghan Communists in Kabul had been, at best, tolerated, if only because they were, after all, Afghans (albeit misguided and irreligious ones). Over thousands of years, Afghanistan had an unmatched track record in expelling foreign rulers. As soon as it became known that the Russians had invaded, a jihad (holy war) was declared against the infidel (atheist, really) foreigners.

But religious fervor was not the key weapon in the Afghan arsenal; deception was. Long before Islam came to Afghanistan, the local tribesmen had used camouflage, trickery, and imaginative deceptions to thwart the designs of foreign invaders and each other. The warrior culture of the Afghan tribes revolved around weapons and raiding. Warfare was considered a sport, and feuds would go on for centuries. Thousands of largely self-sufficient villages raised crops and tended herds to sustain themselves. Although Afghanistan was (and still is) one of the poorest nations in the world, the population was heavily armed and expert in the use of basic infantry weapons. But the Afghans were not suicidal. They were experts at stealth and staying alive in a fight against seemingly hopeless odds. They would need all these skills to deal with a twentieth-century Soviet army.

The physical nature of Afghanistan and its primitive infrastructure was a major ally of the Afghans. Lacking railroads, and possessing few roads, Afghanistan's transportation net could not easily supply a large modern army. The Soviets quickly concluded that because of these logistical limitations, they could not support more than a hundred and fifty thousand troops in Afghanistan without greatly improving the nation's road network. Early on, the decision was made not to expand the transportation system. After all, the Afghans were a primitive people, and the Red Army had an enormous air force to compensate for the relative lack of troops in the country. Moreover, right across the border in the Soviet Union, there were several large air bases and ample railroad access to the rest of the Soviet Union. Ten or so divisions inside Afghanistan and hundreds of aircraft and helicopters in support should be sufficient.

The Soviet calculation was wrong, of course, and by the time they realized it, it was too late to begin building railroads in Afghanistan. Moreover, after the war was under way for a few years, the Soviets had a difficult time even protecting truck convoys from all the tribes now in arms. The added task of guarding hundreds of miles of railroads would be insurmountable.

Afghan resistance fighters used deception to even the odds. They fought a guerrilla war in an area where guerrilla war was thought (by outsiders) to be impossible, or at least very difficult. Afghanistan is largely devoid of forests, or cover of any kind. But the Soviets found out that the traditional Afghan costume of dark-colored (often browns and blacks) clothing made people invisible from the air whenever they simply lay down on the ground and remained still. They made maximum use of whatever cover was available. When Afghan fighters moved with munitions-laden pack animals, they either moved only at night, or always stayed near whatever cover there was (for the mules), and listened carefully for any low-flying aircraft or helicopters. These fighters blended in with local civilians, who rarely betrayed them to the Soviets. When the Russians decided to "clear the countryside" of Afghan civilians, this merely moved millions of Afghans to neighboring Pakistan and Iran, where the Afghan fighters now had safe bases from which to operate.

The biggest Soviet handicap was its self-deception. Well aware that the Soviet Union had conquered most of the Central Asian tribes in the nineteenth century, and reconquered them in the 1920s after the Russian civil war ended, the Soviets felt they knew how to handle the Afghans.

But there was a substantial difference between the Central Asian tribes and those in Afghanistan. The Central Asian peoples didn't have as many mountains to hide in. This made a big difference. Moreover, Soviet operations in the 1920s had the benefit of railroads, which made it much easier to concentrate military forces. The Central Asian peoples were more dependent on trade, and towns with markets. Most of the Afghans were not, being largely self-sufficient in their mountain valleys. Most important, the Afghans were able to find secure sanctuaries in neighboring countries, and extensive support from Western and Islamic patrons. The Central Asian tribes never had this advantage, or at least not to a significant extent. The Soviets, who were so keen on calculation in military matters, made a serious miscalculation in Afghanistan.

Actually, the Soviets' miscalculation was political. The Red Army staff officers who did examine the "Afghanistan problem" concluded that it would take more troops to win than Afghanistan's infrastructure could support. Soviet political leaders overruled their generals and went in anyway. As always, self-deception is the deadliest form of deception.

The basic Afghan tactic was to resist until the invader tired of the fighting and went home. This had worked many times in the past, and the Afghans saw no reason why it would not work this time as well. The Soviets, of course, saw the campaign as a typical bit of warfare in which all they had to do was destroy the enemy. But, as is common in guerrilla warfare, you had to find the little buggers first. The Afghans refused to be found in great numbers. Although the Soviets ended up killing or injuring over 10 percent of the Afghan population, they were never able to put a serious dent in the number of Afghan fighters.

Over the course of the nine-year war, the Soviets tried many different gambits. The most successful was to bring in thousands of their own commandos (*spetznaz*) and several brigades of air mobile troops. But the disorganized nature of the Afghan resistance made it impossible to use these elite Soviet troops effectively. Afghan fighters were generally operating in many small, autonomous groups. In effect, the Afghans set out from their Pakistani camps as tribal "war parties" to inflict whatever mayhem they could on the Soviets. Later in the war, the Afghans received hundreds of portable Stinger antiaircraft missiles. This made life very difficult for the Russian commandos and air mobile troops. The Afghans' favorite form of combat was ambush, and with the Stinger, they could now ambush Soviet aircraft.

The Soviets tried politics. Attempting to buy off many of the Afghan tribes, they found that the Afghans were wily negotiators. Much to the Soviets' chagrin, an Afghan bought would not always stay bought. Lies, subterfuges, and other trickery were also a favorite Afghan sport, and the Soviets were usually the victims. The more brutal the Soviets were, the more determined the Afghans became.

Eventually, the war ended when the deceptions the Soviet government used on its own people to keep the war going became untenable politically. From the beginning, the Soviet propaganda machine portrayed the invasion as an intervention based on an invitation by the Afghans. The official line was that the Afghan people pleaded for the Soviet troops to come in and protect them from foreign invaders, counterrevolutionaries, bandits, or whatever. Since most of the Soviet troops fighting in Afghanistan were conscripts, there was a constant flow of veterans, and body bags, coming back to remind the Soviet people that, as usual, the propaganda was a lie.

Many Soviet citizens began to compare Afghanistan with Vietnam, and the comparison was apt. Both wars involved a superpower interfering in another nation's civil war. And in both cases, the superpower was backing the weaker side. Equally significant, the faction that was supported by the foreigners eventually lost. Worse yet, the Soviet economy was beginning to unravel after twenty years of trying to outbuild the United States in an arms race, and the new political leaders in the mid-1980s were preaching "openness" and "restructuring" and were willing to withdraw from Afghanistan in 1988 to keep the Soviet population from getting more restless. Three years later, the Soviet Union fell apart. Self-deception always catches up with you.

Star Wars, the Deception

Here is a tale of technical deception on a grand scale. And it worked (the deception more so than the technology).

In the early 1980s, the United States embarked on an ambitious program to develop an antiballistic-missile defense system. Exotic technology, much of it placed in orbit, would be at the core of this effort which had the official title of "Strategic Defense Initiative" (SDI). Most people called it "Star Wars" (after the science fiction movie of the same name). Historians of technology pointed out that the scope of the scientific and engineering breakthroughs SDI proposed to achieve were rarely gained without a lot of false starts, detours, budget overruns, and much abject failure. But as the pet project of President Ronald Reagan, SDI went forward.

What was not known for many years was that SDI had another use, as a deception to force the Soviet Union to divert resources from workable technologies to the highly speculative work required to compete with SDI. This was all part of a "competitive strategies" technique and the increasing realization that the Soviet economy was coming apart because of high defense spending.

"Competitive strategies" was partly a deception program, but primarily, it was a plan to emphasize work on weapons and equipment that were easier for the West to develop, and more difficult for the Soviets to make or counter. Emphasis was also placed on developing weapons that the Soviets would have the most trouble dealing with. For example, antitank weapons that used microprocessors (i.e., computers) were given a lot of attention, as were precision aircraft bombing systems (more computers), stealth aircraft, and electronic warfare (still more computers).

Through the 1980s, it became increasingly obvious that the arms race the Soviets began in the late 1960s was doing greater harm to them than to anyone else. Although the official American line was that the Soviet arms buildup was going on unabated and was a great threat to world peace, insiders knew better. American intelligence agencies were compiling a growing mass of evidence indicating that all was not well in the Workers' Paradise.

Many of the new Soviet weapons had serious flaws and were often rushed into production more for political purposes than to give the

Red Army more combat power. The appearance of a new generation of Soviet leaders preaching *glasnost* (openness) and *perestroika* (restructuring) made it obvious to all that something was not working in the Soviet Union. Western economists pointed out that, since the 1970s, the Soviets (according to their own statistics) were diverting increasing amounts of capital from industrial investment (to modernize and keep it competitive) to military spending (which did little but scare money out of Western taxpayers).

Out of this was born a secret, semiofficial plan to purposely expose the Soviets to technical information about American advanced technologies—some of it real and some of it made up—that we knew the Soviets would have a hard time matching. One of the more notable bits of make-believe concerned SDI tests that were rigged (or the announcements of the results were, depending on whom you believe). At the time, some Pentagon officials were afraid that the deception was working better on Congress than it was on the Soviets. However, after the Soviet Union disintegrated, it became known that the Soviets did believe that many of the technical marvels being announced in the West would work. A historian of Soviet affairs would not find this surprising, as the Russians have long had a sometimes irrational awe of Western (and particularly American) technology. The Soviets spent billions of dollars a year on special projects to steal Western technology. Most Soviet computer technology, for example, was nearly an exact (but markedly lower-quality) copy of American equipment. What they couldn't copy, they would steal or (illegally) buy.

The Soviets also projected many of their own characteristics on the Americans. The most important example of this was their belief that, no matter what was published in open sources about American technology, the "good stuff" was probably kept secret. The U.S. Air Force's "stealth" aircraft projects (which were kept pretty secret by any standard) appeared to the Soviets as proof that, as impressive as American technology appeared, there was even better stuff that no one knew about. Moreover, the Soviet counterpart of the "military-industrial complex" had a vested interest in advanced American military technology. Only with this kind of threat could the Soviet weapons builders grab a lion's share of the Soviet Union's wealth. Thus hoist by their own petard, the Soviets were ripe for some technomilitary deceptions of the highest order.

The American deception plan was simple. As billions of dollars were poured into the SDI (Star Wars) project, an enthusiastic public relations campaign in favor of the project was kept going. The ostensible reason for the PR stuff was to keep the pressure on Congress to provide the money needed.

But during the 1980s, many people in the United States government came to understand how the Soviets handled news from the West. The Soviets had a large bureaucracy devoted to studying what was going on in the West, and these experts provided regular (daily, weekly, monthly) summaries of their findings to various levels of Soviet bureaucracy. SDI touts knew that if certain SDI developments were announced in a certain way, it would come across to the Soviet leadership as development potentially leading to a crisis situation.

There were two things about SDI that bothered the Soviet leadership a great deal: one was that SDI might work, and thus make useless the hundreds of billions of dollars the Soviets had invested in ICBMs; the second, and more immediate, danger was that the Americans would get parts of SDI working, which would enable the United States to quickly shut down the Soviet (and anyone else's) satellite system. Without satellites, a nation is blind. If there's one thing the Soviets fear more than being underarmed, it's being uninformed.

The Soviets knew they would have to match the SDI effort to maintain their superpower status. If both nations had a fully (or partially) functioning Star Wars system, neither would have an advantage, and both would retain their superpower status.

An effective SDI might take ten years to install, or (more likely) longer. If it appeared that the Americans were making progress, the Soviet Union would have to spend a lot of money to keep up.

The Soviet tactic since before World War II was to steal a lot of technology from the West, and fill in with their own limited resources to obtained working systems. This had worked with everything from tanks to nuclear weapons, and the Soviets were prepared to do the same with SDI. But the Soviets had been working on SDI since the 1970s, and they knew how formidable technical obstacles were. The American SDI specialists knew of the Soviet work; indeed, it was the alarming reports of the Soviet SDI project that helped get the American one going.

American scientists knew that the Soviets were having trouble with SDI technology, and decided to capitalize on these problems. As the SDI program went forward, U.S. government publicists issued blizzards of information reporting work that was being done, the vast amounts of money and talent involved, and, increasingly, the actual (or promised) results. While American critics picked apart SDI, the

Soviets tended to believe the PR, and became more nervous as SDI's achievements (real or imagined) grew apace.

The American deception effort wasn't planned from the beginning. In fact, the deception sort of just evolved. At first, it was thought that the actual progress on SDI research would scare the Soviets into bankrupting themselves trying to keep up. But it soon became apparent that the technical obstacles were going to be too formidable even for American science to overcome. At that point, it was decided to pass deliberately misleading information to the Soviets. Of course, the American researchers would know what was performing and what was not and would continue to work on the still-stubborn problems. But the Soviets would not be certain how well SDI was doing and would, in their typical fashion, believe the worst. The Soviets were expected to go into techno-panic, and this is precisely what they did. After the fall of the Soviet Union in 1991, one senior Soviet leader was asked what effect the SDI race had, and he replied, "It hastened our collapse by five years."

While this deception was something of a success, it was marred by the failure to clearly tell Congress that it was a deception. One could argue that Congress could not be trusted to keep the secret, but then, certain Congressional committees have for years been privy to top-secret material without leaking choice bits. Or at least, mostly without leaking crucial items. On highly politicized items, there is more risk that a member of Congress or a staffer would leak something. In any event, the deception went forward, had the desired effect, and Congress found out only later. Congress was not happy. But SDI had in the meantime mutated into a more modest program to protect America and its allies against the more primitive weapons of the post–Cold War bad guys.

BDA Blues

The Coalition offensive of the 1990–1991 Gulf War kicked off with a spectacular, and quite successful, six-week bombing campaign in early 1990. The one problem with this operation was the difficulty in discovering precisely what damage had been done. Bomb Damage Assessment (BDA) is what this double-checking is called. For many decades, the normal drill was to have a recon aircraft, often the one that spotted the targets in the first place, return after the bombing to take more pictures to determine what was hit by the attack. This would also settle questions about whether another bombing mission was necessary, and how effective your weapons and techniques were in the first place.

With all the effort that went into precision-bombing systems in the previous twenty years, you would have thought that more attention would have been paid to BDA. But practicing BDA in peacetime is difficult. An enemy who is bombed will (and often does) take actions that make BDA difficult. For example, as soon as the bombers have left (and often even before), the bombed troops will start cleaning up the damage, and moving anything portable before the bombers can return. All this usually takes place before the recon aircraft show up to begin the BDA process. Thus what the recon aircraft see is often not what existed right after the bombs went off. Photos are taken anyway, and the BDA process continues back at a friendly air base, where specialists examine the photos and try to determine what damage was done. Their task is complicated, because they usually have only photos of what the target looked like before the bombing. They have little idea of how much camouflage was present in the target area, or what the enemy troops did in the immediate aftermath of the bombing.

This is the essence of the BDA problem. In all areas of military affairs, it's a constant peacetime problem figuring out what potential enemies will do in wartime. Such was the case with BDA in Iraq and Kuwait in 1991. As much as was known about Iraq before the war, it was insufficient to clarify what the Iraqis would do to the BDA situation. There were ample satellite photos of Iraqi military activities during its 1980–1988 war with Iran, and much data were also collected at ground level. But during the Iran-Iraq war, there was nothing like the bombing campaign the Coalition unleashed in January 1991. Thus there was no prior experience on how the Iraqis could be expected to react once they had to deal with all this bombardment.

The Coalition soon discovered that the Iraqis could be quite resourceful when faced with a new danger from above. In addition to using various forms of camouflage and fortification for their armored vehicles, they eventually went to the extreme of actually burying their tanks. When these buried Iraqi tanks were being bombed, the Coalition quickly discovered that not all the apparent "dug-in" positions contained tanks. They thought that if a bomb hit what appeared to be a dug-in tank, and the next day a black spot appeared in the area, this indicated that a tank had been hit and exploded (the resulting fire having caused the black smudge). But the Iraqis soon realized that a black spot was a form of protection, and it wasn't until after the war that American intelligence analysts discovered that not all black spots on the ground are the same.

The Iraqis proved to be quite adept at camouflage and various forms of deception. Even at the front line, they were quite skillful in making BDA more difficult. False fortifications, multiple fortifications (with not all of them manned), and phony minefields all contributed to the trepidation of Coalition troops as they prepared to attack the Iraqi ground forces.

Not only was BDA a problem, but finding targets in the first place was made more of a chore. This, plus problems with weather, caused 25 percent of the bombers returning without having found anything to attack. Not wanting to return with unused bombs, many pilots stretched the definition of a "target" and hit anything that might be one. The Iraqis quickly discovered what the enemy bombers were looking for, and just as quickly provided suitable "targets" that did not contain Iraqi troops or weapons.

The proof of the Iraqi success in these deceptions was discovered when captured Iraqi unit commanders generally admitted that the vast majority of their vehicle losses were to Coalition ground forces, not aircraft. A Coalition tank could get up close and root out Iraqi troops and vehicles. A Coalition aircraft had to do its work from two or three miles up in the air and while moving at several hundred miles an hour.

Camouflaging targets, and quickly clearing and recamouflaging bombed targets, is a low-tech activity. Anyone can do it. Historically, nearly everyone has. What future opponents will tend to forget is that Coalition bombers did make life decidedly unpleasant by systematically destroying what they could find. Often, the targets that were hit consisted of supply trucks heading for frontline Iraqi troops. The shortage of supply at the front caused most Iraqis to take sick, or at least saw their morale plummet, and many did die from the lack of supply. Morale was eventually completely shattered, causing most Iraqis to surrender as soon as Coalition ground forces came into view.

Air power wasn't decisive by itself, particularly when confronted with Iraqi deceptions, but the bombers and Coalition ground forces proved to be a devastating combination. And Iraq's skill at deceiving the Coalition's BDA experts ultimately only led to even greater devastation.

SCUD Hunt

Iraq was expected to use its SCUD missiles during the 1990–1991 Persian Gulf War. To forestall this, the known SCUD launch sites were bombed during the early days of the six-week bombing campaign. But when the SCUDs later rained down on Saudi Arabia and Israel, it was clear that the Iraqis had run a successful deception campaign to protect their SCUD launch capability. It got worse after that. Not only were Iraqi deceptions able to keep many of their launchers hidden before the war, but they were able to defeat massive efforts by Coalition aircraft to find these launchers during the war. There was nothing high-tech or mysterious about Iraqi deceptions. Like most police states, Iraq was able to keep much information on their SCUDs away from prying eyes before the war. The Coalition thought they knew how many SCUD launchers the Iraqis had from the satellite photos taken before the war and what little firsthand information friendly agents were able to provide. But as the war went on, it became clear that only the Iraqis knew exactly how many launchers existed, and the Iraqis weren't talking.

The additional launchers were largely locally made rigs based on truck-trailer chassis. This configuration had several advantages. For one thing, these launchers still looked like truck trailers, especially from the air. The custom-built launchers that the Russians sold with the SCUDs had a rather distinctive appearance, and these were the launchers Coalition pilots were trained to spot. Once the Coalition became aware of all the homemade launchers, a lot of trucks driving along Iraqi highways became targets. Worse yet, the homemade launchers were cleverly hidden. Their size (smaller than the official Soviet launchers) and appearance made it easy to park a launcher among truck trailers and leave it to the Coalition pilots to figure out which was the launcher. It was an impossible task. The Iraqis were up to the task of playing cat and mouse with enemy pilots. Launchers were hidden during the day. Any form of concealment would do, including highway overpasses, culverts, clumps of trees, or inside towns and villages. Where concealment was insufficient, camouflage was used.

Many civilians and military leaders in the West thought that American satellites would quickly pinpoint the location of Iraqi launchers.

The Iraqis knew better, having received, and used, American satellite intelligence information during their 1980s war with Iran. The satellites only pass over the area a few times a day, and it requires several hours (at least) to process and analyze the resulting information. Some of the satellite data is in the form of photographs; some is "imagery" (infrared and other sensor scans showing heat patterns or the presence of metal objects). While the satellite data could help in finding and identifying large military units (tank battalions, etc.) or bases (especially air bases), it was pretty much a bust in finding Iraqi mobile SCUD launchers. Most of the SCUD launchers bombed early on were the fixed ones, which were rather like miniature Cape Canaverals. These were easy to spot. The Iraqis kept quiet about their efforts to build homemade launchers, although there was information available on this effort. Coalition intelligence agencies did not make as much of Iraqi launcher building as they should have. Not that it would have helped all that much. These mobile, homemade launchers would still have been very difficult to spot.

When the continued Iraqi SCUD launches became a major political problem, extensive efforts were made to stop them. The Coalition military commanders had already announced that they had "destroyed most of the Iraqi SCUD launchers," so they were quite eager to eliminate this embarrassment to their military prowess. Nothing really worked. Even combat aircraft crisscrossing the likely launch areas at night were unable to stop the launches. Bombing raids on likely storage areas for the missiles themselves may have had some success, and a few launchers were apparently hit. But the Iraqi launch crews were no dummies. They had the advantage, and they used it to the hilt.

The continued SCUD launches were the one Iraqi success of the war, and the Iraqis made the most of it in their propaganda. This achievement was a well thought out and applied use of deception.

"Hail Mary"

"Hail Mary" referred to the maneuver used to go around the Iraqi flank during the Persian Gulf War of 1990–1991. The term "Hail Mary" is taken from an old football term for a long pass that you hope will connect (thus you pray, "Hail Mary . . .").

This Gulf War was notable for the amount of active and passive deception that was attempted, and the extent to which a lot of it succeeded. This was unusual, because the side that was most successful was the one that had the least amount of recent combat experience. The American-led Coalition was largely composed of U.S. divisions that had last operated in large campaigns during the early 1970s, some twenty years earlier. Yet, the Americans were able to pull off one major feat of active deception by causing the Iraqis to think that Coalition troops would come straight through Iraqi positions near the Persian Gulf, rather than swinging around through the open desert inland and past the dangling Iraqi flank.

The Iraqis were no amateurs in the deception department. Iraq had recently concluded a major war with Iran in 1988. There was a fair amount of deception used in that conflict, by both sides. It was understood by the Coalition generals that the Iraqis would not be easy to fool. The Iraqis had garnered a lot of experience in defensive fighting, and in the construction of field fortifications. The basic problem was that the Iraqis used their experience in fortification, which had served them so well during the Iran-Iraq war, to promptly throw up an impressive-looking line of defenses along the Kuwait-Saudi-Arabian border. This presented a seemingly formidable obstacle to the Coalition. And everyone knew that the Americans were very chary of suffering high casualties.

While the Coalition soon determined that they could breach these fortifications with not much trouble, they were more interested in running the Iraqis out of Kuwait quickly and with minimal Coalition casualties. To do this required moving half of the Coalition's divisions far inland so they could advance through the Iraqi desert. This would be the classical "go around the flank and roll up the enemy line" maneuver. The Iraqi western flank in Kuwait went out into hundreds of miles of desert. This was too vast an area for the Iraqis to defend with the type of fortified positions they built on the Kuwait border with Saudi Arabia. However, if the Iraqis felt that the Coalition was likely to come through the desert, they could have put more troops out there, and perhaps minefields in key areas. These Iraqi troops could not have stopped the Coalition advance through the desert, but they could have slowed it down and caused more casualties. By doing this, the Iraqis would have had time to get their reserves (the Republican Guard) into position to cause more trouble. Or, worse yet, the Republican Guard would have had time to get out of Kuwait intact.

If the Iraqis could be convinced that the main attack was going to come straight up from Saudi Arabia, they would keep their reserve units due north of the border, ready to hit the attacking force coming straight up from the south, or take a less aggressive option and promptly retreat back into Iraq. The "Hail Mary" would have the Coalition armored divisions charge across the empty desert to the west of the Iraqi lines and surround all of the Iraqi forces in Kuwait. If properly executed, the Iraqis would still be waiting for the Coalition forces coming straight through the fortified line when the Coalition flanking force would come, unexpectedly, storming out of the western wastelands.

For the "Hail Mary" to be successful, the Coalition had to deal with the following items:

1. The Iraqis had some knowledge of where Coalition forces were. Although the Coalition had a screen of Arab troops along the border, there was still a lot of Bedouin traffic across the open desert areas to the west. Some of these "Bedouin" were Iraqi agents. Moreover, the Western media were covering the deployment of troops into Saudi Arabia. Although movements of the press were controlled, a lot of information was getting out via the media, or from people who were simply driving through the area. The Coalition could not assume their movements, and the location of their units, was a secret from the Iraqis.

2. The Iraqis had to be made to think that the flanking maneuver would not be used. The Coalition allowed U.S. Marine divisions to get a lot of press attention, especially when the marines were practicing their breaching tactics (techniques for attacking fortified lines) and their amphibious landings. The British division was also filmed undergoing the same training. Other mobile divisions were kept near the Persian Gulf, as were most of the millions of tons of supplies that were being stockpiled for the coming battle.

3. The Coalition had to be able to move over a hundred thousand troops and twenty thousand vehicles quickly from the vicinity of the Persian Gulf nearly a hundred miles inland in order to execute the flanking movement. This was the key to making the "Hail Mary" work. Right up to the day when the ground battle actually began, extra care was taken to keep the press from seeing the massive movement of troops to the west. Accompanying this movement was an equally large mass of trucks carrying the needed supplies.

For success, the Coalition deception was dependent on two things. First, for weeks, information was sent out indicating that the Coalition was preparing to blast straight through the Iraqi lines. This succeeded, but not so much because of the Coalition's press manipulation. The Iraqis were convinced that any Coalition force advancing into the desert would promptly get lost in the trackless wastes, and thus had no choice but to come straight north. This was curious for several reasons. The Iraqis themselves were not used to operating in the desert. Most Iraqis lived in river valleys or mountains. They considered the desert a dangerous place, and had regarded it as such for thousands of years. While the Iraqis knew that the Americans had the new global positioning system (GPS) navigation devices (that used satellite signals to give a precise location of the receiver), they did not realize that GPS would revolutionize desert navigation. The Iraqis, especially their leader, Saddam Hussein, also did not appreciate the high degree of professionalism among American forces. Saddam was still thinking of the Americans as they were in the demoralizing years after Vietnam. It was this professionalism, as well as the GPS, that enabled U.S. forces to succeed in the desert. This was a case of the Coalition deception working, with some assist from some Iraqi self-deception.

The second deception Coalition forces depended on was a rapid movement of flanking forces at about the same time the attack through the Iraqi defense line took place. This would, in effect, be using the assault on Iraqi fortifications as a feint. The half-dozen divisions swinging around the Iraqi flank would be the main attack and would do most of the damage. This huge mass of over twenty thousand vehicles had to move at the rate of over a hundred miles a day for nearly a week. This was an unprecedented rate of speed for such a large force over trackless desert terrain. The Iraqis felt it couldn't be done. By the time it was done, it was too late for the Iraqis to do anything about it.

As the Coalition "left hook" reached the Tigres River, deep in Iraqi territory, Iraqi forces were just becoming aware that something had gone terribly wrong. Meanwhile, the Coalition assault through Iraqi fortifications had gone better than expected. Most of the Iraqi forces on the Saudi Arabian border were in full retreat. These Iraqi forces reached the Tigres after the Coalition "Hail Mary" forces had already gotten there. Some Iraqi troops, in flight from Coalition forces that had blasted their way through the fortifications, then came upon more Coalition troops who had rapidly advanced through the desert to the west.

The hundred-hour war ended with Iraqi forces in complete disarray. This swift defeat of the Iraqi forces was not a sure thing. It happened because a deception worked.

They Also Serve, Who Only Float and Wait

Sometimes, the perfect deception is one in which you do nothing.

During the six-week bombing campaign of the 1990–1991 Gulf War, much was made of the U.S. Marines aboard their amphibious assault ships in the Persian Gulf. Coalition commanders did nothing to discourage speculation that these marines would be used in the upcoming land battle.

What the Iraqis didn't know was that early on, the decision had been made not to use amphibious assaults. The reasons were twofold. Principally, the U.S. government had decided to put great emphasis on keeping American casualties as low as possible. Amphibious assaults are notorious for being bloody. The marines are trained to push across a defended beach despite high casualties. Indeed, U.S. Marines have never been pushed off a beach on which they have landed, often in the face of horrendous losses. While it was expected that the marines could land and fight through Iraqi beach defenses without exceptionally high losses, experienced marines knew there would be some friendly casualties.

The second reason for avoiding attacking from the sea was the presence of mines in the waters off Kuwait, and the lack of adequate mine-clearing capability in the Coalition fleet. All it took was for one assault ship to hit a mine and there would instantly be many marine casualties. But Coalition commanders knew that the long Kuwaiti coast was, technically, quite vulnerable to an amphibious assault, and so did the Iraqis.

A decision by the United States to put strong emphasis on keeping casualties down was kept secret, although if the Iraqis had studied American domestic politics closely enough, they could have reasonably deduced that this nation's strong aversion to battle losses would have precluded something as risky as an amphibious assault.

The American government played the "casualty issue" to the media in such a way that they acknowledged this fear without letting on that they had already decided to shape their entire strategy around keeping American deaths as low as possible. The Iraqis did assume that Americans would not stomach heavy losses, and featured this assessment in their propaganda. But the Coalition deception plan gave much exposure to the marines offshore, and their landing exercises in the months leading up to the ground war. The press was allowed access to the corps, and the marines did their usual job of playing up their combat-ready image. The Iraqis fell for this also, and deployed several divisions to guard the Kuwaiti coast.

When the ground fighting began, there were over five Iraqi troops guarding the beaches for every marine floating offshore. To make sure that the Iraqis stayed put on the beach as Coalition forces advanced north from Saudi Arabia behind the Iraqis, marine amphibious craft and their supporting warships maneuvered offshore as if getting into position for a landing. During these movements, two American ships did hit mines. Several casualties occurred, although the ships were not sunk. U.S. Marines did not land; the Iraqis guarding the beaches were captured or dispersed by Coalition troops coming from behind. The deception worked.

Stealthy Deception

When aircraft first began to adopt deceptive measures during World War II (1939–1945), they relied largely on electronics or operating at night, and to a lesser extent, on camouflage and flying low to avoid radar. Thirty years after World War II, another way was found to camouflage the presence of aircraft.

In the 1970s, technology had reached the point where it was thought that additional technology could make an attacking aircraft invisible, or nearly so, to radar. This was the concept that became known as "Stealth Aircraft." Ironically, this was not entirely new. The British had a very popular and successful wooden aircraft (the Mosquito) which was used as a reconnaissance plane. Because of its wooden construction, the Mosquito had a lot less metal (mainly the two engines) to reflect radar waves. Ironically, at the same time radar came into use, wood was being replaced by metal for aircraft construction. But early in World War II, when radar technology was still quite primitive, a single Mosquito would get past all but the most alert and expert radar operators. Even as radar became more capable, a single Mosquito was difficult to spot with the electronic devices available.

Ever since World War II, radar has been the biggest threat to aircraft. Stealth aircraft render most current radars much less effective. To appreciate this fact, you have to understand how radar functions. It works much like a flashlight at night. The beam bounces light off an object so that your eye can see it. The light that hits an object at night literally bounces off the object and back into your eyes, where you "see" it. Radar sends out high-frequency beams that bounce off objects. The large dish that is typically portrayed as a radar site "sees" the reflected signal and these days uses a computer to figure out where the object is. As with a flashlight, the color of an object has a lot to do with whether you can see something. For example, a light-colored object will reflect more light; a dark object will absorb light and give you a much less visible "return." While a radar receiver can "see" at a longer range than the human eye, its signals are more easily scrambled. Part of a radar signal can be sent off in another direction when it hits an object at an angle, and doesn't return to the radar dish. This is the primary technique of stealth: lots of rounded surfaces (as well as material that absorbs radar signals). For this reason, a pickup truck gives a radar signal return of 200 (square meters). A Boeing 747 jet, with all the round surfaces typical of an aircraft, only gives a return of 100. The B-52 bomber has even more curves and has a return of 40. A cruise missile or fighter bomber has a return of 10. The B-1B bomber, designed to minimize radar return, has a return of .4. The B-2 and F-117 stealth aircraft go a few steps further and produce a return of. 01, about the same as that of a large bird. These are the radar returns obtained if the aircraft is flying straight into the radar beam. Most aircraft, particularly the "semi-stealth" B-1B, have much larger exposure to radar if they are hit with radar beams from the side or rear, where the radar beams have a lot more surface to bounce off.

The B-2 and F-117 also make extensive use of radar-absorbent materials. Just as dark objects do not reflect light, certain nonmetallic composites tend to absorb radar signals. Thus a radar that could pick up a cruise missile at almost two hundred miles would not see a B-1B until it was only about sixty miles away, and the B-2 would not be picked up until it was about forty miles off. That might not seem like much of a problem for the radar user, except that the B-1B and B-2 carry sensors that tell them where enemy radar signals are going to be weakest or nonexistent, thus allowing the aircraft to dodge and weave around areas in which they are more likely to be spotted. Yes, the B-2 can be detected, but often not soon enough to scramble and direct fighters or launch missiles efficiently.

Stealth is not invulnerable, but it turns aircraft into a-needle-in-the-haystack problem for searching radars. Worse, Stealth bombers carry cruise missiles or short-range attack missiles. They don't even have to pass over the target to attack it. By the time you find a stealth aircraft, your radar may be seconds away from destruction by a radar homing missile. More likely, the Stealth bomber, cruising along at nine miles a minute, will have left your sector without your even being sure it's gone, or that it was there in the first place.

The American 1970s stealth effort resulted in the F-117A. Its primary purpose was to lower U.S. aircraft losses in combat. To appreciate this, you have to understand how air attacks are usually done. Without stealth aircraft, you must send in larger groups of aircraft ("strike packages"), led by expensively equipped, single-purpose, electronics warfare planes (Wild Weasels). This approach will either figure out how to avoid enemy air defenses, or destroy a lot of enemy ground defenses that get in the way. If nothing else, this makes it easier for subsequent raids. Large strike packages were first used during World War II. Then, and now, such a large operation also attracts enemy attention and often develops into a major air battle. The strike-package approach surrenders the element of surprise, and you end up using up to ten support aircraft for every one that is actually going to bomb the target. The Wild Weasels leading these raids have radar detection and jamming equipment which can either hide the group from enemy radar or prevent the enemy from making accurate use of its ground-to-air missiles. You still need fighters to deal with enemy interceptors. Thus the battle can range from ten thousand meters and up as well as down to ground level. The Wild Weasels carry missiles that home in on enemy ground radars. The most dangerous opposition still comes from enemy guns, which often can fire without radar in clear weather. For this reason, raids at night and in bad weather are often preferred. If all goes according to plan, the Weasels will protect the electronically less sophisticated strike aircraft to the targets, where the bombs or missiles are used. Everyone then fights their way home past a thoroughly alerted enemy. The stealth aircraft are well suited to perform the Wild Weasel missions.

But stealth aircraft don't normally lead strike packages; they go in alone to destroy critical targets. In particular, air-defense headquarters and similar targets are popular stealth targets. Stealth aircraft deceive the defender into thinking there's nothing out there. Until the guided bombs hit their targets. In effect, stealth aircraft are flying ambushes. They hide behind darkness and radar "invisibility."

Stealth can be defeated. And now that there are several stealth aircraft in service, and their capabilities are generally known, much is being done to strip away the stealth advantage. This is done by taking advantage of the fact that stealth technology does not make an aircraft "invisible"; it just produces a weaker signal for the radar-receiving equipment to process. It cuts detection range of radars to about 25–30 percent of their normal capability. Thus you can simply add more radars, or adopt a cheaper alternative, such as adding more radar signal receivers. The latter method would allow the many signals scattered by the stealth aircraft's odd shape to be reassembled by a single computer. It's also theoretically possible to use television signals. But the TV approach requires new receivers and powerful computers. This might work within industrialized nations, where there are a lot more TV transmitters, but less-developed nations would have problems with the TV approach. However, even if an exact position is not determined with any radar technique, the defender can get a good idea where the stealth aircraft is and send interceptors to the area for a closer look. But with the introduction of stealth cruise missiles, sixteen of which can be carried by the B-2, this solution becomes more difficult. Problem is, most nations are not too strong in that kind of computer technology.

Whichever solution is adopted, the bill will be expensive. But there are theoretically cheaper ways to increase stealth detectability. Different radar frequencies that are not absorbed by the B-2's special outer skin can be used. Different radar frequencies can double the return. But first, you have to figure out exactly what the B-2 is made of, build the radar, and either test it on an aircraft made of the same materials or bet the farm that your theoretical calculations are accurate.

But wait; you may be able to dispense with radar. Stealth aircraft still use jet engines, which throw off a lot of heat. Infrared (heat-sensing) detectors are becoming much more common. The Russian MiG-29 fighter uses an infrared sensor instead of radar for its cannon and short-range missiles.

Satellites use infrared sensors to detect aircraft, although a handful of industrialized nations are the only ones that have these satellites. Infrared sensors can be placed on the ground, a lot of them, and wired to a central location to indicate the passage of a stealth aircraft. Another visible item is the contrail of the aircraft's jet engines. While a corrosive chemical can be added to the fuel to make the contrail invisible, it can still be seen by ultraviolet detectors. The B-2 also uses some active radar, particularly for its low-altitude navigation system, which can be detected. However, the emissions are brief and low-powered. The principal means of navigating are with an inertial guidance system and position updates from satellites. The radar is used largely to keep track of altitude and upcoming obstacles.

Also keep in mind that for all its apparent efficiency, electronic-equipment performance is a now-and-again thing. Anyone who has experienced the occasionally unstable performance of radio and TV equipment knows what this is all about.

Any electronic gear that sends signals through the atmosphere must also contend with natural interference. As a result, equipment specifications are misleadingly deceptive. A radar with a quoted range of sixty-two miles will not spot every target every time at that range. More detailed equipment specs will reveal that the radar manufacturer warrants that at sixty-two miles, targets of a certain (usually large) size can be spotted 90 percent of the time; at thirty-one miles, 99 percent of the time. In practice, spotting probability may be under 10 percent at sixty-two miles and only 50 percent at thirty-one miles. This is the element that stealth technology exploits.

Stealth does not try to make something invisible to radar, simply harder to detect. Since detection must be continuous to be useful, a target that blinks on and off upon the radar screen is less likely to be tracked and hit by a missile. The electronic battlefield is one of probabilities, not certainties. Victory will go to the side that can best cope when the gadgets don't perform according to the spec sheet.

Stealth aircraft provide the air force generals with another deception tool. Even with the uncertainty of its use, stealth remains a potent weapon because, even if it doesn't work as advertised because of some clever defense the enemy has put up, you are never sure just how ineffective the aircraft is. Normally, aircraft are not used as weapons of deception. The stealth aircraft changes all that. The more uncertainty the enemy has to worry about, the less powerful he is. Because of its use as a deception, and because of its very existence, stealth aircraft have had an impact far beyond the bombs they drop. Such is the power of deception.

Deception at Sea

One would think that deception at sea in the twentieth century would be difficult, if not impossible. Such has not been the case.

Camouflage, concealment, and a host of other deception techniques have been widely used at sea in this country. In fact, all of the tools of deception are used at sea, often in very imaginative ways.

- While it's certainly true that there are few places to hide at sea, concealment is available at times. Capable sailors find and use this concealment when they must. There are two forms of concealment at sea. The more obvious is the weather. Fog and squalls have long provided concealment during naval operations. While these weather conditions are generally looked upon as a hazard, when things are going poorly for you, these same navigation perils become a refuge from enemy observation. Hiding in the weather declined in effectiveness with the introduction of radar during World War II, but it still has its uses. Not all aircraft, for example, have radar that can keep track of ships. One can still expect ships to seek concealment in a nearby fog bank when they encounter hostile aircraft.

  Ships can also make their own concealment. "Laying smoke" is a technique that developed early in this century. This was simultaneous with the development of modern ship classes (battleships, cruisers, and destroyers). Destroyers were the escorts for the larger ships and, if the battle turned against you, your destroyers would be sent forward to "make smoke" in order to cover the retreat of larger ships. The introduction of antiship missiles made the use of smoke less effective, but the technique is still used.

  An ancient, and still effective, concealment method is the use of land to hide behind. This can be a peninsula or island for surface ships, or shallow water for submarines. Actually, surface ships can also "hide" in shallow water if larger ships (with a deeper draft) are hunting for smaller ships (with a shallower draft). Small coastal warships can hide up a shallow river that larger ships cannot enter. Despite the frequent use of the term "high seas," most naval battles have, and continue to, occur close to land. A cagey sailor takes nearby land and water depth into account when planning an operation.

  Submarines use another dimension of aquatic geography when looking for concealment. Submarines are typically searched for using sonar (sort of an underwater radar). Sonar equipment broadcasts noise, and then listens for the sound bounced back after it has hit something (like an enemy submarine). Water, however, does some strange things to the bounced-back sound, depending on water salinity and temperature. In shallow water, the sound will also bounce off the seabed. Moreover, water of different temperature (and sometimes salinity) tends to form layers. There may be several markedly different layers between you and the object you are "pinging" (bouncing noises off). Even with the help of powerful computers, it's often difficult to get a good idea of where, and how far away, the target you have pinged is. Submarines take advantage of these layers, using onboard instruments to keep track of the layers they are passing through. If they want to hide, they will get several different layers of water behind themselves and their pursuer. This is often sufficient to make it impossible for the sonar to locate its target with any accuracy.

- Camouflage is another deception tool at sea. Modern ships have always used as a form of camouflage the nearly ubiquitous "battleship gray" paint job they wear. This gray color serves to make ships less visible under most visibility conditions. The sea is a fairly monotonous place when it comes to color. Much of the time, it's just shades of gray and blue.

  More attention was paid to more elaborate camouflage paint jobs during World War I, when ships had to face the widespread use of submarines for the first time. Because subs often attacked while submerged, the mere shape of a ship stood out on the horizon. The sub commander was looking at its targets from a periscope that was only a few feet above the water level. In effect, a periscope view was looking up at its targets, and not seeing much under the best of conditions. Sailors on the receiving end of these torpedoes soon realized that, while a new paint job would not hide them, it could throw off the aim of the sub captain. World War I torpedoes were not as accurate as later models, and the sub captain had to look through the periscope and estimate range and speed of the target ship before he ordered speed and direction settings for the torpedoes about to be launched. A paint job on the target ship could make it difficult to estimate distance or speed for the target and thus there was a good chance of aiming the torpedo incorrectly. This concept led to various "dazzle" paint schemes, which broke up the silhouette of the ship, suggested it was moving in the opposite direction, that it was moving at high speed, and so forth. The extent to which this approach was effective has never been demonstrated, but it appears to have some value. At the same time that merchantmen and warships were being repainted, better antisubmarine warfare techniques were being developed. All agreed that the new paint jobs didn't make it any easier for the torpedoes to hit something. And the obvious effort expended in repainting ships improved the shaky morale of civilian crews on the merchantmen.

Similar paint schemes were tried on combat ships. Two are worthy of note, and both were intended to make life harder for enemy submarines. One was to paint "waves" on the bow. This would make a submarine captain viewing the ship through his periscope think that the target was going faster than it actually was. This was an important consideration, as the sub captain had to estimate the speed of the target and fire the torpedoes "ahead" of the target to make sure torpedo and ship courses intersected. Another painting trick to bedevil submarine skippers was to paint a smaller ship (like a destroyer) on the side of a carrier or battleship. This deception only had a chance of working if visibility was bad. If it did work, it might cause the sub captain not to fire, as he would not want to "waste" torpedoes on something like a destroyer. If nothing else, the phantom destroyer would cause a delay in the firing of torpedoes, in the hope that the "destroyer" alongside the larger ship would move out of the way. Neither of these paint schemes was particularly popular, as there was no way of knowing if it was worth the trouble. But given the number of carriers and battleships that got hit by sub torpedoes during the two world wars, any protective measure seemed reasonable.

When World War II came around, the bizarre paint schemes of World War I were not used much by the Americans. They felt that the effort could be better spent on antisubmarine warfare techniques. Moreover, the United States noted that the most dangerous actions were by groups ("wolfpacks") of U-boats attacking at night while on the surface (the subs were hard to see at night, so they surfaced to fire their torpedoes with greater accuracy). The British stuck with the World War I zigzag paint schemes on their merchant ships, if only to give crew morale a boost, as did the Japanese. In fact, the Japanese tried a few interesting wrinkles to the scheme, even painting aircraft-carrier decks so that, from the air, they suggested a battleship, in the hope of deceiving aerial reconnaissance.

But this was not the end of camouflage for ships. The Japanese soon found themselves under constant attack by Allied aircraft in the Pacific. While Japanese warships could shoot back, and were designed to evade attacks quickly, or survive a fair amount of damage, the slower, unarmed, and less robustly constructed cargo ships needed all the help they could get.

The Japanese came up with a rather clever concept. They turned some of their cargo ships into what appeared to be little islands, or simply part of the shoreline. In the Solomons and New Guinea area, where fighting went on into 1945, the Japanese were desperate to supply their troops. By 1943, Allied aircraft controlled the air most of the time, and any Japanese cargo ship caught at sea in daylight was almost certain to be sunk by aircraft. By 1944, the Allies were constantly patrolling the area's waters with ships as well. The Japanese ships could move at night and escape aerial bombardment, but come daylight—there they were. Or were they? The Japanese first sought locations near the shore where they could hide their supply ships. Most Japanese cargo ships in general, and more so in the New Guinea area, were quite small (three hundred to a thousand tons). Some were simply motorized barges. The Japanese called their numerous small cargo craft, aptly enough, "sea trucks." These shallow draft vessels could get close to shore, or even up the larger streams on some of the islands. Many of them were capable of being hauled onto the beach. From this, it was but a small step to festooning the ships with camouflage netting which would be "garnished" with foliage. This done, a ship close inshore would, from the air (and sometimes the sea), look like part of the shore. If a ship could not get very close to shore by daylight, it would lay on even more foliage so that it appeared like one of the many small islands that were sometimes found off the larger ones. Since these small ships were slow, they could only cover about eighty miles a night. Cheap and often crudely built, the "sea trucks" needed a lot of maintenance. To take care of this, and to make the most of those areas that were easiest to hide in, the Japanese set up small "truck stops" for their nocturnal supply ships. Fuel, spare parts, tools, and mechanics were stationed at these stops, enabling the Japanese to keep the maximum number of sea trucks in operation. Despite the fact that the Allies increasingly dominated the air and sea, the Japanese were able to keep troops fighting in areas hundreds of miles behind the forwardmost Allied units. At the end of the war, Japanese troops were still holding out in New Guinea and the Solomons, in no small part because of the use of the camouflaged sea trucks.

It took a while before the Allies caught on to this Japanese camouflage effort. It was only a matter of time before a sharp-eyed Allied pilot caught sight of a Japanese boat with a sloppy camouflage job. There were also reports about cleverly camouflaged ships as told by the civilian coast watchers hiding out on Japanese-controlled islands. Even with their secret revealed, it was not always possible to spot every camouflaged Japanese ship. But the Allies tried, and many patches of island jungle were bombed and shot up simply because they looked suspiciously like a hidden boat.

During the Vietnam War, the Communists reinvented the Japanese "sea truck" as a way to avoid the seeming ever-present American aircraft. So, like any good deception technique, you can expect to see this one revived, or reinvented, when needed.

- False and planted information is another very popular deception technique for naval forces, because ships at war spend most of their time "out there somewhere." Armies are large and sprawl all over the place, making them easier to keep track of, and making it more difficult to pretend that they are where they aren't. Once a group of warships departs from its port, it's relatively easy to make the enemy believe that the ships are going where they aren't. The introduction of aircraft and space satellites made it easier to keep track of ships. But the oceans are still vast, and even with modern technology, it is still possible to convince the enemy that your ships are where they aren't. Most of these deceptions are accomplished with electronic trickery. As navies become increasingly dependent on long-range sensors, the opportunities to "plant" false information via counterfeit electronic transmissions grew. As early as World War I, when warships first carried wireless telegraph equipment into action, the potential for electronic deception was recognized, and used. This tactic accelerated during World War II and continues to grow up to the present. Since there have been no major naval campaigns since 1945, this use of false and planted information at sea has not seen much use. But the naval tricksters are ready, and waiting.

- Ruses in naval warfare are another matter. At sea, even under benign conditions, it is often difficult to identify ships until they are relatively close. Ships' lookouts and aircrew are trained to identify ships accurately, but even with these often strenuous efforts, misidentifications are common. With all that, ruses are not that often used by warships. Ship captains will take advantage of opportunities to employ ruses and make the enemy think he is looking at a friendly ship. But the opportunities are relatively rare. It's difficult enough to keep track of who's who, even when ruses are not being played out. This is particularly true during combat, when "friendly fire" damage is quite common. Always has been, still is. Even when there is no shooting going on, ships at sea still blunder into each other at night or in foul weather. Normally, warships are eager to be identified quickly and accurately.

Commerce raiders are a different story. Early in World War I, German commerce raiders (armed merchant ships) had much success pretending they were Allied merchantmen, until they revealed their guns and demanded surrender. The Germans used some simple techniques, like simply flying an Allied flag and painting the name of a known Allied merchantman on their hull. This did not always work, because there were ship-identification books available, and many ship captains kept one handy. If a wary Allied sailor checked one of these books and found that the name on the German ship didn't match the description of the real ship of that name, they would attempt to flee. The Germans overcame this problem by carrying equipment that allowed them to raise a false funnel and otherwise change their superstructure to roughly match the description of the Allied ship whose name they were hiding behind. But the reason commerce raiders were not a huge success, despite the effectiveness of their methods, was that they were slow merchant ships being pursued by swifter warships. Moreover, the oceans were largely controlled by Allied ports. Even neutral ports had pro-Allied people around who would quickly report (over the new-fangled wireless telegraph) the raider's location. The Germans could run (slowly), but they couldn't hide. Within a few months, all were hunted down and put out of service.

- Another deception at sea is the use of displays. It wasn't until World War I, and the widespread naval use of wireless telegraphy, that one could make convincing displays. World War I saw the beginning of "traffic analysis," the monitoring of the volume and source of enemy radio messages. Most of these messages were coded, but you didn't have to break the code to derive useful information from traffic analysis. This was a serviceable technique, but it wasn't long before it was realized that radio traffic could be faked. For example, say you want to take your fleet to sea in order to surprise the enemy. Arrange for a number of radio operators to stay in port and continue to send the radio traffic the enemy was used to picking up while the fleet was in port. Meanwhile, your warships steamed off under conditions of radio silence (no transmissions). This trick often worked, or at least kept the enemy guessing. During World War II, the use of radio-based displays continued. Traffic analysts knew about the risk of an enemy display, and this simply made their job more of a sporting proposition. After World War II, with the proliferation of antiship missiles, a new form of display came into use: the electronic decoy. The target ship, on being alerted that a missile was on its way, would fire off some decoy rockets that would use heat or electronic transmissions to make these decoys appear, to the missiles' guidance system, as a ship. Decoys had to look more like the ship than the ship itself, so that the missile would head for the more valuable (larger) target. Knowledge of how the missiles' guidance system worked was necessary, as several target-finding systems could be used. Some homed on heat, others on electronic transmissions, while some used radar. All of these methods could be "foxed" (deceived), but you had to know which method to use. Some missiles use all known methods. This is more expensive, but so is the ship. Missile designers, in return, keep making their guidance systems more "intelligent," enabling the guidance system to figure out what is a decoy rocket and what is a ship by the relative speed of the two "targets." This can be overcome with helicopters. Some ships equipped with helicopters can send out a chopper with a decoy device dangling from a long cable. This is generally more effective than decoy rockets, because the guidance system can be programmed to ignore something that "looks" like a ship but is moving too fast to be a ship (like a rocket carrying a decoy device).

- Demonstrations constitute yet another ancient naval technique. In this century, they were often combined with displays using phony radio traffic. Twentieth-century fleets tend to consist of several separate groups of ships. If the enemy saw a group of destroyers or cruisers, he would think that these were the "scouting group" for a larger fleet of battleships or carriers. As time went on, carrier aircraft replaced the scouting group of surface ships, so the presence of a few carrier aircraft would indicate the presence of a larger force. But both of these assumptions could be false. The scouting group could have been sent out to distract the enemy main force. This is a classic use of demonstration. Later in World War II, the presence of a few carrier aircraft could be just as deceptive. But those carrier aircraft could have come from one lone carrier (often a smaller and more expendable one), from a land base or, in the case of the Japanese, one of a handful of their large submarines that could carry a float plane.

  The World War II carrier battles were often a duel of demonstrations, as each side tried to get the other side to send most of their aircraft after a deceptive target. It was especially true in the first year of World War II in the Pacific, when the carriers were especially vulnerable to air attack. It was only later that carrier anti-aircraft defenses and damage-control procedures were markedly improved. Thus during these early carrier battles, whoever got the first shot in at the carriers did the most damage and often won the battle. But even late in the war, demonstrations were used to good effect. The Japanese used a series of demonstrations to bring about one of the few instances of surface ships getting at enemy carriers (during the 1944 invasion of the Philippines).

- A feint, in land combat, is a smaller attack that you want to convince the enemy is your major attack, so that you can make your major push somewhere else (where you hope it is unexpected). Naval warfare in this century, with its use of radio and aircraft, has depended more on demonstrations. One could make a case that feints and demonstrations rather merge in modern naval warfare. That said, it is true that admirals prefer demonstrations over feints, if only because at sea the former is easier to pull off than the latter. Nevertheless, feints have been used, as smaller forces are sacrificed to enhance the chances of success for a larger force. And feints at sea will no doubt continue to be used.

- Lies are used at sea, even though, unlike land commanders, twentieth-century admirals rarely have an opportunity to speak with

> or communicate with each other as much as land combat commanders do. If you can't talk to your opposite number, you can't lie to him. This does not mean that admirals don't lie to each other, just that they do their most productive lying before war breaks out. Most of the lying has to do with the capabilities of their ships. Land warfare uses a great number of different weapons and pieces of sundry equipment that are more accessible to public view than the innards of warships. Moreover, when ships practice with their weapons and equipment, they generally do it at sea, far from the curious and discerning view of potential enemies. Navies have often been quite touchy about outsiders looking in on their naval exercises. With aircraft and satellites, it is easier to keep an eye on enemy training exercises. So it followed that one could hold misleading exercises. In other words, lie to the recon aircraft and satellites.

Naval officials can spread all manner of lies about their ships' capabilities, as well as saying nothing at all and leaving performance estimates to the vivid imaginations of their foes. Before World War I, there wasn't much sense of secrecy, but there were only a few nations building large fleets, and each thought its stuff was the best and didn't pay much attention to what other warship-building powers were doing. World War I was full of surprises emanating from these prewar attitudes. It turned out that some nations did build noticeably better ships (the Germans, for example), and technology advanced at a rapid pace while the fighting went on. By the 1920s (even though a naval arms-control treaty had been signed by the major nations), lying admirals were common. By the end of World War II, secrecy had become a mania in naval circles. The lies, of course, don't get revealed and resolved until the shooting starts.

Zapping Deception

In the last thirty years, many time-tested deception methods have been compromised by developments in electronic sensors. This is not a radical development, as technology has been gradually defeating camouflage and deception efforts for over a century. The telescope allowed soldiers to get a closer, and more revealing, look at enemy deceptions. Telegraph and radio allowed commanders to exchange information more rapidly and thus sort out enemy deception efforts more quickly. Aircraft and photography enabled armies to look farther and in more detail to uncover deceptions.

All of these gadgets were present in great numbers during World War II, yet deceptions of all types proliferated and succeeded. But today, we have sensors that can see heat, magnetic fields, electrical activity, and different types of metal. Cheap and powerful computers can take fragmentary information from all manner of sensors and make a clever guess at (and reconstruct a picture of) what is out there. The new sensors can see people and vehicles hiding in forests, or even underground. Worse yet, missiles can home in on this sort of information, vastly shortening the time between detection and attack.

This, the second generation of antideception devices, has had more impact on warfare than the first. When telescopes, telegraph, and aircraft went into use, it required relatively simple countermeasures to maintain concealment and deception. The troops had to work harder, but they needed little more than shovels, sweat, and their wits to defeat this first generation of sensors.

As always, the troops' incentive was self-preservation, a marvelously effective inducement to improvise. The second generation of sensors requires more than shovels and smarts. Hiding heat is very difficult, particularly when those heat sensors are backed up by computers that can quickly sort through a large number of possible countermeasures. Hiding large chunks of metal or electrical activity is also difficult.

There are ways to fox the new sensors, but the best of these involve the use of more technology (electronic or heat devices). Many armed forces can't afford to buy a lot of new technology to defeat the latest sensors. This gives the rich guys a larger edge than they had the first time around when they were the ones with all the aircraft, radios, and fancy binoculars.

But this does not give the wealthy and well-equipped armies an insurmountable advantage. While the less well-equipped troops will be at a disadvantage, the underdog will adapt. This has always been so, and woe to the high-tech warrior who believes that the fighting and danger will be trivial simply because of equipment superiority. Sensors are largely developed in peacetime, when there is no life-and-death wartime incentive to perfect them, and they adapt to enemy countermeasures. When combat use does arrive, it does so with a new set of rules for what works and what doesn't. The Gulf War and Somalia are but two of the latest examples of how low-tech adversaries quickly learn to survive under the gaze of high-tech sensors.

Another downside of the new generation of sensors is that the side that can afford the most of them will find that these new gadgets can also be used against them. Even the poorest armies can afford some of this high tech. Heat sensors and night-vision devices are relatively cheap and are even sold to civilians. Low-tech armed forces usually lose any ability to fly their aircraft (if they had any in the first place) early in a conflict with a major power. Thus the little guys will have to be more concerned with the dreaded "unseen death from the night skies."

Depending on how determined and resourceful the low-budget troops are, they can obtain and use many of the modern sensors. Although the Cold War is over and the Soviets are no longer around to equip ragtag revolutionary movements with high-tech gear, their successors, the Russians, are still willing to sell the stuff. While the Russians offer bargains, many other industrialized nations sell the same gear at higher prices (and sometimes higher quality). So the high-tech trooper can expect to encounter equally equipped opponents no matter where, or who, he is fighting.

Quantity of technology will be as important as quality. American infantry running into the foe *du jour* during a 1990s war somewhere can expect to encounter an opponent who has at least a few starlight scopes that can see through the murkiest evening. But the force with the greatest number of these gadgets will have the larger advantage.

As with the previous generation of antideception gear, the better equipped side is prudent to assume that its low-tech opponent may have some of the latest stuff. In the many guerrilla wars since World War II, well-equipped government forces were constantly surprised to discover that their ragged opponents were quick to use radios and whatever other electronic devices they could get their hands on. All this goes under the ancient wisdom of: "Never underestimate the enemy"—no matter how wretched he may appear.

While the latest antideception gear is relatively available to anyone with cash and a strong desire to get the stuff, the means to counter these items is more difficult to obtain. For example, special paints and camouflage mats that defeat modern sensors are strictly a military item and more difficult to obtain. As always, fortune favors the bigger (and richer) battalions. But the less well-equipped trooper still has his wits and desire for self-preservation. This combination has often produced countermeasures that well-financed scientists developing new sensors never thought of. Such surprises are deadly on the battlefield, and often demoralizing to the high-tech troopers who thought they had an unstoppable edge. While many developers and users of these sensors are aware of this situation, there's only so much you can do about it without benefit of a live foe. No one has yet developed a method to dream up new countermeasures more effectively than an enemy soldier facing a life-and-death duel with a new sensor.

The one thing the underfinanced have going for them these days is worldwide television media. While rich nations can afford a lot of spiffy gadgets for their troops, they also get mightily upset when their boys get killed and it is displayed on the tube. But that's another subject.

Trust, but Verify

The phrase, "Trust, but verify," was often used by President Ronald Reagan when he was negotiating arms reductions with the Soviets in the 1980s. Reagan would speak the term in Russian, referring to it as an old Russian proverb. That it was, and the Soviets had long extended its use to their own armed forces.

Even before the Bolsheviks took over in 1917, Russia's army was full of paid informers who reported to the Czar's secret police. But the Communists took the concept of "keeping an eye on the troops" to new heights. What they did, in effect, was run what amounted to an ongoing deception on their own soldiers. While the troops were constantly lectured on their duties and responsibilities, what really maintained discipline was the realization that there were several different organizations of informers (KGB and military) constantly on the lookout for disloyal activities. No soldier was ever sure if the fellow next to him was the one sending in reports for somebody.

The knowledge that the informer system existed made it nearly impossible for Soviet units to organize mutinies or rebellions (although some did occur). And this was exactly what was intended. The Communists had overthrown the Czar with an armed rebellion led by disaffected soldiers. In fact, most of the Communist troops were the Czar's soldiers who had been convinced to rebel. The Communists wanted to ensure that they did not end up like the Czar—on the wrong end of Russian bayonets.

This system of controls on the troops was not nearly as efficient as it purported to be. That was the deception. It was a deception both on the troops (who believed they were being carefully watched) and the Communist government (who thought it had a finger on whatever was going on in the ranks). Most of the informers were either forced into the job (in return for not being punished for some infraction) or were men who sought to "get back" at enemies in their unit by turning them in for often fictitious transgressions. There were also a few true believers in the Communist system, but these were much the minority.

The officers who controlled informers would routinely fabricate their reports of "all is well." While senior officials wanted loyal troops, they were most unhappy with their subordinates if there were reports of skullduggery. So the informer system became largely a fiction at all levels. Yet, it worked, mainly because the troops knew of its existence and were deceived into believing that the system worked and that it would be prudent to avoid any troublesome behavior.

There were actually three different organizations that might be involved in infiltrating and keeping an eye on a unit (as small as a hundred troops). Every unit had a deputy commander who was the "political officer." This fellow represented the Communist party, and kept an eye on troop morale and doctrinal purity. He also recruited his informers, to provide him with information about who was doing, and saying, what when outside the view of the unit's officers. In addition, the KGB (Soviet secret police) recruited its own informers, or had its agents assigned to a unit that was suspected of disloyal (or simply illegal) behavior. And on top of that, the military police might get involved with investigators if they suspected something was amiss.

The reality was that none of these security organizations was particularly efficient. The political officer was basically a politician in uniform. His principal job was to avoid any embarrassing (to the Communist party) incidents. His informers were usually known, and as a result, they were often useless. The whole routine turned into a big game, with everyone pretending they were doing something useful.

Just how meaningful the system was can be appreciated when you realize that it was largely done away with during World War II. Defeating the Germans was more important than maintaining any pretense of doctrinal purity in the ranks. Early in that war, it was realized that the political officer/informer system was harmful to the efficient functioning of combat units. So the system was suspended, but reinstated after the war was over, the Germans were defeated, and the Communists again were more concerned with the loyalty of their troops than their combat ability.

The KGB's main job was assuring the loyalty of everyone, including Communist party officials. The KGB, in effect, "guarded the guards." While the top 10 percent of KGB personnel were quite bright and capable, the rest varied widely in ability and dedication. The KGB folks assigned to watching the troops were not as pervasive as the political officers and their informers, but they did have the advantage of greater powers.

Although watching the troops was not the highest priority task, the KGB still had an edge in the quality of people assigned to this area. Moreover, a KGB official could be arrested only by another KGB officer. While political officers were actually army officers, KGB types were strictly KGB and did not have to deal with military commanders, unless they wanted to. Thus the KGB could simply walk into an army headquarters and announce that they wanted some of their people assigned (as army personnel) to such-and-such unit and it better be kept secret or else. The KGB agent would then arrive in army uniform and would be well-trained in army procedures. The agent would fit right in, make his observations, send his report, and arrests might or might not follow, depending on what was found.

Actually, KGB agents could always find someone breaking one rule or another. This is true in every organization, and Soviet organizations had more rules than most. It was impossible to get your job done as an army officer without transgressing one regulation or another. The KGB would enforce the rules if they felt it would be good for the KGB. For example, if Soviet leadership was upset with reports of soldiers stealing and selling army equipment (a common practice), then the KGB would go out and make some arrests. But the KGB also had the future to worry about, so they would often confront the guilty soldiers or officers and give them a choice: arrest or becoming a KGB informer. The newly recruited informer might never be used, but the KGB now had the poor wretch in its pocket, to be used in the future if the need arose.

In wartime, the wealth of information the KGB had on army personnel was expected to prove useful in running deceptions on the enemy. KGB files would show who was really capable and who was simply able to play the army game well enough to satisfy the peacetime generals. When the situation got serious, and the right people had to be transferred or promoted to undertake tricky operations (such as deceptions), the KGB often had a better knowledge of who was hot than the army leadership had. The KGB also possessed its own elite units of hand-picked troops. These were to be used to enforce discipline among mutinous army troops, and to carry out especially critical actions. Again, deceptions called for extraordinarily reliable troops, and the KGB had them.

But the KGB's main game was control, and the information that enabled them to gain and maintain that control. While the army used its informers and political officers as watchdogs over its own members, the KGB used its powers and informers to deceive, as deception was the tool the KGB was created to forge and wield. Even after the Soviet Union fell apart, the KGB survived. An organization that inspired fear and uncertainty throughout the world was too valuable to destroy. But then, no one was willing to risk the fallout resulting from scattering KGB personnel, or of allowing KGB files to fall into the wrong hands. The KGB is an example of a deception machine grown so large and unfathomable that no one is willing to risk the aftereffects of its demise.

Most important, with the Communist informers and political officers gone, no one can guarantee the reliability of the Russian armed forces better than the KGB. Its ability to be aware of what exactly is going on in the armed forces is invaluable, and will be for some time to come. And the officers and troops are still operating under the shadow of the KGB. The fear and uncertainty of the troops may not have as much factual basis as it once had, but that is the mark of a good deception. What the KGB might know is still sufficient to keep the troops in line.

Great Deceptions of the Cold War

When the Cold War finally ended in the early 1990s, many of its secrets were revealed. There were many surprises in these revelations, as members of the Soviet general staff talked, and hordes of formerly secret files from East Germany were perused.

The major deception was the Soviet's ability to make themselves look more formidable than they actually were. Moreover, the Soviets were also able to conceal just how combat-ready they were in Central Europe. This may seem a strange combination, but it turned out to be the case, creating a situation where the West overestimated the global power of the Soviet Union, while underestimating the Red Menace in its own backyard.

Secrecy and deception were always primary concerns of the Soviet state. Part of this was ancient Russian custom, carried over (like so many other things) from the czarist monarchy to the Communist state. But the Communists also brought with them a fervent belief in "science" and "the ends justify the means." Put another way, the Communists were willing to try anything new, and were completely amoral about using it. With this mind-set, the Soviets quickly outstripped the rest of the world in developing ways to keep their secrets and to run deceptions. The Soviet police state was awesome in its degree of control over the population. This created a lot of dissatisfied citizens who would be ideal candidates for Western intelligence agencies to recruit as spies. But so tight was security in the Soviet Union, the West ended up getting most of their information from satellite and electronic reconnaissance. The Soviets backed up their mania for secrecy by spreading a lot of false information.

The classic example of Soviet deception was maps. Publicly available maps of Soviet cities had deliberate errors and misinformation on them, just to confuse Soviet citizens and, especially, foreigners. Accurate maps were considered state secrets and were available to only a small circle of trusted Soviet officials. To assist American diplomats, the CIA eventually created its own accurate map of Moscow. This particular item became a prized possession of foreign tourists fortunate enough to get a copy. The Soviets were not amused, and the United States found it prudent to keep this accurate Moscow street map something of a secret. Thus the Soviet devotion to secrecy even infected foreigners in the Soviet Union.

Recruiting Soviet citizens as spies was no easy task. The USSR security system went to extremes to keep Soviet citizens in general, and key Soviet officials in particular, away from foreigners. The more secret information a Soviet official had access to, the more likely he was to be watched. Senior members of the Soviet government had their homes bugged and were periodically put under surveillance. The most senior members of the Soviet elite (even KGB—Soviet secret police officers) were forbidden contact with foreigners unless accompanied by a member of the KGB. But any Soviet citizen coming into contact with foreigners was likely to be spotted by one of the millions of paid informers, and reported. Such consorting with foreigners soon led to surveillance and a grueling interrogation (or worse).

Even though the Soviet state became increasingly corrupt, you could never be sure that anyone you were making a deal with was not an informer. This had a profound effect on Western efforts to figure out what the capabilities of the Soviet armed forces were. While the satellites could count armored vehicles and barracks, they could not determine how effective the vehicles or the troops were.

Worse yet, Soviet secrecy was so impenetrable and Soviet leaders so reticent in public, it was impossible to figure out what the Soviet leadership's intentions were. In the West, there arose a minor industry called "Kremlinology" that studied such things as which Soviet leader stood closest to the head guy at major celebrations. This was thought to indicate who was more powerful than whom (and it was later revealed that who stood where meant nothing).

In such a murky atmosphere, it was no surprise that the Soviets were able to run some very successful large-scale deceptions. The major one was the idea that the Soviet armed forces were more than they actually were. The Soviets had always considered foreigners possessing misinformation about Soviet armed forces to be a major asset. Enemies could not make decent military plans for action against the Soviet Union if they weren't sure what they would be up against.

Because the Cold War sought to avoid conflict between the American and Soviet superpowers (after all, both had nuclear weapons), the battles were "fought" through the media, public opinion, and, from time to time, via third parties who did the actual shooting. In this situation, where talk and impressions were more important than actual firepower, it was most crucial to appear militarily powerful. Both sides grasped this principle and played it for all they could.

The Soviets had a major advantage in the media campaign, as unfettered media in the West kept everyone informed about what the West had. Better yet, a free press was often a very critical press and the media on both sides of the Iron Curtain stressed the scandalous situations that could be revealed (or invented) about Western armed forces and governments. The Western military had a major problem in that they were held up to criticism and ridicule by everyone's media. The press was tightly controlled in Communist countries. Communist armed forces were to be praised, and those of the West, denigrated. More important, nothing about the Soviet military got into the media without official permission. While the Communists allowed a lot of pure PR-type stuff to appear about the troops, they made sure that there was little that Western intelligence officers could make much use of. To make matters worse, Soviet intelligence officers would deliberately place military information into their media that was calculated to deceive Western observers, or at least confuse them.

The Soviets were also well aware that there were special-interest groups (military and industrial) in the United States that had a vested interest in the Soviet Union being perceived as a mighty military power. The Western military-industrial complex had its own staffs of experts who studied the fragments of information available on the Soviet Union, and issued reports "proving" that the Red Army was a dire menace and was getting stronger. Oddly enough, this same game was played in the Soviet Union by senior military leaders and the managers of the arms industries. While the Soviets publicly proclaimed the superior quality and quantity of their armed forces, privately, they thought otherwise.

Part of this came from the way experts on both sides of the Iron Curtain counted the forces arrayed in both camps. American analysts tended just to compare the strength of the United States against that of the Soviet Union. The Soviets were more realistic and counted the numerous and well-armed allies of the United States, as well as potentially hostile nations on the USSR's borders (such as China). The Soviet balance of forces was much less favorable to the USSR than the American calculation of the same situation.

Basically, both superpowers had developed powerful special-interest groups in their respective armed forces and arms industries. These groups had a vested interest in making the other side look formidable enough to justify ever-increasing defense budgets. This exaggeration-of-enemy-strength syndrome had repercussions throughout every corner of both nations. One of the more successful deceptions of the Cold War was the way in which the military-industrial complexes of both nations managed to keep their little inflated numbers game out of the spotlight. People tend to believe the worst, and the experts in both nations tended to present dire analyses of the military situation versus the other superpower.

While reports released to the public in the West made the Soviets look numerous and ten feet tall, assessment within the intelligence agencies was more mixed. Starting in the 1970s, more attention was paid to the qualitative factor, an area in which the Soviets appeared to be continually slipping as the Cold War ground on toward its conclusion in the late 1980s. "Quality" proved to be a slippery object to get a handle on. While it was pretty straightforward to show satellite photos of military equipment (and the reports that gave tallies and totals), applying a qualitative value led to major ongoing disagreements.

One school of thought (to which the authors of this book belonged) looked at the situation in a historical context. Seen from this perspective, indications were that as the Soviet armed forces grew larger in the 1970s and 1980s, they were also becoming less capable on a man-for-man basis. All armies, and the Russians in particular, tend to lose their fighting edge during many years of peace. The Soviets had not demonstrated any capability to overcome this problem and, despite the constant stream of Communist propaganda to the contrary, it became increasingly apparent that the Red Army was developing feet of clay.

The counterargument sometimes made was, "So what—they have so many modern weapons." The performance of Soviet weapons, where they were used in Third World conflicts, cast a cloud over the implied capabilities of Soviet equipment. The response to this was, "Well, what do you expect from a bunch of ill-trained and -led Third World soldiers." Then came Afghanistan in 1979. Through the mid-1980s, Soviet troops were in action against poorly armed, organized, and trained Afghan irregulars. It was not a pretty picture, and the Red Army did not distinguish itself.

The counterargument this time was, "The Soviet forces were not trained for guerrilla warfare, and that kind of combat is tough no matter who you are." As a coda, Afghanistan was dubbed "the Soviet Vietnam." But if you talked to Afghans (especially former Afghan Army officers who went over to the rebels after having been trained by the Russians) who had met the Soviets in combat, you got the impression that Soviet military training was as ragtag as everything else in the Soviet Union.

The Soviets had their own problems maintaining their illusion of military power. They had more faith in the lessons of history and, increasingly, they tripped over them. A frequent exercise within the Soviet general staff was to have a new analyst (usually a young officer with the equivalent of a Ph.D.) do an analytical study of the current military balance. As always, the study began with the results of World War II battles, with logical modifications for changes in training and equipment during the intervening decades. Most new analysts knew that the generals wanted the numbers "corrected" to show that the Red Army was still able to crush NATO forces without too much trouble.

But from time to time, a rather more honest appraisal was forthcoming. This was not appreciated by the generals, and the offending analyst was sent off to some distant research institute and his overly realistic study hidden from view. These honest studies showed that the declining quality of Soviet troops, plus the better training of Western troops, was making the "correlation of forces" (a favorite Soviet term for the likely odds of one side or the other prevailing in a battle) increasingly less favorable for the Soviets. No one (outside of the Soviet general staff) knew about these unflattering studies until the early 1990s, when the banished analysts (who were often the most capable researchers in Russia) were able to talk to Westerners.

At about the same time, it was revealed (via the secret Warsaw Pact war plans discovered in East Germany) that the Soviet general staff had listened to its dissident, and overly frank, analysts. The Soviets decided that the only way to deal with aggression from NATO was to launch a nuclear attack, followed by several dozen mechanized divisions. But the prospect of a nuclear wasteland in Western Europe eventually proved too grim for the Soviets, and in the 1980s, they scaled back the planned use of nuclear weapons.

In the midst of all these preparations in East Germany, the Soviets did manage to make their forces far more capable of launching a surprise attack on Western Europe than NATO had anticipated. This did come as quite a shock to Western intelligence agencies, although some allowed as how they knew but could not permit this information to become publicly known for fear of endangering their sources. Perhaps, but conceivably, massive Soviet efforts in the secrecy and deception areas had worked. Apparently, they had.

But deception is like a bullet; it hits whoever is in the way. Deception is often accompanied by self-deception, sometimes because those who are running deceptions get taken in by their own creations. This was certainly the case with the Soviets, where some members of the senior military leadership were shocked at the sorry performance of their troops in Afghanistan. Of course, other generals were not surprised, but this just demonstrated how these two sets of beliefs (one realistic, the other a false front created for the West) could coexist in peacetime.

Even in the West, there was a deception of sorts played on the American public regarding the capabilities of the United States armed forces. The glow of respect the armed forces had gained during World War II was extinguished by Vietnam. The decade after Vietnam did see American armed forces sink to a low ebb of effectiveness.

Since bad news plays well in the press, these conditions were well publicized. But the media didn't catch on to the reforms going on in the military, and continued to cover the military as if it were hopelessly mired in decline and disintegration. Thus the results of the Gulf War came as a shock to all concerned.

The Iraqis had gotten most of their information about the U.S. military from the American press, as did the rest of the world. This unrealistic view of American military capabilities was one reason why Iraq invaded Kuwait in the first place. In the months leading up to the Coalition counterattack, the CNN broadcasts seen in Baghdad were full of U.S. "military experts" essentially agreeing with Saddam Hussein's prediction that the coming battle would result in thousands of Americans being slaughtered to no effect.

The efficient manner in which the mainly American troops rolled over the Iraqis in 1991 came as a surprise to everyone except those who had followed the progress of American military reforms in the 1970s and 1980s. Even many of the troops were surprised at their professionalism in the face of an armed enemy. The troops had been exposed to a constant deluge of negative reporting from the media regarding their capabilities. They stood a little taller after Desert Storm, which just goes to show that the military really can't fight media coverage during peacetime. Without a war to sort out all the opinions from the abundant peacetime pundits, the bad (real or invented) news tends to get more play, and plausibility.

The Cold War ended without a war to determine what the trillions of dollars America and the Soviet Union had spent on arms would have produced on the battlefield. The results in Afghanistan and Iraq did leave an impression on many military professionals and civilian analysts. The impression was that the Soviets would have lost, unless they resorted to nuclear weapons, in which case, everyone would have lost, big time.

But the Soviets did win in the deception sweepstakes. Throughout the Cold War, the Red Army (and navy and air force) was seen as a foe of towering capabilities. The reality was that Soviet forces were large and a mess. Many historians pointed out that the Cold War Soviet armed forces was similar to the equally impressive Red Army just before World War II. In 1940, the Soviets had the largest armed forces in the world. They had more tanks, aircraft, and submarines than anyone else. They also possessed some of the most modern weapons and equipment in the world in 1940. Yet, the Soviets embarrassed themselves when they invaded Finland in late 1939.

Less than two years later, the Red Army fell to pieces during the German invasion. Yes, the Soviets won their war with Germany, but only because they could afford to lose thirty million people. The extent of these losses was unknown until after the Cold War ended, and the actual Soviet statistics of its World War II experience were published. By Soviet reckoning, eleven million were combat losses, the other nineteen million were "sanitary" losses (disease, privation, etc.). The Soviet Union came out of World War II in worse shape than any other combatant. Yet, the Soviets managed to conceal the true extent of their losses and steadily shaped an image of a mighty military and economic superpower. But it was all a false front, which was not revealed until the early 1990s.

The Cold War was fought more with deceptions than with bullets, even though America lost half as many dead during the Cold War as it did during World War II. Yet, the Soviets lost fewer troops during the same period, and pulled off far more deceptions. Thus, while the Soviet Union lost the Cold War, it did, in retrospect, win most of the battles.

Somalia, Land of Deception

Somalia in 1993 was the site of numerous deceptions, not all of which were accurately reported at the time.

In late 1992, over a year's worth of civil strife and famine led to UN intervention. This action was led by American troops. Most of the Americans were withdrawn in a few months, and a polyglot force from numerous other nations came in to keep the peace. But there was no peace, not least because of a number of deceptions.

But first, a little background. Somalia is one of the many areas in Africa that were never nations. Until it was colonized in the nineteenth century, it had never before even been united. It was turned into an independent nation by the colonial powers (Britain and Italy) after World War II.

The coastal areas of Somalia were first settled thirteen hundred years ago by Arab and Persian traders. Three hundred years later, there was a major movement of "Cushite" peoples from Ethiopia into what is now known as Somalia. The Cushites are a distinctive African people who speak Hamitic languages (related to Arabic and Hebrew). The Somalis became Somalis largely by developing a distinct (but related to other Ethiopian) language and intermarrying with the Arabs and Persians on the coast. To this day, the Somalis consider themselves Arab, not "African."

This linguistic homogeneity was the only unity the Somalis had ever known. Split into dozens of clans, subclans, and tribes, the Somalis were further divided by geography and function. Along the coast were a series of towns and cities that had long served as trading posts. Here, seagoing Arab, Persian, and Indian merchants met and traded with Somalis. The Somalis regularly offered slaves, which were usually obtained from among non-Somali peoples to the west and south. But most Somalis were nomads living away from the coast. Mounted on horseback, these Somalis have long been the scourge of the region because of their raiding for slaves, livestock, and anything else that could be carried away. This raiding-and-trading background made the Somalis expert at deception, a trait the UN was to underestimate initially.

The Somalis never got along with their neighbors, or each other. The British and Italians gradually took over Somalia in the nineteenth century; first the coastal towns, and then the interior. But in the late nineteenth century, there arose a Somali preacher who, by sheer force of personality, led an armed resistance to British rule.

Despite the overwhelming British military superiority, the Somali rebels kept at it. The British had dealt with fanatic Islamic rebels before, but the Somalis kept on coming year after year, oblivious to losses. What was more telling was that the Somalis generally stayed out in the bush, ceding control of the towns to the British. The Somalis openly boasted that the British were fools for coming out into the back country to kill (and be killed by) a few Somalis. This lasted until the "Mad Mullah" Mohammed ibn Addoellah died of natural causes in 1920.

This affair should have been a wake-up call for those who stormed into Somalia in 1992. The Somalis play by their own rules, and, seemingly, laugh in the face of death. Moreover, the Somalis use this psychological advantage to run some imaginative deceptions on the foreigners. The British experience with the Somalis earlier in the century had to be relearned all over again.

The Somalia intervention by the United Nations began because of a deception. While Somalia has never been a particularly well-organized place, it did not generally suffer famine and civil disorder of the magnitude encountered in 1990.

When Somalia was "created" in 1960, the two colonial powers (Italy in the south, Britain in the north) left behind a nation that didn't even have a written language. A small cadre of educated Somalis (literate in Arabic, English, Italian, or other languages) formed a government.

In 1969, clan politics came to a boil when Mohammed Barre, the head of one of the smaller clans and the head of the army, took over the country. There are six major clans (Dir, Hawiye, Darod, Rahanwin, Digil, and Issaq) in Somalia. Each of these clan families has subclans which often serve as a greater focus of loyalty than the major clans.

Barre, belonging to the Darod clan, was faced with prolonged clan warfare. So he did what Somali strongmen have been doing for centuries: he cut a deal with a powerful foreign nation. In this case, it was the Soviet Union that came to his aid, in 1974. The Soviets provided weapons (to overawe Barre's opponents), money (to pay off Barre's allies), and, most important, the technology of a police state (something the Soviets had perfected since the 1920s).

Barre kept the other clans at bay, but even he knew this would not last forever. And it didn't. In the same year Barre cut his deal with the Soviets, the ancient aristocracy of Ethiopia was overthrown by Communist rebels. The Soviets saw the opportunity to acquire another client in the region, but in a typical bit of deception, the Soviets didn't keep Barre apprised of their plans for Ethiopia.

Neither did Barre tell his Soviet patron what Somalia planned for Ethiopia.

In 1977, he ordered an invasion of Ethiopia. It seems that when Somalia was created in 1960, the national boundaries did not encompass all the Somali people. Parts of neighboring Kenya (to the south), Ethiopia (to the west), and Djibouti (to the north) all contained sizable Somali populations.

In Ethiopia, there was a large desert area called the Ogaden, which was populated (thinly at that) by Somali nomads. Somalia had long claimed the area, and now Barre saw an opportunity to improve his standing among the other clans by staging a little invasion against the disorganized revolutionary Ethiopian government.

This turned out to be a disaster. Forced to choose between their Somali and Ethiopian clients, the Soviets dropped Somalia and took up with the much larger Ethiopia. Soviet military aid was quickly forthcoming (mainly in the form of Cuban mercenaries), and the Somali troops were promptly cleared out of the Ogaden.

Soviet military advisors actually ran this campaign, and pulled off some imaginative battlefield and diplomatic deceptions while doing it. The Somalis didn't know what hit them (militarily or diplomatically) until after they were undone. The Soviets used secrecy, feints, and speed to rapidly reverse the situation in the Ogaden.

By 1978, Barre was in bad shape. With his Soviet meal ticket gone, and the disgrace of the failed Ogaden invasion hanging over him, he now began getting more significant opposition from the other clans. Initially, America was not willing to provide much support unless Somalia renounced their claims on neighboring territory. Barre would not do this, as it would make him appear weaker to the other clans.

Barre's hold on power was largely a deception. His Darod clan was vastly outnumbered by all the others. But by using bribes, propaganda, and a ruthless secret police, he gave most Somalis a convincing illusion of unassailable power.

In 1979, Somalia was able to obtain American support because of the Iranian seizure of the U.S. embassy in Tehran that year. America now needed local bases, and this Somalia could provide. The United States was still unwilling to provide much in the way of weapons (to discourage further Somali attacks on neighboring countries), but a lot more economic aid was forthcoming. Barre used the economic aid to keep his supporters happy, and to buy additional weapons from other nations.

Using foreign-aid funds to buy weapons was one of the deceptions that brought about the crises in Somalia during the early 1990s. While American economic aid was to be used for the benefit of all Somalis, Barre saw to it that only members of his clan, and key individuals from other clans, were taken care of. Most of the money went to purchasing weapons and other equipment needed to keep Barre in power.

This particular deception was not unique to Somalia, but was (and is) quite common in nations receiving foreign aid. The ruling clique that gets the money generally hangs on to it, or at the very least, sees to it that very little "trickles down" to the starving masses that appear on American television news and prompt the aid in the first place. But while a government in power is doing this, it is customary for other nations (including those giving the aid) not to make a big stink. A little stink, perhaps, and often a lot of private (diplomat-to-diplomat) complaining.

In 1991, the Somali government in power disappeared. Barre first encountered organized resistance in 1988, when the northern clans rose up in rebellion. While this was brutally suppressed, there were two other clans preparing to resist the government. The armed unrest in the country became too much for Barre's gunmen, and in early 1991, the government collapsed.

This, in itself, was not a disaster. The northern clans promptly seceded and formed "Somaliland" and have been relatively peaceful ever since. But in the south, there were three major (and several minor) groups contending for control of what was left of the central government. This was the root cause of the ensuing disaster.

None of these factions was strong enough to defeat the others. Since the country was broke and the constant fighting made normal commerce impossible, the only major source of income was stealing foreign aid. There were still plenty of guns around, and most Somalis knew how to use them. Like Afghanistan (and the United States), Somalia was a "gun culture." It was accepted that adult males were often armed (or have access to guns) and experienced in the use of weapons. While United Nations (and other) relief workers were trying to import food for the starving population, each clan sought to make sure its people got taken care of first. This was often guaranteed at gunpoint. The clans that controlled the major ports (particularly the capital of Mogadishu) stole all the food and other foreign aid they could. What they didn't eat, they sold (for whatever they could get) to other Somalis (or across the border in Ethiopia and Kenya).

The various armed clans managed to convince the outside world that there was a civil war going on, at least for a while. By the end of 1992, it was pretty obvious that there was a stalemated civil war going on, financed by plundering famine relief and the foreigners who were bringing it in. The TV pictures of starving Somalis (who did not belong to a well-armed clan), plus the growing complaints of put-upon relief workers led to America leading (with thirty thousand troops) a major military operation in December 1992, to "neutralize" the armed clans and allow the food aid to move freely. Awed by the considerable American firepower (and the high reputation of American troops after the Gulf War), the clans backed off.

The Somalis knew that the United Nations could be lulled into believing peace had broken out, so the clans hid their weapons and bided their time. Sure enough, by early summer, most of the American troops were withdrawn. The Americans were replaced by a nearly equal number of soldiers from many nations, including Pakistan, Italy, Bangladesh, and Malaysia. These troops were stationed all over the country, with the largest concentration in the capital, Mogadishu.

The capital contained over 10 percent of Somalia's population, as well as the best port and the symbolism that goes with being the largest city in the country. Mogadishu also contained the most heavily armed clan armies.

In June, a force of Pakistani troops was led into a trap by a clever deception, and twenty-four Pakistani soldiers were killed and mutilated. The deception that made this possible was rather gruesome. A crowd of women and children swarmed around the Pakistani troops. Knowing that the United Nations troops would not (or were at least reluctant to) fire at women and children, the Somali gunmen were able to get close enough to shoot up the Pakistanis before the United Nations troops could defend themselves. Of course, some Somali women and children were caught in the subsequent cross fire, but that was a risk the Somalis were willing to accept. It got them into position to kill the "enemy," and all those dead women and children played well in the Western press.

Including women and children in combat operations is an ancient practice, one which has survived into this century. It is not considered "civilized," but if a group feels threatened enough, they will do it. The practice was frequently seen during World War II, as well as the many small wars that followed. It took several incidents like this before the UN forces realized this use of women and children as concealment for gunmen was not an accident, but use of an ancient deception.

United Nations forces were not above some useful deceptions. When America sent in its rangers in the wake of the Pakistani deaths, a detachment of the elite Delta Force (antiterrorist commandos) also went in. The rangers were to assist in finding the Somali leaders responsible for the Pakistani incident. American troops were aware that whatever they did would be subject to constant scrutiny by the press. To get around the possibility of the element of surprise being compromised, some operations were made easier to spot, while truly vital ones were performed in a much more low-key fashion. In effect, some of the American activity was to distract the press, in order to protect the missions that would be even riskier if exposed to the real-time glare of press attention.

Although most American journalists got out of Somalia during the summer of 1993, they left behind a number of Somali journalists to get the job done for them. There was never any shortage of enterprising cameramen prowling the streets of Mogadishu, looking for something newsworthy. The United Nations command knew that this kind of journalism could be dangerous to the troops (as well as to the Somalis) and learned to work around it as best they could. It's difficult to use surprise when your preparations are liable to show up on the evening news for all to see.

There was more deception going on in Somalia than most people realized. But that's just what one can expect from well-run deception.

The Balkans: Child of Byzantium

When Yugoslavia finally came apart in the early 1990s, the world was treated to an epic demonstration of deceptions. Some of these deceptions are not yet general knowledge. But given the history of the area, it's not surprising how often deception was used.

Long a bailiwick of the Byzantines (who themselves disappeared in 1453), it was in the Balkans that dozens of local ethnic groups were long kept under control with a rich menu of deceptions by the Byzantines, and later the Turks. The Balkan peoples were generally the victims, but the concept of using deception when sufficient force was not available became accepted as a fundamental concept. The world took a while to realize that the Balkan peoples were playing by a different set of rules when it came to what was real and what was not.

We can trace the current crop of deceptions back to World War II. Early in this conflict (1941), the king of Yugoslavia was pro-German. The reasons for this stance were manifold, but the principal one was that the king sat on a very shaky throne. Then, as now, the Balkans were populated by a lot of different peoples, none of whom liked one another very much. As with the present, the Serbs were the largest group and, again like the present, they controlled the army. The Serbs also saw the Germans as an ancient enemy, and likely to favor the other ethnic groups over the Serbs. Seeing the king's action as a blow to Serbian interests, the Serb-dominated army promptly overthrew the king's ministers and declared Yugoslavia anti-German. The Germans invaded shortly thereafter and occupied the Balkans until near the end of the war. Soon after the German invasion, the Yugoslavs organized partisans and went to war with their occupiers, and with one another.

Until late 1943, most of the occupying troops were Italian. Since the Italians weren't as bloody-minded as the Germans, they often arranged informal truces with the partisans. Thus the partisans were free to fight one another, which they did with great ferocity throughout the occupation. When Italy surrendered to the Allies in late 1943, the Italian troops left (although many, not being great fans of the Germans, joined the partisans).

There were many different local armies operating in Yugoslavia during the war. The principal groups were:

- The Croats: These really weren't partisans, as Croatia was pro-German and was set up as an independent country by the Germans. In return, the Croats were expected to help in fighting the partisans. This the Croats did with great enthusiasm, organizing more troops per capita for the Nazi cause than any other German ally.

- The Chetniks: These were "loyalist" partisans, seeking a return of the old government (albeit not its pro-German attitudes). Formed under the leadership of a Yugoslav (actually, Serbian) army colonel (Draza Mihajlović), these were the first partisans in Europe to take up arms against the Germans. The Allies initially supported the Chetniks, until they discovered the Chetniks were more interested in fighting the Croats and Communist partisans than they were the Germans. In fact, it was the frequent Chetnik truces with the Germans that put off the Allies more than anything else.

- The Communists: Led by "Tito" (Josip Broz), these partisans were most representative of all the ethnic groups in Yugoslavia (including Serbs), and the most determined in their struggle against the Germans. The Communists not only wanted to expel the Germans, they also wanted to establish a Communist government in Yugoslavia. The Communists were also known to make a deal with the Germans from time to time in order to take a shot at the Chetniks or Croats. But it was the Chetniks who sinned the most in this department. In 1944, the Communists became the principal recipient of Allied aid (air-dropped supplies and agents for the most part). This, plus their superior organization and singleness of purpose, enabled them to grab control of the country when the Germans evacuated in late 1944. Tito then became dictator of a Communist-style government and ruled with a firm (and sometimes blood-soaked) hand until his death in 1980.

- In addition to the Croats and partisans, there were also local troops recruited by the Germans to fight the partisans. The Germans took advantage of ethnic animosities to do this. There were several Waffen SS divisions raised from among the Bosnian Muslim population. These fellows were told to go get those Christian partisans, and any women or children who got in their way. This tended to get very ugly.

As you can see, the ethnic rivalries that rend the Balkans today were also at work during World War II. All manner of deceptions were used then, and many of the same ones are used today. One of the nastier deceptions used then and now is to commit some atrocity against your own people, and then blame it on the enemy. This served the dual purpose of uniting your own people behind the current strongman, while allowing you to make the other side look bad to some external power (the Allies then; the UN in the 1990s). This "hurting your own" gambit was also used to punish followers who had gotten out of line. By making it look like the enemy did it, you got your message across without starting a feud among your own troops.

Saying one thing and then blatantly (or not so blatantly) doing another was a favorite deception then, and still is now. During World War II, partisan groups would promise the Allies just about anything to get those airdrops of weapons and ammunition. But these same partisans would then go ahead with their hidden agenda anyway (like using the airdropped weapons to attack other Yugoslavs instead of the Germans).

Tito even managed to pull off a masterful deception on Josef Stalin. Throughout the war, Tito paid lip service to "Communist unity." To Stalin, this meant Russian troops and secret police would march into the Balkans to ensure that a pro-Soviet government would be set up after the war. To Tito, this meant that when the Soviets reached the Yugoslav border, they would be greeted with heavily armed (with tanks and artillery) partisans who would keep the Russians out. Tito won this one, and Stalin was never able to get his revenge.

The biggest deception Tito perpetuated was the creation of Yugoslavia itself. As formed in the wake of World War I (as the "Kingdom of Serbs, Croats, and Slovenes," changed to "Yugoslavia" in 1929), the nation never jelled. The majority Serbs wanted their own state, and domination of the other minorities. The Croats and Slovenes not only wanted independence, but preferred alliance with Germany (and "the West") rather than the traditional Balkan "protector," Russia. Since the Chetniks were mainly Serbs, they suffered the most, and their political power (as well as their numbers) was much reduced after the war. Tito used the carrot as well as the stick, giving out the choice jobs and other favors to those who went along with his idea of a unified Yugoslavia. But the Serbs didn't forget. When Tito died in 1980, there was no one else of his stature or skill to keep the illusion going.

Just as he did in wartime, Tito had a knack for dividing his enemies during his long peacetime rule. He rearranged the provincial borders of Yugoslavia so that the Serbs were spread all over the place. With their power thus diluted, he felt safe to appoint loyal (to him) Serbs to key jobs. No Serb leader was allowed to appeal to Serb nationalism. It was a virtuoso performance that none of his successors were able (or willing) to match. Serb nationalists soon arose, as did like-minded men among the other nationalities, and Yugoslavia came apart in an orgy of violence.

The deceptions used during this war (or wars) depended heavily on doing one thing and making it appear as something else to the outside world. As Saddam Hussein realized, Cable News Network (CNN) can be a significant ally if you project the right image.

The unraveling of Yugoslavia began a few years after Tito died, as the secret police and intricate political and financial relationships Tito had built up began to weaken. Local politicians soon found that the path to power was via championing historical grievances of one ethnic group against another. The Croats and Slovenes in particular wanted to separate themselves from the rest of the "south Slavs" (as they had been trying to do for most of the twentieth century).

The Croats went first, declaring independence in June of 1991, followed shortly by the Slovenes and the Bosnians. The Croats and Slovenes relied on support from their historical patrons, the Germans, and they got it. The Slovenes managed the transition rather peacefully, being a homogenous province. Not so the Croats. The Croats painted a rosy picture to the outside world, and carefully hid what was being said inside Croatia regarding their large Serbian minority.

Although the Serbs had lost about 10 percent of their population (mostly to Croats and other nationalities rather than the Germans) during World War II, they had been made to feel safe during the time of Tito's police state. But now the Croats were declaring independence and using the same terms and symbols Croats used during World War II when the Serbs in Croatia were being massacred. The Serbian Croats had good reason to be terrified, and soon the guns came out into the open. It's still difficult to sort out who started shooting first, but Serbs from Serbia were promptly on the move to "protect" the Serbs in Croatia.

Serbia initially justified its warlike actions as an attempt to protect Serbs living among other ethnic groups. Some 30 percent of the Serbs in the Balkans live outside the borders of what is considered "Serbia." The other nationalities saw the Serb operations as an attempt to expand Serbian territories. Both sides were right, but no one would admit it. While the truth may sometimes require a bodyguard of lies, in the Balkans, the truth gets trampled by an army of deceit. Slovenia soon got involved in the fighting, as Serb troops crossed the border to protect the Serbs of Slovenia.

Soon we saw the phrase "ethnic cleansing" being added to our vocabulary. Removing people of another ethnic group by terror and massacre is nothing new in human history. It is, unfortunately, quite common. The Nazi racial policies during World War II were, basically, "ethnic cleansing." The Nazis preferred to exterminate some groups (like Jews, gypsies, and political opponents) while the Serbs and other "ethnic cleansers" of the 1990s preferred to turn their ethnic opponent populations into refugees. This played better on CNN than piles of dead bodies.

Ironically, during World War II, it was the Serbs who were most often on the receiving end of the "ethnic cleansing" at the hands of Croats and Bosnian Muslims. It wasn't pretty then, but there were no television camera crews around either. But the memories remained fixed in Serbian minds. From this you get the "What's the problem?" response from Serbs when they are accused by Western media of being murderous thugs. Revenge, in some minds, justifies all sorts of horrors. And deceptions are used to make the mess more palatable to outside eyes.

Although Tito proclaimed Yugoslavia to be ethnically "integrated," most of the intermarriage took place in the cities, not in the countryside. If there ever were any real Yugoslavs, they lived in the cities. The ancient hatreds survived, and thrived, out in the rural areas. It was there that most of the ancient fighting and killing took place.

The cities could remake themselves; the rest of the country could not. Through most of Yugoslavia you could go from a wholly Serbian town to one down the road populated exclusively by some other group.

Ironically, the different ethnic groups of the Balkans speak the same language. There are different dialects, but everyone can understand one another. What makes the difference is religion (Orthodox and Roman Catholicism, plus Islam) and alphabets (Roman and Cyrillic). These differences, in all their combination, created half-a-dozen hostile tribes willing to die for their differences.

All sides in this conflict attempted to play down to the outside world the extent of their animosities with one another. All sides realized that foreigners just wouldn't understand. This was true with the Americans, and that was important. The Europeans were more knowledgeable about Balkan history and the vitality of ethnic hatreds. This, in a nutshell, is why Europeans were reluctant to get too involved with Balkan peacemaking.

The United States was the only superpower, and could lead, or veto, any Balkan intervention. The warring groups also knew that they were all guilty of some pretty hideous atrocities. It really was a replay of World War II, and local leaders knew that this ugliness would not play well on the outside. In this situation, Serbia was at a major disadvantage. With the most soldiers, the most territory to fight over, and the most people outside of their traditional territory, the Serbs were going to commit the most atrocities and most frequently get caught by the media.

The Croats, Slovenes, and, especially, the Bosnians, made the most of this. They played up the Serb atrocities and played down their own. In several cases, the Bosnians were caught committing some of these atrocities against Bosnians. At first, when the Serbs pointed this out, no one believed them. As the evidence piled up that the Bosnians were not loath to attack some of their own people to gain an edge on the evening news, the moral landscape became pretty murky. But the Serbs were still identified as the major villains.

To overcome their image problem, the Serbs, or at least the Serbs from Serbia, began to conceal and camouflage their support of the Serbs in Bosnia (where they comprise about 40 percent of the population). The Serbs also played an effective game of diplomatic deception in Bosnia, as did the Croats. Both groups were eager to carve up Bosnia without triggering foreign intervention. So they played up the desire of the Bosnian Serbs to have autonomy within "Bosnia" rather than the real goal of all Serbs (unification into a "Greater Serbia"). The same attitude was assumed by the Bosnian Croats.

The Bosnian Muslims (plus Bosnian Serbs and Croats who considered themselves Bosnian first), who comprised a little less than half the "nation's" population, had no trouble presenting themselves as embattled. It took a bit more deception to present themselves as "the first United Nations member to be wiped off the map." The Bosnians had to wage a war of press releases and contorted history to make a case that they were a nation.

None of the new nations in the Balkans had much experience as independent entities in the last thousand years. The area had always been dominated by empires that, at best, let the different ethnic groups exist as provinces. In the past, the United Nations had only admitted countries that were already generally accepted as nations and were functioning as such. The case(s) in Yugoslavia were a lot less clear-cut. First, several component parts of Yugoslavia declared themselves independent in 1991. Then these new "nations" fell to arguing and fighting over where the borders would be drawn. The United Nations jumped in and admitted several (but not all) of them before the border disputes had been settled (by diplomats or soldiers).

Bosnia's principal claim to nationhood was its United Nations recognition, as all the other normally accepted symbols of statehood were buried by the chaos of the war with Serbs and Croats. There was no functioning government, and the borders changed from month to month. Should the Bosnians lose their United Nations recognition, and the hope of foreign recognition it implied, there would be nothing left. The Bosnians were outnumbered and surrounded. Deception was the most powerful weapon they had. The extent to which they used that weapon will not be known for years. But if the past history of deception in the area is any indication, deception was used early and often.

One of the more notable deceptions was the ever-popular self-deception. Many educated people in the West increasingly called for armed action against the Serbs who were ravaging Bosnia. One of the arguments for intervention was that the Serbs wouldn't resist if faced with a more powerful opponent. What is amazing about this is that it flies in the face of Serb history in the twentieth century. Consider the record of Serb combativeness. The Serbs fought two wars just before World War I. Then they fought through World War I (mainly against the Austrians). There was a lot of civil disorder between the two world wars. Then the Serbs were overrun by the Germans in 1941. Yet, the Serbs fought as partisans throughout the war (belonging to one faction or another). Yugoslavia was the only nation in World War II that liberated itself from German occupation. The nation (and especially the Serbs) lost over 10 percent of its population. To keep the Serbs down after the war, Tito killed and jailed thousands more. And ten years after Tito's death, the Serbs were at it again. What is amazing is that anyone who knows all of this can deceive himself into believing that the Serbs will back down in the face of superior force. The fact of the matter is that in most of Serbia's twentieth-century wars, they have been up against a stronger opponent and never backed off.

Self-deception is the most tragic form of deception, because it is the one form that is the most easily avoided. All you have to do is check a history book.

The only problem with studying deception in ongoing events is the success of the deceptions themselves. If they work, you won't know a deception was going on, at least for a while. The 1990s wars in Bosnia used just about every military and diplomatic deception known to man. The details will follow, although some may never be known.

Deception Via the Front Page and CNN

The onset of the electronic "Global Village" has created new fields for deception, and new dangers to their success. The instantaneous transmission of news, and particularly pictures, has opened up a whole new dimension for prospective tricksters to explore. There were many examples of this in the 1991 Gulf War.

Coalition forces (i.e., the United States), engaged in a massive manipulation of the media, primarily to fool Saddam Hussein, but also to prevent some of the more unpleasant realities of war from disrupting political support for the operation. The Iraqis, meanwhile, played the media as if it were a major military asset (which it was). After Kuwait was invaded, Saddam Hussein accomplished more via Cable News Network (CNN) appearances than he did subsequently on the battlefield.

An incident during the Gulf War may serve to illustrate the possibilities. The landing of the first Iraqi SCUD missiles in Tel Aviv was covered live by all international telecommunications organizations. The media's reaction was almost instantaneous, and it was hysterical. Within minutes of the attack, telejournalists were proclaiming, "live on television," that: The Iraqi missiles carried poison gas; and the Israeli Air Force was already on its way to deliver a retaliatory strike.

Within a few more minutes, it was clear that both reports were false. But for those few minutes, there existed the possibility of a disastrous development, one which would certainly have caused the collapse of the carefully constructed anti-Iraq Coalition: active Israeli participation in the war. That one or more of the Arab partners in the Coalition did not opt out as soon as these erroneous reports began circulating was due primarily to the speed with which they were corrected.

Now consider the possibilities inherent in a carefully prepared deception that grabs instant international attention at a critical moment, thus provoking a reaction likely to have undesirable political side effects. Sound absurd? It isn't. Consider one of the dirty little secrets of modern diplomacy: that national leaders and their staffs usually get late-breaking news first from television and radio, not from their enormous intelligence staffs.

In recognition of this, the CIA set up its own closed-circuit TV news operation. In this way, they could try to compete with the timeliness of CNN while still providing their specialty (secret information). Ironically, the CIA, with a larger "news budget" than CNN (or any other network), still has to hustle to keep up.

But only American officials have access to the CIA network; the vast majority of the world's diplomats and government leaders must rely on the likes of CNN for most of their information during a fast-breaking situation. Moreover, even if network news is false or misleading, the politicians and generals have to deal with it. While what appears on a television screen can be quite deceptive, it is also most compelling. The TV screen often becomes an alternate reality, and one that cannot be ignored. Democracies, in particular, are subject to considerable public pressure generated by television coverage.

Until about a century ago, government had the edge in news technology. Only governments could afford the vast numbers of informants, as well as fast riders and fast ships to speed the news to kings and prime ministers. Then came the telegraph and a proliferation of daily newspapers. News, and diplomacy, haven't been the same since. But it was the introduction of satellite-based communications and twenty-four-hour television news that really changed the situation. On January 17, the air war against Iraq began, live, on CNN. Suitcase-size satellite dishes and even smaller cameras allow anyone, anywhere, to record an event and simultaneously broadcast it worldwide.

The military is not pleased with the newfound ability of anyone to broadcast images (and sound) worldwide using equipment that could be carried by one person. What has not been mentioned thus far is the deception opportunities now possible via satellite news reporting. Consider the possibilities. "Live" news can be faked or slanted to support a deception. This was done during the 1991 Gulf War, where the Coalition press officers only let the TV cameramen see what the military wanted seen (especially by Saddam Hussein and his advisors).

Faking the news is nothing new; telejournalists get caught doing it periodically, and those within the TV news business are aware that there is a great deal of flexibility in what is shown and how it is presented. The truth is a nebulous concept, while the need to fill the screen with "newsworthy" images is compelling. While it is difficult for the American military to control the press, such is not the case in many other nations. Moreover, the ever-falling cost (and size) of satellite broadcast equipment means that it's only a matter of time before the many freelance telejournalists are so equipped. At that point, TV news editors will be faced with some tricky decisions, as they view some juicy footage coming in live over the satellite from a freelancer (who may be working for more than one paymaster). Such exciting footage may be a deception, but if it is compelling (some atrocity or another), the competitive nature of television news will cause someone to show it without double-checking its provenance. At that point, the other news broadcasters will be under considerable pressure to follow. While that item may later be proven false, the initial image has more impact than subsequent retractions. The Iraqis play this game with great gusto, and success, as do other nations that have tight control over the media within their borders.

Even in situations where the local government doesn't control the domestic media, they have used fabricated events to shape world opinion. Somali warlords deliberately sent mobs of women and children against armed (and nervous) United Nations troops. The warlords knew that images of dead women and children would aid their cause. In Bosnia, several of the "other side's atrocities" have been traced back to the same group the victims belonged to. This is nothing new in Balkan-style politics, but it garners greater benefits, courtesy of worldwide TV coverage.

The combination of instant (but difficult to quickly verify) news images and the dependence of national leaders on this information is a volatile mixture. The urge to use instant news as a form of deception is a temptation that many cannot resist. The professional purveyors of deception are only beginning to appreciate and exercise the potential of these opportunities. You can expect to see more television news that is not all it appears to be.

For the soldiers, it isn't deception on the evening news that worries them, but rather the truth being broadcast before its time.

Since the relatively independent mass media began sending reporters to battlefields in the nineteenth century, the military has seen some ugly side effects. The reporters often have a different agenda from the troops'. Moreover, the reporters usually don't fully understand the consequences of what they report. This was first seen during the American Civil War (1861–1865). Here, we found Union reporters eager for a scoop, and the means to obtain one. Stories were often sent back by telegraph and appeared in print within hours. Within twenty-four hours, those newspapers could be (and often were) in enemy hands, providing the Confederates with valuable information they could not otherwise have obtained.

Some Union generals banned reporters from their area of operations, but this was not easy to enforce. At other times, reporters were threatened with execution if they were caught providing (knowingly or otherwise) military information to the enemy. This proved legally and politically impossible.

By the twentieth century, the military had gotten better organized, so they were able to seal off the combat zone from unauthorized journalists. But more important, the military made strenuous efforts to educate reporters about what was dangerous to report and what wasn't. Moreover, dispatches going back to the newspapers and radio stations could be reviewed by officers, and dangerous items could be pointed out and excised.

Vietnam was a new era, not because there was now TV exposure to worry about, but because it was not a declared war, and the military did not have the authority to control the press tightly. Moreover, the war was unpopular back home, and the press had to respond to these attitudes in its reporting. The American government tried to project a more optimistic Vietnam reality than actually existed. The media and the military developed mutual bad feelings as a result of their Vietnam experiences.

Vietnam-era bad feelings between media and military have abated considerably, but now, there are more problems for military deception and security. The biggest problem is that, for the foreseeable future, there will be no big, declared, wars that will enable the military to control the media tightly. Moreover, these future wars will inevitably create fiercely pro and anti factions back home. The media have come to depend on exploiting adversarial relationships in their programming. Thus reporters will be searching more diligently for "the other side of the story," even if it doesn't exist (or, worse yet, is mundane). Scandal and outrage get better ratings than reporting banal reality. This is made worse by the ever-shrinking pool of reporters who have any understanding of what military affairs are all about. Most current reporters have no military experience, and few of the people they hang out with do, either.

The great fear among military commanders during the 1991 Gulf War was that an eager reporter with a TV camera and portable satellite dish would "scoop the world" by showing details of the initial Coalition ground attacks. This would have resulted in greater Coalition casualties, as Iraqi generals took notes from CNN broadcasts and issued more effective orders to their troops as a result. This situation is still with us, and will become easier to pull off as satellite broadcasting equipment becomes cheaper and more portable.

During the Gulf War, only major news organizations could afford the suitcase-size satellite gear. The military could afford to have press officers work with network editors to warn them about what not to show. But in the next few years, this equipment will become cheap enough for many more reporters to carry. This is especially the case with freelancers, who can always get their "scoop" on some foreign TV network, and from there, most of the world's TV nets will be under enormous pressure to show the scoop also. The cat will be out of the bag, and some risky deception will be blown (along with many of the participating troops).

Such is one of the dangers of military operations in the Balkans in the 1990s. Fast, secret, and decisive military action is the kind that will accomplish peacekeeping goals most effectively. But the danger of a television crew catching you in the act (or preparations), and putting the pictures onto a satellite feed gives the most able and forceful military commanders pause.

It was only during the Cold War that deception via the media became big business. It was only during the 1980s that the American military began working out ways to deal with its own media as an integral aspect of military planning. The results were seen in the Gulf War, and the American media haven't quite gotten over the shock. But the military is still at a disadvantage when dealing with the press. The military uses secrecy as one of its principal weapons, while the media strive to unmask secrets to maintain their position in a fiercely competitive business.

Remember this well: The media and the military have a historically new relationship. It's not necessarily an unfriendly one. But the media can hurt the troops, and the brass have to constantly come up with new ways to avoid this from happening.

New Directions in Deception

Electronics have been changing the practice of deception throughout this century, and this trend will gather steam as we enter a second hundred years of high tech. Not just the electronic weapons that have been around for over fifty years, but increasingly, the media as well. As we enter the twenty-first century, we will see electronic and media deception merging somewhat, particularly in the smaller wars that have historically filled the decades (and sometimes centuries) between total conflagrations like World War I and World War II.

The last period of smaller wars, the late nineteenth century, was also the time in which the media (in the form of newspapers) were becoming a force in political affairs. Before the nineteenth century, there were no mass media in the twentieth-century sense. There were some newspapers, but most of them were in the pay of one power broker or another. But even during the nineteenth century, the media, and media deception, played a role in military affairs. This was seen in the American Civil War, as well as in some of the conflicts that occurred up to World War I. In the late 1890s, for example, the Cuban revolutionaries actively supplied the American "Yellow Press" with phony atrocity tales to inflame public opinion, in the ultimately successful hope that the United States would go to war with Spain. Military deception also played a role in this period, as it always had, and the generals had a hard time coming up with ways to deal with the mass media. It wasn't until World War II that the military began to see the media as just another means of running deceptions. As history goes, we can expect to see history repeat itself in the coming decades.

But there are sharp differences between the late nineteenth century and the late twentieth. Automation in military affairs has reached the point where combat robots are becoming a reality. The cruise missile is literally a robot bomber, with some of the current models being capable of bombing one or more targets before running themselves into one more. On the ground, the U.S. Army is ready to deploy a series of robot antivehicle weapons. These devices, once placed and turned on, literally listen for enemy vehicles or helicopters. Once detected, homing missiles are launched. From the beginning, these robotic antivehicle weapons were designed with deception in mind. It will be a while before robot soldiers are as bright and resourceful as human troops, so antideception had to be built into the robots' computers. Of course, this computer intelligence is not expected to give human brains an excessive workout. But on the battlefield, randomness counts for more than braininess, and every bit of cleverness counts, even for robot warriors.

Yet, in some cases we already have robots fighting robots. Modern aircraft must contend with air defense systems that are largely computer controlled. To counter this, fighters and bombers carry their own computers that endeavor to keep the human pilot one step ahead of the computer-controlled radars and computers trying to shoot them down. Especially in the air, it is the speed of combat that is giving the machines more decision-making power. Particularly when trying to evade enemy missiles, the total time available to evaluate the situation and act is often measured in seconds. Humans are increasingly too slow for this, and it's becoming more common for the computers to be put in control of those combat functions that require immense calculation and very rapid decision making. With situations like this, the deception techniques must be carefully thought out ahead of time and built into the aircraft and missile computers. Humans are still making the decisions, but as time goes by, humans are making fewer of them.

While the battles are becoming faster, and more frequently controlled by machine intelligence, the higher-level decision making is now spread among a far larger number of people. So even here, the Art of Deception is changing.

There was a time, not so long ago, when decisions about how to fight a war were made by a small group of people. Now, the appearance of mass electronic media has expanded the decision-making group considerably. In effect, the population of an industrialized nation makes decisions regularly via opinion polls. This intensifies media pressure on political leaders. It's no accident that the first targets rebels go after, when staging a coup d'état, are television and radio stations.

More and more often, political decisions are made based on what populations see about the battlefield situation through the mass media. This has brought about an increased temptation for governments to run deceptions on their own populations as well as on the armed enemy. Governments deceiving their own people is nothing new, but mass media, opinion polls, and democracy have created more incentive for national leaders to seek relief from the chorus of contradictions.

Public opinion is a fickle thing, especially when driven by the critical and sensational tendencies of the media. While molding public opinion is somewhat looked down upon, this is precisely what governments are supposed to do in order to lead. America was the first nation to develop this form of ongoing public debate, a style of government that was present when the nation was formed and has merely become speedier and more immediate as media became more pervasive and the population better educated.

The "American System" is now found in most industrialized nations and many developing ones. Thus when these nations go to war, the citizens tend to get deceived along with the enemy. If anything, the deception of your own voters is more important than those run on the enemy. Your own people can usually run you out of office faster than the armed enemy can. Thus it is obvious which deception is more important. Fortunately, victory on the battlefield usually brings with it a de facto pardon for wartime deceptions against your own people. At least this is the case in major wars. But the many smaller military actions leave the public, and the media, in a less forgiving mood.

With all that, the easiest deception to pull on your own constituents is the "I didn't know, I went forward with the best information available" gambit. The many meetings national leaders have with their experts, diplomats, and generals are done in private. Exactly who said what to whom, and what all went into the final decisions, is largely lost, or deliberately hidden, in the murk. Incomplete information becomes the leaders' friend and provides a plethora of options.

There is still danger, for modern government bureaucracies are vast, and many of the civil servants are willing to leak information that would compromise a government deception. While the risk of leaks is always there, so many leaks ooze out of the bureaucracy that one can put additional leaks into play to counteract the original damaging leak, on the theory that you can bury the real info in a mass of false data. So even if your opponents have access to accurate information, they have so much to sort through that they gain no benefit from it.

An interesting offshoot of televised warfare is the shock effect the normal battlefield carnage has on an audience unfamiliar with such mayhem. While movies have long gone out of their way to portray gruesome combat mutilations, people are still able to separate Hollywood make-believe from the real thing on their TV sets.

It has reached the point where a population caught up in a patriotic war fever of blood lust is turned into instant pacifists when they view their countrymen in uniform being shot to pieces. Aside from making it difficult to pursue a war to its conclusion, the televised slaughter has made a minor industry (and religion) out of "nonlethal weapons."

The oldest form of nonlethal weapons was the use of clubs and whips rather than swords and spears to subdue a civil disturbance. Modern forms of nonlethal force are tear gas (and related nonlethal gasses) and rubber bullets. There are also water cannon, cattle prods, attack dogs, horsemen, and a new generation of high-tech "nonlethal" weapons.

Actually, any of these devices can (and often do) kill. But they are much less likely to kill than the weapons they replace. The problem is, no one has come up with a nonlethal weapon that will work effectively against an enemy who is armed with lethal weapons. Send in troops armed with tear gas, clubs, and rubber bullets and they will not do well against a foe equipped with assault rifles.

It has become fashionable to postulate a not-too-distant future when "sonic disruptors" (directed high-decibel sound that disables), "flash disruptors" (bright lights that do the same), and sundry new chemicals will allow an armed enemy to be subdued without the normal bloodshed. The prognosis for this future, at least in the immediate sense, is not good. These devices often get a workout with police first, and the success rate has not been high. An armed opponent is going to be quite resourceful, and will find ways to use his old-fashioned "lethal" weapon to injure his less bloody-minded (but more modishly equipped) opponent.

Moreover, a policy of using nonlethal weapons lacking a solid track record of success will deny the use of some classic nonlethal deceptions. For example, the Soviet Union always maintained a bloody-minded attitude toward terrorists. The Soviets said, in effect, "What you do to me, we'll do to you, the same way, but more so." An example of this occurred in Lebanon during the 1980s. One of the many Lebanese factions seized a Soviet agent and held him hostage. The Soviets dispatched some of their KGB people to seize someone belonging to the offending faction. The Soviets then began sending body parts of their captive to the faction leaders. The message was, "Release our man, or we'll up the ante." Now, that might normally just get the faction to escalate the affair. But the Soviets had a reputation for not retreating from the use of such savage tit for tat. The Soviets may not have even had the means to push the affair any further than they already had. But their reputation for ruthlessness made the deception work. And it did work. It was a classic "Byzantine technique," and the Russians have long been admirers of the Byzantine way of taking care of such affairs.

Deception, particularly in little wars and struggles with terrorist organizations, consists of a lot of bluff and bluster. Your reputation counts for a lot. If you get tagged as a Goody Two-Shoes who believes in "nonlethal force," the meat eaters out there will simply lick their chops and plan accordingly.

A similar situation took place in Somalia during 1993. The Adid faction slaughtered twenty-four Pakistani UN peacekeeping troops. The United Nations put a price on Mr. Adid's head, and the United States contributed commandos to help in the hunt. One of the commando operations turned into a large-scale shoot-out. The United States suffered seventy-six casualties (including eighteen dead). But the Somalis lost ten times as many people. If America had turned around and said, "You want some more?" Mr. Adid would have been hard-pressed to maintain the morale of his gunmen.

The Somalis were aware of their past, but the UN peacekeepers were not. British soldiers early in this century developed an effective tactic for dealing with hostile Somalis: "Shoot early, shoot on sight, shoot often." The other useful observation of the British regarding the Somalis was that most Somalis are hostile, particularly if they have an opportunity to act without too much fear of retribution.

But the shoot-out (and its immediate aftermath) between the Americans and Somalis was partially televised; the American public was horrified; and popular opinion turned an Adid defeat into a victory. The fighting (among the Somalis) continued, and United Nations troops died in vain. All for lack of will to establish a reputation for ruthlessness in the face of lawlessness. Had some willpower been demonstrated, it would have been possible to use the "Shape up or you know what we're capable of" deception to pacify the country with a lot less bloodshed. This technique has worked many times in the past, even in Somalia. But to do it, you have to know what you're doing, and perhaps have a little privacy while you're doing it.

The age of troops doing violent (even if effective) "peacekeeping" is over. The terrorists and minor-league despots know this and are no longer susceptible to the threats and deceptions that commonly accompanied the traditional military/political cure for the Somalias and Bosnias in times gone by. Many a time in the past, local diplomats could run a very effective deception (a bluff) on the bully of the moment. If your troops were known to be effective, and capable of getting where they were needed, the bad guys would often back down before facing hostile bayonets.

One should pay careful attention to the fact that tyrants were the first to note the new influence of mass media on deception. Hitler, Stalin, and, of late, Saddam Hussein, all used the media to run deceptions. But modern midget moguls learned to use the media not only to deceive, but also to disarm their more powerful opponents. Thus the media have been turned into a "nonlethal weapon," but one that cannot be used readily by a nation with a free press.

That is not the end of the story; simply the current chapter. Deception has constantly evolved, and adapted to new technologies and conditions. You may not recognize it when you first see it, but you will be seeing new forms of deception. Cleverness, limited means, and a strong sense of self-preservation will make it so.

By Way of a Conclusion

Deception looks to be a growth industry in the future. And not merely essentially military deceptions. Indeed, in "the new world order," political and diplomatic deceptions (for domestic as well as foreign consumption) will probably become more important. As the likelihood of major war diminishes, the chances of smaller wars increases. Political and diplomatic trickery can be important tools in dealing with minor threats to world peace, whether through averting them entirely or by bringing them to a successful conclusion. But such deceptions may also serve to bring about little wars, or prolong them.

In the future, we will certainly need political and military leaders who are at least as clever, if not more so, than those in the past. After all, the world will be a much trickier place, and they will have to be able to spot ruses and deceptions perpetrated by foreign leaders and be able to perpetrate a few of their own when necessary.

When will it be necessary to employ deception, tricks, and ruses? When you are weak and when you are strong; when you are attacking and when you are retreating; when you are winning and when you are losing; when you are at war and when you are at peace. In short, to recall the ancient Chinese adage which we quoted at the beginning of this book: "There can never be enough deception . . ."

RECOMMENDED READING

A surprising number of books deal with deception, trickery, and ruses in war. What follows is a mere sampler, for the reader who may be interested in pursuing the topic further. Remember, however, that very often the cleverest deceptions can be found only by careful reading of quite ordinary historical treatments of wars, campaigns, and battles.

A few hoary standbys: Most of these are available in various translations.

The Bible, notably, Old Testament historical books
Sextus Frontinus, *Strategematicon*
Herodotus, *The History*
Flavius Josephus, *History of the Jewish War*
Titus Livy, *Roman History*
Niccolò Machiavelli, *The Prince*
Polyaenus, *Strategica*
Thucydides, *History of the Peloponnesian War*

Classic Chinese works dealing with warfare, and deception in war, are best consulted in *The Seven Military Classics of Ancient China*, translated by Ralph D. Sawyer (Boulder: Westview Press, 1993), which has some of the wiliest deceptions on record, and a good deal on the philosophy of deception, as well.

More modern works on the subject, which stress the Second World War, are:

William B. Breuer, *Hoodwinking Hitler: The Normandy Deception* (Westport, CT: Praeger, 1993).
Anthony Cave Brown, *Bodyguard of Lies* (New York: Harper &

Row, 1975) is still the best single treatment of the deceptions that helped to protect the secret of D-Day.

Charles Cruickshank, *Deception in World War II* (New York: Oxford University Press, 1979) concentrates on the European Theater, and usefully supplements *Bodyguard of Lies*.

John B. Dwyer, *Seaborne Deception: The History of U.S. Beach Jumpers* (New York: Praeger, 1992).

David Glantz, *Soviet Military Deception in the Second World War* (Totawa, NJ: Frank Cass, 1989) is an exhaustive and informative look at some of the most masterful deceptions in World War II.

Michael I. Handel, *Military Deception in Peace and War* (Jerusalem: Magnes Press, 1985) is a good, serious if short treatment of the art of deception through the ages.

J. C. Masterman, *The Double-Cross System in the War of 1939–1945* (New Haven: Yale, 1972).

Seymour Reit, *Masquerade: The Amazing Camouflage Deceptions of World War II* (New York: Hawthorne, 1978).

INDEX

INDEX

INDEX

INDEX

INDEX

INDEX